Birds of
MADAGASCAR AND
THE INDIAN OCEAN ISLANDS

Seychelles, Comoros, Mauritius,
Reunion and Rodrigues

HELM FIELD GUIDES

Birds of MADAGASCAR AND THE INDIAN OCEAN ISLANDS

Seychelles, Comoros, Mauritius, Reunion and Rodrigues

Frank Hawkins, Roger Safford and Adrian Skerrett

Illustrated by
John Gale and Brian Small

CHRISTOPHER HELM
LONDON

Christopher Helm
An imprint of Bloomsbury Publishing Plc

50 Bedford Square
London
WC1B 3DP
UK

1385 Broadway
New York
NY 10018
USA

www.bloomsbury.com

BLOOMSBURY, CHRISTOPHER HELM and the Helm logo are trademarks
of Bloomsbury Publishing Plc

First published 2015

© text by Frank Hawkins, Roger Safford
and Adrian Skerrett 2015
© artwork by John Gale and Brian Small 2015

Frank Hawkins, Roger Safford and Adrian Skerrett have asserted their right under the
Copyright, Designs and Patents Act, 1988, to be identified as Authors of this work.

All rights reserved. No part of this publication may be reproduced or transmitted in any form or by
any means, electronic or mechanical, including photocopying, recording, or any information storage
or retrieval system, without prior permission in writing from the publishers.

No responsibility for loss caused to any individual or organisation acting on or refraining from action
as a result of the material in this publication can be accepted by Bloomsbury or the authors.

A catalogue record for this book is available from the British Library.

Library of Congress Cataloguing-in-Publication data has been applied for.

PB: 978-1-4729-2409-4
ePub: 978-1-4729-2410-0
ePDF: 978-1-4729-2411-7

2 4 6 8 10 9 7 5 3 1

Designed by Julie Dando, Fluke Art
Printed and bound in China

Cover artwork by John Gale
Front: Pitta-like Ground-roller
Back (top to bottom): Seychelles Paradise Flycatcher;
Cuckoo-roller; Yellow-bellied Sunbird Asity; Subdesert Mesite

To find out more about our authors and books visit www.bloomsbury.com
Here you will find extracts, author interviews, details of forthcoming events and the option to sign up for our newsletters.

CONTENTS

	Plate	Page
ACKNOWLEDGEMENTS		7
ABOUT THIS BOOK		9
Species accounts		9
Maps and distribution		10
Names and classification		11
Abbreviations		13
Glossary of terms		13
Topography		16
THE MALAGASY REGION AND ITS BIRDS		18
GEOGRAPHY, CLIMATE AND HABITATS		20
Madagascar		20
Seychelles		27
Comoros		31
Mascarenes (Mauritius, Reunion, Rodrigues)		35
Outer Islands		40
BIOGEOGRAPHY: COLONISATION, DIFFERENTIATION AND SURVIVAL		40
ENDEMIC FAMILIES AND GENERA		45
LIST OF BIRD ORGANISATIONS		57
PLATE SECTION		59
Part 1: Larger birds (non-passerines): Malagasy Region		
Albatrosses DIOMEDEIDAE	1	60
Petrels and shearwaters PROCELLARIIDAE	1–4	60–66
Storm-petrels HYDROBATIDAE	5	68
Tropicbirds PHAETHONTIDAE	6	70
Boobies SULIDAE	6	70
Cormorants PHALACROCORACIDAE	7	72
Darters ANHINGIDAE	7	72
Frigatebirds FREGATIDAE	7	72
Herons, egrets and bitterns ARDEIDAE	8–10	74–78
Hamerkop SCOPIDAE	11	80
Storks CICONIIDAE	11	80
Flamingos PHOENICOPTERIDAE	11	80
Ibises and spoonbills THRESKIORNITHIDAE	12	82
Grebes PODICIPEDIDAE	13	84

	Plate	Page
Ducks, geese and swans ANATIDAE	13–15	84–88
Kites, hawks and eagles ACCIPITRIDAE	16–18	90–94
Falcons FALCONIDAE	19–20	96–98
Francolins, partridges, pheasants and quails PHASIANIDAE	21–22	100–102
Guineafowl NUMIDIDAE	21	100
Buttonquails TURNICIDAE	22	102
Mesites MESITORNITHIDAE	23	104
Rails, crakes and gallinules RALLIDAE	24–26	106–110
Jacanas JACANIDAE	26	110
Painted-snipes ROSTRATULIDAE	27	112
Crab-plover DROMADIDAE	27	112
Stilts and avocets RECURVIROSTRIDAE	27	122
Pratincoles GLAREOLIDAE	27	112
Plovers and lapwings CHARADRIIDAE	28–29	114–116
Sandpipers and snipes SCOLOPACIDAE	27, 30–31	112, 118–120
Skuas STERCORARIIDAE	32	122
Gulls LARIDAE	33	124
Terns STERNIDAE	33–36	124–130

Part 2: Landbirds ('near-passerines' and passerines)

	Plate	Page
Madagascar	37–61	132–180
Granitic Seychelles	62–65	182–188
Coralline Seychelles	66–69	190–196
Comoros	70–79	198–216
Mauritius and Rodrigues	80–84	218–226
Reunion	85–89	228–236
Outer Islands: Agalega and Mozambique Channel	90–92	238–242

Part 3: Vagrants: Malagasy region   93–124   244–306

APPENDIX 1: CHECKLIST OF BIRDS OF THE MALAGASY REGION   308
APPENDIX 2: A NEW CLASSIFICATION OF BIRD ORDERS AND FAMILIES
  NATIVE TO THE MALAGASY REGION   324
REFERENCES   328
INDEX   329

ACKNOWLEDGEMENTS

This book follows the publication in 2013 of *The Birds of Africa, Volume 8: The Malagasy Region* (*BoA 8*: Safford & Hawkins 2013), and draws on the research that went into that larger volume. Although *BoA 8* was a multi-authored work, all the texts on identification and voice therein were prepared by two of the authors of this field guide (Roger Safford and Frank Hawkins). Identification and voice texts for vagrants were not provided in *BoA 8*. Therefore, the species texts in this book were prepared by Hawkins and Safford, building on content they had written for *BoA 8*, and by Adrian Skerrett, who drafted new or revised text for Seychelles species and for all vagrants. In addition, the section in the Introduction on *Biogeography: colonisation, differentiation and survival* has been adopted (and much shortened) from the equivalent *BoA 8* section by Ben. H. Warren, Roger Safford, Dominique Strasberg and Christophe Thébaud.

We thank the commissioning editor at Bloomsbury (and its imprint, Christopher Helm), Nigel Redman, for steering the work to completion, together with Julie Dando at Fluke Art, who prepared the material for publication. Nigel Redman was preceded in his role by Andy Richford at Academic Press, with whom the idea of producing both *BoA 8* and this field guide was originally discussed. The whole text was edited by Guy M. Kirwan.

The authors of *BoA 8*, in addition to ourselves, were Rado Andriamasimanana, Aristide Andrianarimisa, Steve Bachman, Ruth M. Brown, Nancy Bunbury, Malcolm D. Burgess, Aurélie Chowrimootoo, N. J. Collar, Richard Dale, J. M. M. Ekstrom, C. J. Feare, Damien Fouillot, Yannick Giloux, Steven M. Goodman, Valérie Grondin, Jim J. Groombridge, Marion Gschweng, Carl G. Jones, Fabien Jan, Victorin Laboudallon, Olivier Langrand, Michel Louette, Rob Lucking, Gulo Mauror, Justin Moat, Raoul Mulder, Malcolm A. C. Nicoll, David J. Pearson, Marc Rabenandrasana, Claire Raisin, Hajanirina Rakotomanana, Anna Reuleaux, Lily-Arison Réné de Roland, Heather Richards, D. S. Richardson, Martin Riethmuller, Gérard Rocamora, Marc Salamolard, Autumn Sartain, Tom Schulenberg, N. Seddon, Dominique Strasberg, Kirsty J. Swinnerton, Tamas Székely, Joseph Taylor, J. A. Tobias, Christophe Thébaud, Simon Tollington, Luciana Vega, Ross Wanless, Ben H. Warren, H. Glyn Young, Stuart Young, Sama Zefania and Nicolas Zuel.

In addition to the work they did on their species or themes, many of them provided very valuable assistance in the preparation of other parts of *BoA 8* and hence this book. For this, we particularly thank Rado Andriamasimanana, Aristide Andrianarimisa, Nancy Bunbury, Nigel Collar, Chris Feare, Steven Goodman, Carl Jones, Olivier Langrand, Michel Louette, Marc Rabenandrasana, Hajanirina Rakotomanana, Gerard Rocamora, Marc Salamolard, Autumn Sartain, Vikash Tatayah, Ben Warren and Glyn Young.

The following contacts provided particularly valuable support with respect to particular species or groups of species: Neil Baker (East African migrants); Nick Block (passerines of Madagascar, especially tetrakas); Simon Bruslund (on Malagasy birds, notably herons, ibises and Long-tailed Ground-rollers, held at Weltvogelpark Walsrode); Matthieu Le Corre (seabirds); Andrew Cristinacce, Rina Nichols and Jannie Fries Linnebjerg (Mascarene passerines); Guy Kirwan (numerous aspects); Christopher Lever (introduced species) and Dieter Oschadleus (weavers). Several photographers provided new insights into various species, and we specifically recognise Solohery Rasamison, Bruno Andriandraotomalaza Raveloson and Andriamandranto Ravoahangy for their remarkable work in Madagascar, which deserves wider attention; also Paul Desnousse and Catherina Onezia for Seychelles material.

Many other people helped the editors, artists and authors to complete this book in countless ways, including providing unpublished information, photographs, references and manuscripts: Roland Albignac, Gary Allport, Charles Anderson, Luciano Andriamaro, Rado Andriamasimanana, Aristide Andrianarimisa, Voahirana Andriantsalama, the late Fleurette Andriatsilavo, W. R. P. Bourne, Vincent Bretagnolle, Rachel Bristol, Nancy Bunbury, Brigitte Carr-Dirick, Anthony Cheke, Salvatore Cerchio, Claude Chappuis, Neil Cheshire, Alice Cibois, Philippe Clergeau, Mark Cocker, Callan Cohen, Andrew Cooke, Blaise Cooke, François-Xavier Couzi, Helen Crowley, Marianne Damholdt, Normand David, Ron Demey, Edward Dickinson, Tim Dodman, Hugh Doulton, Robert J. Dowsett, Françoise Dowsett-Lemaire, Gerald Driessens, Will Duckworth, Kevin Duffy, Joanna Durbin, Lee Durrell, Jonathan C. Eames, Mike Evans, Mark Fenn, Brian Finch, Lincoln Fishpool, Damien Fouillot, Jerôme Fuchs, Jörg Ganzhorn, Nick Garbutt, Charlie Gardner, Justin Gerlach, Steve Goodman, Katie Green, Phil Gregory, Steven Gregory, Tom Gray, Jim Groombridge, Dominique Halleux, Faramalala Miadana Harisoa, Matthew Hatchwell, Clare Hawkins, Marc Herremans, Erik Hirschfeld, Philip Hockey, Josep del Hoyo, Pierre Huguet, Julian Hume, David Hyrenbach, Louise Jasper, Michael Jennings, Knud Jønsson, Peter Kappeler, Sarah Karpanty, Chris Kehoe, Jan Komdeur, Joris Komen, Samit Kundu, Frank Lambert, Olivier Langrand, David Lees, Richard Lewis, Jeremy Lindsell and the East African Rarities Committee, Sam Lloyd, the Madagascar

Pond Heron discussion group, James Mackinnon, Klaus Malling Olsen, Clive Mann, Etienne Marais, Charles Marsh, Greg Middleton, Russ Mittermeier, David Meyers, Pete Morris, Raoul Mulder, Thomas Mutschler, Sheila O'Connor, Oliver Nasirwa, Martin Nicoll, David Notton, Michel Ottaviani, the late Mario Perschke, Mark Pidgeon, Oliver Pierson, Jean-Michel Probst, Rick Prum, Rivo Rabarisoa, the late Rabemavaza, Marc Rabenandrasana, Sahondra Radilofe, Marie Jeanne Raherilalao, Sylvère Rakotofiringa, Michel Rakotombololona, Daniel Rakotondravony, Odon Rakotonomenjanahary, Leon Rajaobelina, Serge Rajaobelina, Julien Ramanampamonjy, Olga Ramiarison, Ramanitra Narisoa, Voninavoko Raminoarisoa, Jaime Ramos, Jean-Jacques Randriamanindry, Harison Randrianasolo, Patrick Rasamoela, Joelisoa Ratsirarson, Richard Ranft, Pamela Rasmussen, Jean-Claude Razafimahaimodison, Sushma Reddy, Don Reid, Adam Riley, Lily-Arison René de Roland, C. J. R. Robertson, Peter Robertson, Kees Roselaar, Jonathan Rossouw, Peter Ryan, Beau Rowlands, Richard Schodde, Derek Schuurman, Hadoram Shirihai, David Showler, Ian (J. C.) Sinclair, Iben Hove Sørensen, Michael Sorenson, Claire Spottiswoode, Eleanor Sterling, Tim Stewart, Richard Switzer, Colin Taylor, Sam The Seing, Russell Thorstrom, Tony Tree, Clémentine Virginie, Rick Watson, Roger Wilkinson, Lucienne Wilmé, Michael Wink, Sébastien Wohlhauser, Lance Woolaver, Satoshi Yamagishi, Andy Young and Sama Zefania. We also thank those, too many to list, who provided their published material or contacts, and sincerely apologise to any whose names we may have accidentally omitted.

We thank members of Seychelles Bird Records Committee for their contributions in assessing all vagrants to Seychelles and sometimes the wider region, Michael Betts, Ian Bullock, John Bowler, David Fisher, John Phillips and Rob Lucking.

Other conservation and research organisations in the Malagasy region whose staff contributed in many ways include the Mauritian Wildlife Foundation, Société d'Études Ornithologiques de La Réunion, Island Conservation Society (Seychelles), Seychelles Islands Foundation, Nature Seychelles, Asity Madagascar, Madagascar National Parks, The Peregrine Fund, Durrell Wildlife Conservation Trust, Fanamby, Conservation International, WWF, Wildlife Conservation Society, Parc Botanique et Zoologique de Tsimbazaza, and University of Antananarivo.

For access to specimens, we thank Robert Prŷs-Jones, Mark Adams, Hein van Grouw, and the other staff of the Bird Group at the Natural History Museum, in Tring, UK; Michael Brooke, Ray Symonds and Mathew Lowe at the University Museum of Zoology, Cambridge, UK; and Jacques Cuisin, Marie Portas, Anne Préviato and Jean-François Voisin at the Muséum National d'Histoire Naturelle, Paris, France. Thanks also to Friederike Woog of the Staatliches Museum für Naturkunde, Stuttgart, Germany, for an exceptionally varied range of help and advice. Other assistance was gratefully received from librarians at BirdLife International: Christine Alder, Jeremy Speck, Janet Chow, John Sherwell, Rohan Holley and Lizzie Atkinson.

We are very grateful to the two artists, John Gale and Brian Small, for their superb illustrations. However, our thanks go far beyond this, for their ornithological advice and especially their willingness to take on the very many changes and additions, compared to our original plans, which we asked of them. Similar thanks go to David Pearson, who wrote the detailed plumage descriptions for *BoA 8* – like the artists, often 'above and beyond the call of duty' – which revealed many of the characters described and illustrated here, and shared his immense knowledge of African birds.

Finally, we thank our families for their support, understanding, encouragement and above all patience.

ABOUT THIS BOOK

This volume is intended to enable observers to identify every bird that they might reasonably expect to see in the tropical south-western Indian Ocean away from the African mainland. This geographical region includes Madagascar and the archipelagos of Seychelles, the Comoros (Grande Comore, Anjouan, Mohéli and Mayotte), the Mascarenes (Reunion, Mauritius, Rodrigues) and certain associated but remoter islets (the 'Outer Islands'). It is often called the Malagasy Region. It does not cover the archipelagos of the Chagos (British Indian Ocean Territory), the Maldives and Lakshadweep (the Laccadives), Pemba, Socotra and other islands close to East Africa, or the southern Indian Ocean islands.

Species accounts

Each part of the region has a bird community that is fairly distinct from that of other areas. The book has been designed for use in the field so that, as far as possible, observers in a specific part of the region will not be overwhelmed by information on bird species they are unlikely to see. The species accounts are therefore divided into three colour-coded sections, with Part 2 further subdivided as detailed below.

section colour	
	PART 1 deals with larger birds, many found around water, often occurring over large parts of the region: these are the grebes, seabirds, ducks, herons and allies, raptors, gamebirds, rails, waders/shorebirds, skuas, gulls, and terns; in other words, waterbirds, raptors and gamebirds, which many readers will recognise as the 'non-passerines'.
	PART 2 covers all of the landbirds (passerines and so-called 'near-passerines') and is arranged by island or archipelago:
	Madagascar
	Granitic Seychelles
	Coralline Seychelles
	Comoros
	Mauritius and Rodrigues
	Reunion
	Outer Islands
	PART 3 covers all vagrants to the region.

Thus, observers visiting one part of the region (say, Madagascar) will usually be able to find the birds they encounter in two parts of the book: Part 1 and the relevant section of Part 2. Part 3 exclusively covers vagrants to the entire region. Within each of these sections, the sequence is broadly taxonomic, with some similar but unrelated species grouped together for convenience.

The reason we have organised the book in this way is that a very high proportion of landbirds (covered mainly in Part 2) are restricted to one island or archipelago. Therefore, in this book the birds of any given island or group are less likely to be 'lost' among many species the visitor will not see, as would be the case if all the region's species were simply presented in a single sequence. A consequence of this structure is that the few passerines and 'near-passerines' that occur in more than one part of the region appear in the book more than once. We hope that the inconvenience of a few extra pages is more than compensated for by the ease of use.

Similarly, as native island avifaunas are usually comparatively impoverished, there is a disproportionately large number of vagrants to the region which, for some groups, might easily swamp the native and regularly occurring species on the plates. Even though about 28% of the entire avifauna of the region comprises vagrant species, the casual visitor is rather unlikely to encounter such extreme rarities, and so it makes sense to illustrate these species in a separate plate section (Part 3).

Identification To identify a bird, first decide whether it is more likely to be in Part 1 or 2. If it appears to be a waterbird, raptor or gamebird, go to the Plate section of the book (Part 1) and consult the appropriate plate (broadly arranged taxonomically). If a landbird, go to the section of Part 2 that covers the archipelago or island you are visiting. If you can eliminate all species in these sections, you may be watching a vagrant: see Part 3.

Each species text begins with the English and scientific name (see Sequence and taxonomy, below, for more detail), and approximate total length from bill tip to tail tip. Sizes refer to the races that usually occur in the region, and have been simplified to a single figure, useful for comparative purposes, unless a species is exceptionally variable, when a range or different figures for male and female may be given. This is followed by an account of the main identification features, especially coloration and structure (see below for Voice, and Status, habitat and habits). We have focused the text and illustrations to be of maximum use for identification, concentrating on the plumages that are most likely to be encountered, and on features and behaviour that allow similar species to be distinguished. Where there is detailed text on the separation of two species on the same page, we have usually not repeated the same text for both species, but instead provide cross-references as necessary.

Voice Vocalisations are often the key to both finding and identifying birds, and so we describe voice in as much detail as the medium and space allow. Phonetic descriptions of vocalisations were compiled by Frank Hawkins based on published recordings (a list is provided in the References, p. 328) and the personal experience of all of the authors, where necessary checked against unpublished recordings such as those on the website Xeno-canto. Published descriptions of vocalisations are in some cases also used. Typical categories of description include a general indication of frequency that covers the range of most bird songs (low, e.g. a female duck's quack or a pigeon's coo; medium, e.g. a bulbul's chattering; high, e.g. a sunbird's chipping call; and very high, e.g. portions of a fody song), the length of each phrase, the interval between the phrases, and the frequency with which the song is given. The accounts generally start with the most commonly heard vocalisation. Mechanical sounds (e.g. bill snaps, wing noises) are included here, where useful for detection or identification. Sonograms are recommended for further study but are not reproduced here for reasons of space and the difficulty of selecting representative recordings; they can be produced easily from digital recordings.

Status, habitat and habits (SHH) Status includes whether a species is endemic to the region or part of it, and information on distribution that supplements or replaces a map; this often includes altitudinal preferences. Comments on the species' abundance across, or in different parts of, the region, appear here, sometimes in relation to races (subspecies); the latter are introduced under Identification where relevant. A brief description of favoured habitats is followed by any habits that help an observer to locate or identify the species. Biological information is, however, kept to a minimum unless it is of exceptional interest. If a species was designated by BirdLife International in 2014 as globally threatened, the section ends with the BirdLife/IUCN Red List Category given as a two-letter code (for further information, see the BirdLife website www.birdlife.org):

EX: Extinct.
CR: Critically Endangered: species considered to be facing an extremely high risk of extinction in the wild.
EN: Endangered: species considered to be facing a very high risk of extinction in the wild.
VU: Vulnerable: species considered to be facing a high risk of extinction in the wild.
Species considered to be at less risk are designated as follows:
NT: Near Threatened: species close to qualifying, or likely to qualify for a threatened category in the near future.
The remainder are sometimes referred to as 'Least Concern' or 'not globally threatened'; we simply omit any threat category for these species.

Maps and distribution

We provide distribution maps in Part 1 (waterbirds, raptors and gamebirds) for all species, and in Part 2 (landbirds: 'near-passerines' and passerines) for those species that occur in Madagascar. In the remainder of Part 2 (the sections covering the smaller islands), distribution is indicated firstly by the section one is looking at (e.g. the Comoros) and secondly by the text (e.g. which islands on the Comoros). However, for completeness, a map is also provided for each species in these sections, primarily to show the entire distribution of more widespread species in the region. Species known only as vagrants are not mapped.

In all species except those with extremely restricted ranges, the maps show predicted range based on known records and habitat availability (often a combination of vegetation type and altitude) within the species' geographical limits. For the rarest species, known to be absent from apparently suitable areas, only habitat known to be occupied is mapped, and this is stated in the text. For several seabirds, satellite-tracking studies have provided the basis for the non-breeding ranges mapped. All maps were prepared as new for *BoA 8*, with a few revisions since, except for several seabirds for which non-breeding ranges were based on the maps in the *Marine IBA* [Important Bird Area] *e-atlas* published online by BirdLife International; even in these cases, original studies were always checked where published separately. The range limits of many, perhaps most, widespread species are not precisely known, and this introduces an element of uncertainty into range maps; they must always be interpreted with caution.

Dashed ellipse (any colour): found or likely to be found regularly on all islands within the ellipse, although many are too small to map clearly; this is used for the four main archipelagos of Seychelles (the granitic islands, Amirantes, Farquhars and Aldabra archipelago).

Green: resident, breeding or presumed breeding

Orange: breeding visitor

Blue: non-breeding visitor

Hatched (any colour): present but at significantly lower density than in solidly shaded areas

Where necessary, an arrow is used to highlight a small population that might otherwise be overlooked.

? is used where a species may occur but this is not certain; this may be where occurrence is predicted but has not been confirmed, or where one or a few records exist but it is not clear whether the species's presence there is normal or exceptional; the text will indicate which.

× is used for isolated distributional records, usually outside the normal range. One symbol may indicate more than one record. For some seabirds away from breeding colonies, this symbol may appear in the normal non-breeding range, because sightings are so few that individual records merit inclusion.

Illustrations

All the major plumages and distinctive races of all regularly occurring species are illustrated. In Part 2 (landbirds), each section shows only the plumages or races that occur in the part of region being treated (e.g. for Madagascar Turtle Dove, the section on the granitic Seychelles shows only the endemic race *rostrata* and the introduced nominate race). The paintings are based on field experience, examination of specimens at the Natural History Museum, in Tring, UK, a wide selection of references and photographs of skins from other collections, as well as live birds. Emphasis is placed on the identification characteristics of locally endemic or poorly known species.

Names and classification

Sequence and taxonomy This guide follows the taxonomic sequence and scientific nomenclature adopted by *BoA 8*, in turn based on the original *Handbook of Birds of the World* (HBW) volumes 1–16 (not the more recent *HBW and BirdLife International Illustrated Checklist of the Birds of the World*, which so far only covers non-passerines). Two factors led to modifications to the HBW sequence and taxonomy.

First, we introduced a relatively small number of clearly defined exceptions in cases where there are well-supported, more recent findings of key importance in the Malagasy context. These included the expansion of the endemic vanga family (Vangidae) by inclusion of Ward's Vanga *Pseudobias wardi*, newtonias *Newtonia* and Crossley's

Vanga *Mystacornis crossleyi*, and recognition of four 'warbler' families not in HBW, resulting from reorganisation of the huge Old World warbler (former Sylviidae) assemblage: Bernieridae (endemic), Locustellidae, Acrocephalidae and Phylloscopidae (the latter represented in the Malagasy region only by vagrants).

Second, consistent with the book's main function to help with identification, certain pairs or groups of similar-looking species are placed together even if they are not closely related; for example, the sunbird asities (genus *Neodrepanis* in the family Philepittidae) are placed with the members of the true sunbird family (Nectariniidae).

Species and races (subspecies) The difficulties of classifying species whose ranges do not overlap are well known. Not surprisingly, for a region composed of numerous islands, examples abound in the Malagasy region. Many variable species have generally been treated as being composed of two or more races confined to separate islands, or in the case of Madagascar, parts of that vast landmass. Some authorities are increasingly recognising certain (or many) of these forms as species, a process often called 'splitting'; the converse process, 'lumping', is currently rarer. In *BoA 8*, we took a conservative view, avoiding new 'splits' unless published evidence, preferably both molecular and morphological (and sometimes vocal), pointed that way. In this field guide, we maintain that position but point out in the species texts what we believe are the most likely potential 'splits'; a box on p. 55 presents a more comprehensive list of the relatively distinctive races. For many vagrants, we point out either races known to have occurred, or considered most likely to have done so based on field observations or known range. All races are listed in the main checklist (Appendix 1).

English names We normally follow the names recommended by the International Ornithologists' Union (Gill & Wright 2006, and updates published online) with, in most cases, consensus also with the fourth edition of the *Howard and Moore Complete Checklist of Birds of the World* (Dickinson & Remsen 2013, Dickinson & Christidis 2014). It therefore seems unnecessary to repeat here the rules on which these authorities base the names they use. We do not aim to provide an exhaustive list of English names, and present only a few alternatives in cases where two or more quite different names are used regularly and either seems appropriate.

New high-level taxonomy In recent decades, molecular research has been and continues to be revolutionising our understanding of avian (and other) systematics, and it is clear that 'traditional' sequences (used here) do not reflect the phylogeny (evolutionary history) of birds; their main advantage is in terms of stability and familiarity, both being necessities in a field guide. No new, stable sequence has yet emerged, and in the main body of this book, we do not specify orders or families. However, later in the introduction, we provide descriptions of endemic families and genera. Following the completion of the fourth edition of the Howard and Moore Checklist, and a subsequent high-level classification of bird orders (Jarvis *et al.* 2014) readers may be interested to view the high-level groupings (families and orders) of Malagasy birds placed in a newly proposed sequence; see Appendix 2.

Place names and language Names used are normally geographical rather than political. Thus, 'Comoros' refers to all four islands and associated islets (comprising the Union of the Comoros together with the French *Département* of Mayotte), and 'Mauritius' means the island of Mauritius, not the Republic that includes Rodrigues and other islets. Accents are normally retained as in local usage, except in well-known names, most obviously Reunion, where a widely accepted English form is available (in French, it should be called *La Réunion*).

Place names and spellings in Madagascar follow the National Institute of Geodesy and Cartography (Foiben-Taosarintanin'i Madagasikara, FTM) single-sheet 1:2,000,000 or 11-sheet 1:500,000 series. This includes using the modern Malagasy names for the major cities of Antananarivo (Tananarive), Mahajanga (Majunga), Toamasina (Tamatave), Toliara (Tuléar) and Tolagnaro (Fort Dauphin). One of Madagascar's best-known areas for wildlife is around the town of Andasibe, east of Antananarivo, including the rainforests of Mantadia National Park, Analamazaotra Special Reserve and Maromizaha forest, which area is most often referred to as Périnet.

For the Comoros, including Mayotte, we follow the French Institut Géographique National IGN 1:50,000 single map for each island, retaining the well-known French names for the islands: Grande Comore, Mohéli, Anjouan and Mayotte: their local names, used in some literature, are (respectively) Ngazidja, Mwali, Ndzuani and Maoré. We avoid the semi-anglicised 'Grand Comoro' for the first-named. Place names on Mauritius, Reunion and Seychelles are relatively stable, as similar names are generally found on British, French and national maps.

We avoid the adjective 'Madagascan' in favour of 'Malagasy', which is almost universally used by people of all nationalities working in the region, and of course by the Malagasy people themselves. Both are used to mean 'of or relating to Madagascar' (*Oxford English Dictionary*, 2012), the latter being longer established because it is the adjective used in the national language (also called Malagasy); to us, its spelling does not seem so removed from that of the island that it requires abandoning. However, either adjective needs to be used with care in a regional work such as this, to avoid confusion between the region and the island or country, either of which may be referred to as Malagasy.

Abbreviations

♂	male	L	Lake	N	North
♀	female	Mt	Mount	S	South
ad	adult	NP	National Park	E	East
br	breeding	non-br	non-breeding	W	West
C	Central	R	River		
c.	approximately	s	seconds		
cf.	compare with				
I	Island; Is Islands				
imm	immature				
juv	juvenile				

Glossary of terms

Adult: a bird that has achieved the fullest stage of plumage development.
Allopatric: occurring in separate non-overlapping geographical areas.
Alula: the 'thumb' of the wing (sometimes known as the 'bastard wing'), comprising a few short, stiff feathers at the carpal joint.
Austral: relating to southern regions of the world.
Australasia: the zoogeographical region encompassing Australia, New Zealand and New Guinea.
Axillaries: the inner wing-lining feathers between the wing and body (the 'armpit').
Bare parts: the collective name for unfeathered areas, namely the bill, legs, feet, eyes and any unfeathered skin.
Boreal: relating to northern regions of the world.
Carpal: the forward-pointing bend of the folded wing (the 'wrist').
Carpal bar: a line of feathering along and just behind the leading edge of the upperwing.
Caruncle: an outgrowth of bare skin, often loosely referred to as a 'wattle'.
Cere: the soft, fleshy covering at the base of the upper **mandible** in some birds (e.g. a **raptor**).
Conspecific: two or more races belonging to the same species.
Coverts: small (contour) feathers that cover the quills of the tail and **flight feathers** and also the ear.
Crepuscular: active at dawn and dusk.
Crown: the top of the head.
Culmen: the upper ridge of the bill.
Decurved: curved downward.
Dichromatism: sexes of a given species that differ in coloration.
Dimorphic: having two distinct colour **morphs**.
Ear-coverts: feathers covering the external ear opening.
Eclipse: the dull post-breeding plumage that occurs for a short period in some male ducks and some gamebirds while they undergo wing moult; the term is sometimes used for other species with a dull non-breeding plumage, such as weavers (including fodies) and sunbirds, although for these we use 'non-breeding plumage'.
Endemic: a species or race confined to a particular country or region.
Eye: for the purposes of this book includes the coloured iris.
Eye-ring: differently coloured feathers or skin encircling the eye.
Eyestripe: a coloured line (usually dark) appearing to run horizontally through the eye.
Facial disc: a circle of facial feathers radiating around the eye.

Flank: the side of the body below the folded wing.
Flight feathers: the long **primaries** and **secondaries** of the wing.
Flycatch: the habit of launching forth in pursuit of aerial insect prey and returning to a perch or the ground.
Frontal shield: a flat, horn-like extension at the base of the upper **mandible** covering part of the forehead (e.g. in Common Moorhen).
Gape: the mouth opening, formed in the angle between the two mandibles when the lower jaw is dropped.
Gonydeal angle: a visible 'kink' or angle in the gonys, which is the ridge along the centre of the lower mandible; often present in large gulls and terns, just behind the bill tip.
Graduated: refers to a tail in which the central feathers are longest, the other feathers becoming progressively shorter away from the centre.
Greater coverts: see **Wing-coverts**.
Gular: referring to the throat.
Hepatic: a dark brownish-red **morph** (e.g. in some cuckoos).
Hind-collar: a coloured band of feathers around the **hindneck**.
Hindneck: the back of the neck.
Immature: in this volume, all **plumages** prior to full **adult**.
Jizz: a combination of field characters that collectively help identify a species.
Juvenile: a bird in its first full plumage after the nestling or downy stage.
Kleptoparasitism: where food collected by one bird is stolen by another.
Lek: where several males gather prior to or during the breeding season to engage in competitive displays that may entice visiting females.
Lesser coverts: see **Wing-coverts**.
Lore: the area between the bill and the eye.
Malar: the side of the throat immediately below the lower **mandible,** sometimes marked by a **malar stripe**.
Malar stripe: a coloured streak bordering the throat (e.g. in some pipits).
Mandible: in this volume, refers to either the upper half (the upper mandible) or the lower half (the lower mandible) of the bill, the plural being used for both together.
Mantle: the upper back.
Median coverts: see **Wing-coverts**.
Mirror: a white spot in the black wingtip of some gulls.
Monospecific: in reference to a genus, one with just one recognised species.
Monotypic: in reference to a species, one with no recognised races.
Morph: a plumage colour variant which has a genetic basis and is not merely sexual, seasonal or developmental plumage variation (see **Dimorphic** and **Polymorphic**). In this book we avoid the less precise term 'phase'.
Moult: the process by which old feathers are shed and replaced by newer ones.
Moustachial stripe: a stripe running back from the base of the bill, sometimes above a second streak, the submoustachial stripe.
Nape: the rear of the head between the **crown** and the **hindneck**.
Nectarivory: where a bird consumes the sugar-rich nectar produced by flowering plants.
Nocturnal: active at night.
Nominate race: the first race of a species to have been described (where two or more races exist), which has the same racial name as the species e.g. *Nesoenas picturata picturata*.
Oscine: a **passerine** of the suborder Oscines characterised by vocal organs that are highly specialised for singing.
Palaearctic: relating to the zoogeographical region comprising Eurasia north of the Himalayas, together with North Africa and the temperate part of the Arabian peninsula.
Palaeontology: the study of life existent prior to, but sometimes including, the start of the Holocene (current) epoch.
Passerine: a perching bird of the order Passeriformes.
Pectoral: pertaining to the breast.
Pelagic: pertaining to the open sea.
Phylogeny: the evolutionary history of a species or taxonomic group of organisms.

Plumage: a bird's feathers, used sometimes for a particular seasonal, sexual or age difference.
Polyandrous: a breeding strategy where a single female mates and lays eggs with multiple males over the course of a breeding season, leaving the males to incubate the eggs and raise the chicks.
Polygynandrous: a breeding strategy where two or more males mate exclusively with two or more females (but not necessarily equal in number by sex).
Polygynous: a breeding strategy where a single male mates with multiple females, but each female mates with only one male.
Polymorphic: having several distinct colour **morphs**.
Primaries: the outer **flight feathers** (attached to the 'hand' and 'digits' of the wing).
Primary projection: the length by which the **primaries** extend beyond the **tertials** in the folded wing.
Race: a population of a species that is geographically and morphologically distinct from other populations of the same species; here, synonymous with 'subspecies' (but see specialist literature for more detailed discussion).
Raptor: a bird of prey.
Rump: the area between the lower back and the **uppertail-coverts**.
Scapulars: feathers between the **wing-coverts** and the **mantle**.
Sclerophyllous: plants (often of warm dry climates or high altitudes) with hard leaves stiffened by woody tissue and a relatively short distance between leaf nodes.
Secondaries: the inner **flight feathers** (attached to the 'forearm' of the wing).
Sedentary: making only small, local, or no movements between seasons (i.e. not migratory).
Species: a population or group of populations that do not normally interbreed with other such populations.
Speculum: a bright, often iridescent patch on the **secondaries** in many ducks.
Streamers: elongated slender tail-feathers that may be the innermost (e.g. in tropicbirds) or the outermost (e.g. in some terns).
Subadult: the latter stages of development of an **immature** bird that requires more than one year to reach maturity, prior to achieving full **adult** plumage.
Submoustachial stripe: a stripe running back from the base of the bill below a **moustachial stripe**.
Suboscine: a **passerine** of the suborder Suboscines, supposedly more primitive members of the order, with less well-developed vocal organs than the **oscine** birds.
Subsong: a quieter, less emphatic version of a bird's song.
Sulcus: a narrow groove along the inside edge of the lower mandible, the colour of which is diagnostic in separating some species of albatross.
Supercilium: a stripe immediately above the eye, usually pale.
Sympatric: occurring in the same geographical area, in contrast to **allopatric**.
Tarsus: the 'ankle' bone of the leg.
Tertials: a term applied to the innermost **secondaries**, or secondaries nearest the body in some groups (e.g. waders), which are often conspicuously coloured; true tertials attached to the upper arm bone are found only in some large, long-winged birds (e.g. albatrosses).
Tibia (plural **tibiae**): the 'drumstick' bone of the leg.
Undertail-coverts: the feathers behind the **vent** covering the bases of the tail feathers.
Uppertail-coverts: the feathers behind the **rump** covering the upper surface of the bases of the tail feathers.
Wattle: a fleshy unfeathered (or mostly unfeathered) appendage on the head or neck; a type of **caruncle**.
Web: the flexible, broad side of the feather; also, the loose flap of skin between the toes (e.g. in seabirds and ducks).
Wingbar: a band across the wing, paler in colour than the surrounding feathers.
Wing-coverts: the feathers of the upperwing covering the bases of the **flight feathers**, comprising the very short **lesser coverts** (nearest to the leading edge), the shortish **median coverts** and the longer **greater coverts** (furthest from the leading edge).
Vent: the feathers at the base of the **underparts** forward of the **undertail-coverts**.

Topography

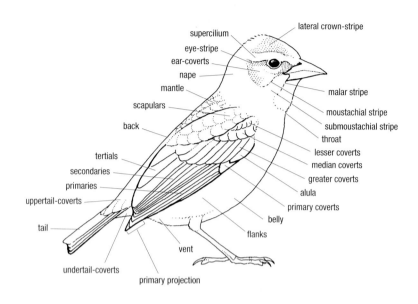

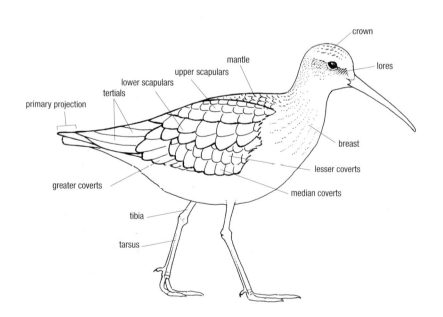

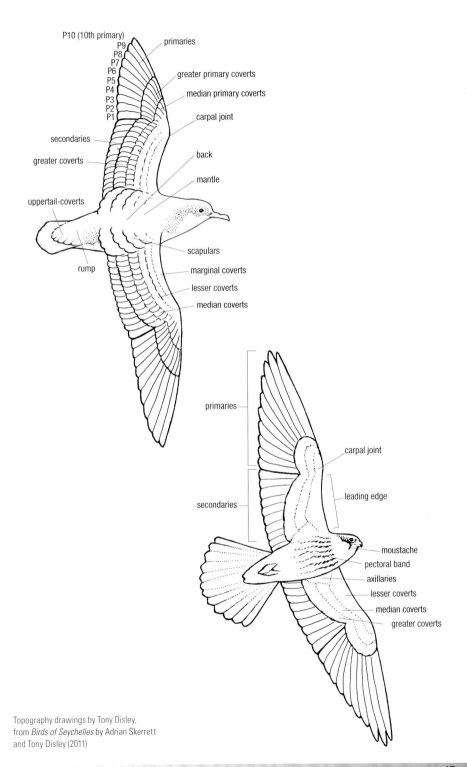

Topography drawings by Tony Disley, from *Birds of Seychelles* by Adrian Skerrett and Tony Disley (2011)

THE MALAGASY REGION AND ITS BIRDS

The Malagasy region contains one of the most extraordinary concentrations of wildlife in the world. It holds the most distinct assemblage of vertebrates, in terms of evolutionary history, after the Australian region, despite being the smallest of all of the world's faunal regions. The hallmark of the flora and fauna of the Malagasy region is not necessarily their diversity (although this is high in some groups of organisms, particularly given the islands' size), but their remarkable levels of endemism, which is marked not only at the species level, but also at higher taxonomic levels; the islands support eight plant families, six bird families (two being also endemic orders) and five primate families that live nowhere else on Earth.

In this section, we describe the region, and attempt to account for its incredible wildlife and in particular birdlife. This is of course far from being the final word on these subjects, but recent advances have led to a level of understanding that did not exist even a few years ago. We do not include a 'where to watch birds' section in the form of a guide to birdwatching sites and localities for the most sought-after species, because nowadays such information is easily found, and kept up to date, on the internet. However, we introduce the main habitats and the birds of most interest to birdwatchers, and this may help the visitor to decide which areas to concentrate on. We strongly encourage observers to submit records to Ebird (www.ebird.org) in order that their records can more easily be integrated into the scientific record. We also encourage observers to send records of rarities seen in Seychelles, with documentation, to the Seychelles Bird Records Committee (www.seychellesbirdrecordscommittee.com).

The region, its islands and its territories

The Malagasy region (Map 1) consists of Madagascar, the archipelagos of Seychelles, the Comoros and the Mascarenes (Mauritius, Reunion and Rodrigues), six more isolated islands or small archipelagos (the Outer Islands), and the sea areas of the accompanying Exclusive Economic Zones (up to 200 nautical miles or 370 km from shore). The region contains four entire countries: Madagascar, Seychelles, Union of the Comoros (Grande Comore, Anjouan and Mohéli) and Mauritius (including Mauritius itself, together with Rodrigues, St Brandon and Agalega, the two latter being part of the Outer Islands). The remaining land is under French administration: Reunion, Mayotte (geographically part of the Comoro archipelago) and the remaining four Outer Islands, known as the Iles Éparses (Europa, Juan de Nova, Iles Glorieuses and Tromelin). The great majority of the human population (88%) lives in Madagascar, although the islands are far more densely settled.

Table 1. Sizes and human populations of islands in the Malagasy region.

Island or island group	Surface area (km²)	Population[a]
Madagascar	587,040	22,005,222
Seychelles	451	92,000
granitic	235	91,500
coralline	216	500
Comoros	2,033	903,645
Grand Comore	1,024	c.346,000
Mohéli	211	c.40,000
Anjouan	424	c.305,000
Mayotte	374	212,645
Mascarenes	4,486	2,129,035
Mauritius	1,865	1,253,000
Reunion	2,512	837,868
Rodrigues	109	38,167
Outer Islands	73	350

[a] Population data from 2011 or 2012, except Seychelles, which is from 2014; see Safford & Hawkins (2013) for sources.

Map 1. The Malagasy region.

Climate: general features

For most of the year (April–October), the Malagasy region's weather is dominated by south-east trade winds (shifting to north-east in the south of the region), pushing ocean currents westwards that split north and south once they approach Madagascar. These steady, warm winds produce mists over high ground but less precipitation in lowland areas, and temperatures of 20–25°C during the day over much of the lowlands. During November–March, very humid, warm easterly winds bring the summer rains. At this season, temperatures in the lowlands increase to 25–30°C over much of the region, with some areas exceeding 35°C during the day, especially inland.

There is a gradient of increased seasonality from north to south through the region, with temperatures in the granitic Seychelles varying only 2–5°C between the coolest and warmest months, but inland, sea level regions of southern Madagascar can have a range of 20°C or more.

Increased ocean temperatures during the summer (rising from 25°C to 28°C) generate cyclones that move from the centre of the Indian Ocean generally north-west, with the trade winds, across the Mascarenes and towards the northern half of Madagascar; the Comoros and Seychelles lie mostly outside the cyclone belt and are rarely hit. These cyclones have produced some of the most severe climatic conditions ever recorded, including winds of over 325 km/h (Cyclone Hudah, Madagascar, 2000) and rainfall of 1.8 m in 24 hours (Reunion, 1966) or 4.9 m in a week (Reunion, 1980). Conditions such as these cause substantial environmental damage through floods or wind-throw, and may have been the cause of bird extinctions through the ages.

Topography influences precipitation and temperature greatly. The mountains of the Comoros, Mascarenes and the east coast of Madagascar receive more rainfall than the lowlands at all seasons, especially in the cooler winter, as clouds form between 500 and 1,500 m on mountain slopes. This cloud layer extends up to about 1,500–2,500 m depending on the day and season, and supports a thick growth of forest in this zone. Above this, the terrain emerges from the cool season cloud and is consequently somewhat drier, often with ericaceous or low grassy and shrubby vegetation.

The summits of the mountains of the region can experience extremes of temperature. The coldest record for Madagascar, on the summit of Andringitra (2,500 m) is −12°C; snow has been recorded on this summit. Sub-zero temperatures have also been recorded on the summits of Tsaratanana (Madagascar), Piton des Neiges (Reunion) and Mt Karthala (Grande Comore).

The rainshadow of these massifs, to their west, is markedly drier. Rainfall can change very significantly over small distances; from east to west of the Andohahela massif, over a distance of little more than 2 km, annual rainfall declines from 2.0 m to about 0.6 m. Rainfall gradients from east to west, and in relation to altitude, on all high islands of significant size in the region may be even more significant.

Vegetation

The flora of islands in the Malagasy region consists of the ancient endemic flora, found in the granitic Seychelles and in Madagascar, a more modern but still remarkably divergent vegetation found nowadays particularly on the mountains of the Mascarenes (and to some extent the Comoros), and the adventitious and agricultural plant communities found in the lowlands of most of the smaller islands of the region. In general, the resident bird fauna occurs mostly in the first two of these categories.

GEOGRAPHY, CLIMATE AND HABITATS

Madagascar

Geography

Madagascar (Maps 2 and 3) is $c.$1,600 km long and a maximum of 580 km wide, covers approximately 587,000 km^2 and is separated from the African continent by the Mozambique Channel, which is $c.$300 km wide at its narrowest point. The high mountains along the east and in the centre are volcanic, some dating from around 88 million years ago, but others are much more recent: Amber Mountain in the far north dates from only a few million years ago. The western and southern plains are mostly of sedimentary origin and more recent. Compared to African and Asian mountains, and even those of Reunion, the mountains of Madagascar are not very high, with few peaks over 2,000 m. The highest is Tsaratanana in the central north at 2,876 m, with (from north to south) Marojejy at 2,173 m, Anjanaharibe Sud at 2,064 m, Ankaratra at 2,643 m, the Ibity–Itremo region at 2,240 m and Andringitra at 2,656 m. Other important montane massifs provide habitat outliers of humid forest into what are otherwise dry areas, including Amber Mountain in the far north (1,475 m), Manongarivo at the western end of the Sambirano region (1,876 m), Andohahela at the far southern end of the eastern rainforests (1,956 m) and Analavelona, north-east of Toliara (1,348 m).

Most rivers in Madagascar run from the high eastern mountains west to the coast, including (from north to south), the Sambirano, Sofia, Mahajamba, Betsiboka, Mahavavy, Manambolo, Tsiribihina, Fiheranana, Onilahy, Linta, Menarandra and Mandrare. The largest lakes are Lake Alaotra (22,000 ha), Lake Kinkony (10,000 ha), Lake Ihotry (6,800 ha) and Lake Itasy (3,500 ha). In the east the coastal Pangalanes saline channel, a lagoon behind a sand barrier linked by man-made canals, extends for over 700 km along the coast. The flatter west is much richer in wetlands, including the lake complex around Bealanana, nestling at the foot of the Tsaratanana Mountains, Lake Kinkony and its satellites in the north-west, Manampape–Mandrozo near Maintirano, the Manambolomaty–Antsalova complex in the central west, Lake Ihotry and satellites, and Lake Tsimanampetsotsa south of Toliara. The southernmost lakes are relics of much larger wetland systems that in drying (since the last warm period, around 4,000–5,000 years ago) achieved high solute content: salt in Ihotry and carbonates and sulphates in Tsimanampetsotsa.

Climate

The central highlands of Madagascar (a plain at about 1,000–1,800 m elevation, mostly 1,300–1,600 m) has a marked dry season between May and November, with temperatures of 11–18°C, followed by a wet season with thunderstorms and occasional cyclones and temperatures of 16–24°C. Further west, temperatures increase with the same basic rhythm of cooler dry winters and warm wet summers; rainfall ranges from 800 to 1,200 mm/year, with more in the north. The cool nights in centre-west Madagascar cause mists and fog to descend during the winter, increasing incidental precipitation at this otherwise very dry season and supporting the growth of fire-resistant forests. Most western forest birds concentrate their breeding season between October and January, when insects are most abundant.

The east experiences in general less seasonal variation in temperature than the west, especially at sea level, with significantly more mist and cloud during the cool season, especially at mid-elevations. Rainfall varies from over 6,000 mm per year on the Masoala Peninsula (north-east) to around 1,500 mm in the far south-east; there is usually a brief dry season in October–November, between the end of the period of mist and cloud, and the start of monsoonal thunderstorms and cyclones in December–January. Many rainforest birds have a very protracted breeding season, from August to March, possibly as a result of this year-round precipitation.

Both the east and west become progressively drier the further you go southwards, with the driest parts of the south receiving less than 300 mm/year.

Vegetation

Madagascar's plant diversity is extraordinary: it is home to an estimated 14,000 plant species, over 90% of which occur nowhere else in the world. Madagascar's endemic flora occurs in a range of habitats, from humid forest to dry spiny bush, and extends to low herbaceous vegetation on rocky substrates and in wetlands. In contrast, the open-country habitats that dominate most of the island contain typically widespread or introduced species.

Most birds are restricted to native forest, and open habitats (except wetlands) possess very few species. The table below shows the relationship between the major habitat types and the typical birds of each one.

Habitat type and best-known or most visited sites (not all species occur at all of them)	Characteristic birds or target species (species in bold type are restricted to this habitat, or nearly so)
Lowland rainforest (0–700 m): Masoala National Park, Andohahela National Park (Parcel 1)	Madagascar Serpent Eagle, Brown Mesite, Red-breasted Coua, Madagascar Red Owl, Short-legged Ground-roller, Scaly Ground-roller, **Dusky Tetraka** [no regular sites currently known], **Red-tailed Newtonia**, **Helmet Vanga**, Bernier's Vanga
Mid-altitude rainforest (700–1,200 m): Andasibe/Périnet and Mantadia National Park, Ranomafana National Park	Madagascar Wood Rail, Blue Coua, Red-fronted Coua, Collared Nightjar, Short-legged Ground-roller, Pitta-like Ground-roller, **Common Sunbird Asity**, Velvet Asity, Green Jery, Spectacled Tetraka, **Wedge-tailed Tetraka**, Rand's Warbler, White-throated Oxylabes, Dark Newtonia, Red-tailed Vanga, Tylas Vanga, Nuthatch Vanga, Crossley's Vanga, Ward's Vanga, Madagascar Starling, Nelicourvi Weaver, Forest Fody
Higher-elevation rainforest (1,200–2,000 m): Marojejy and Ranomafana National Parks, Anjozorobe, Maromizaha (marginal)	**Rufous-headed Ground-roller**, **Yellow-bellied Sunbird Asity**, Forest Rock Thrush, Brown Emu-tail, Grey-crowned Tetraka, Madagascar Yellowbrow, **Cryptic Warbler**
Eastern mid-altitude wetlands: Matsaborimena Lake (Bemanevika), Torotorofotsy, Mantadia National Park, Anjozorobe, Vohiparara (Ranomafana National Park)	Madagascar Grebe, **Meller's Duck**, Madagascar Pochard, Madagascar Rail, **Slender-billed Flufftail**, Madagascar Snipe, **Grey Emu-tail**
Western dry forest: Ampijoroa (Ankarafantsika National Park), Bemaraha National Park, Kirindy, Zombitse–Vohibasia National Park	Banded Kestrel, **White-breasted Mesite**, **Tsingy Wood Rail**, Coquerel's Coua, Crested Coua, Madagascar Pygmy Kingfisher, Madagascar Hoopoe, **Schlegel's Asity**, **Appert's Tetraka**, Long-billed Tetraka, Common Newtonia, Rufous Vanga, **Van Dam's Vanga**, Sickle-billed Vanga, Hook-billed Vanga, White-headed Vanga

Southern spiny forest and (locally) adjoining gallery forest: Ifaty (km 14, Mora-mora), La Table–Sarodrano–St Augustin, Anakao, Tsimanampetsotsa National Park, Cap Ste Marie Special Reserve, Andohahela National Park (Parcel 2), Berenty	Madagascar Cuckoo-hawk, **Subdesert Mesite**, **Long-tailed Ground-roller**, Madagascar Sandgrouse, **Verreaux's Coua**, **Running Coua**, Littoral Rock Thrush, **Thamnornis**, **Archbold's Newtonia**, **Subdesert Brush Warbler**, Sickle-billed Vanga, **Red-shouldered Vanga**, **Lafresnaye's Vanga**
Western wetlands: Ampijoroa (Ankarafantsika National Park), Betsiboka Estuary, Lake Kinkony, Lake Bedo, wetlands around Ifaty, Lake Tsimanampetsotsa	Madagascar Grebe, Madagascar Heron, Madagascar Pond Heron, **Madagascar Sacred Ibis**, **Madagascar Teal**, **Madagascar Fish Eagle**, **Sakalava Rail**, Madagascar Jacana
Highland grassland and savanna: Horombe Plateau, Ambohitantely	Madagascar Harrier (rare), Madagascar Partridge, Madagascar Buttonquail, Madagascar Lark

Approximately one-fifth of Madagascar is covered with primary forest. Of this, just under half is rainforest (i.e. humid forest), western dry forest covers about one-third and the remainder is southern spiny forest. The rest of the island is covered by a mixture of grassland, secondary forest, plantations, urban areas and tree savanna.

Rainforest (humid forest) covers $c.47,000$ km^2, along the length of the eastern scarp (Map 2), with outliers in the centre, north and west, and an extension almost to the west coast in the north-west, sometimes called the Sambirano. Rainforest is found from sea level to almost 2,750 m in the east and in some areas of the central plateau, and is the centre of Malagasy plant species diversity – most of its plant species are found nowhere else. It also has the richest bird fauna in the region, with 193 regularly occurring species, 44 of which are endemic to it, and 16 of these belong to endemic genera. Most members of the endemic families are restricted to rainforest.

Lowland rainforest, Tsitongambarika, SE Madagascar, showing shifting cultivation for rain-fed rice. This is the main cause of habitat destruction in rainforest in Madagascar. The appearance of the forest is similar to (the very few) lowland rainforests elsewhere in the region (Roger Safford).

Mid-altitude and higher-elevation forest, Marojejy National Park, NE Madagascar, showing the wide variation in canopy height with exposure. Again, the appearance of the forest is similar in Madagascar, the Comoros, Seychelles, Mauritius and Reunion (Nick Garbutt).

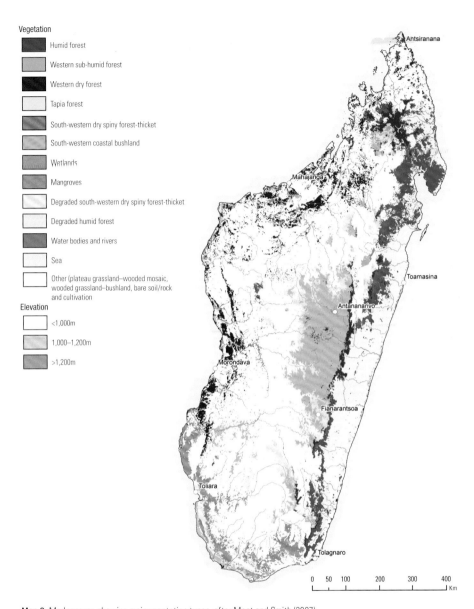

Map 2. Madagascar, showing main vegetation types, after Moat and Smith (2007).

With few exceptions, the primary forest is evergreen, with a closed canopy and multi-layered understorey structure. Canopy height can reach 30–35 m at low to mid-elevations, but declines with increasing elevation; on summits and ridges it can be just 3–4 m.

Lowland rainforest (0–700 m) occurs mostly in the northern half of the east, with outliers around Tolagnaro in the south-east (Andohahela and Tsitongambarika). Mid-altitude forest (mainly 700–1,200 m), the majority of the remaining rainforest, extends along most of the length of the eastern scarp. The canopy is lower (20–25 m), but the forest is as rich as lowland forest in terms of species diversity. This is the richest habitat for birds in Madagascar, as many species occur only at lower and mid-altitudes, some occur only at mid-altitudes and some at mid- to high altitudes, thus overlapping in the middle. The higher-elevation humid forest (mainly 1,200–2,000 m but locally lower

Western deciduous forest, Andranomena Special Reserve, central W Madagascar. This forest type is under high pressure for conversion to agriculture. It is typically very dense, with many small shrubs and trees, but there are many tall species too, including ebonies and rosewood. In the central west, the canopy is often overshadowed by baobabs *Adansonia* spp. (Louise Jasper).

or higher) has a lower canopy and a great abundance of epiphytes (mainly orchids and ferns), with moss on the ground and coating the lower trunks of trees. It occurs only in small areas on the higher parts of the eastern scarp, with notable gaps in the north-east and south-east.

Western dry forest covers approximately 32,000 km^2 and occurs from the northern tip at Cap d'Ambre to about the Mangoky River. It covers altitudes from sea level to *c.*800 m. Plant species diversity is very high and includes many of the famous baobabs of Madagascar, such as Grandidier's Baobab in the Avenue of Baobabs near Morondava. Annual rainfall varies from 600 to 1,500 mm, occurring mainly in summer, with a dry season lasting around six months. The region holds about 172 regularly occurring bird species of which nine are endemic to it.

Southern spiny forest occurs from about the Mangoky River in the south-west throughout the south almost to Tolagnaro in the south-east. It ranges from forest (mostly north of Toliara, including around Ifaty) to impenetrable thicket on permeable limestone plateaux, for example near St Augustin and at Tsimanampetsotsa, to bushland and low scrub

Southern spiny forest, Ranobe, SW Madagascar. The plant community is frequently dominated by members of the Didiereaceae, here on the left of the photo. There are also many members of the Euphorbiaceae, and the understorey is often dense and impenetrable (Louise Jasper).

on sandstone or sand, for instance at Anakao. Plant species diversity is high, with many species being endemic; the flora is dominated by the famous spiny Didiereaceae and *Euphorbia* species. The region holds 159 regularly occurring bird species, so is not species-rich, but ten (and three genera) are endemic to it.

Coastal bushland of the southern spiny forest, near Sarondrano, SW Madagascar. Even more dense and impenetrable than classical spiny forest typically found further inland, this forest type has fewer large Didiereaceae and is dominated by spiny Euphorbiaceae (Louise Jasper).

Wetlands include marshes, lagoons, peat bogs, swamp forests, streams and rivers, and temporary or permanent lakes, which may be fresh, brackish or saline. Marshes, lakes and rivers are found throughout the island, although most abundantly in the west (particularly lakes) and much less commonly in the drier south. Brackish or saline lagoons are frequent along the east coast from Tolagnaro to Fenoarivo, but they are also found in the west. Marshes occur

Eastern wetlands. Shallow-water marshes have relatively low vegetation, dominated by short sedges and grasses, with taller reeds and papyrus-like sedges where water is deeper. This view shows one of the largest areas of this habitat, at Didy in central E Madagascar, although the current condition of the area is uncertain (Roger Safford).

Western wetlands often include shallow pools dominated by water-lilies and abundant underwater vegetation, and are often surrounded by forest with baobabs, as here near Andranomena, central W Madagascar. Many are seasonal, drying out entirely between August and November, while others are deeper and permanent (Frank Hawkins).

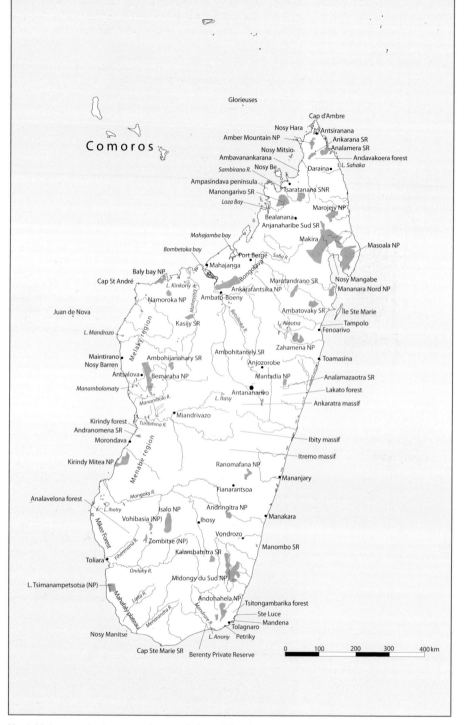

Map 3. Madagascar, showing protected areas and other main locations mentioned in the text.

widely around lakes in the west, and also along narrow valleys or around lakes in the east, mostly at mid-altitudes (1,000–1,500 m). Western wetlands often possess extensive reedbeds, whereas in the east marshes are dominated by papyrus beds and low sedges. Flooded forest with dense stands of *Pandanus* is characteristic of some eastern rainforest areas, for instance in Makira (north-east).

Savanna and grasslands cover much of the central plateau and west of Madagascar. They support few species of plant, are frequently burned to promote regrowth for cattle, and are generally very poor for birds.

The Central Plateau of Madagascar, typified by this scene near Fianarantsoa, often contains villages and agricultural fields in valleys surrounded by bare grassy hills, which may be burned every year. In narrow valleys sheltered from fire, vestiges of primary forest may remain (Frank Hawkins).

Seychelles

Geography

Seychelles consists of four archipelagos and two isolated coral islands scattered across some 1,300 km of the western Indian Ocean, running roughly north-east to south-west, starting 420 km north of Madagascar. The granitic islands (Map 4) occupying the submerged Seychelles Bank in the north are the world's oldest oceanic islands. These are home to most of the human population, with $c.$40 islands of which Mahé is the largest (153 km^2) and highest (Morne Seychellois, 905 m). The sand cays of Bird and Denis are often grouped with the granitic islands due to their proximity and position on the edge of the Seychelles Bank.

The remaining islands of Seychelles are referred to as the coralline Seychelles (Map 5). The Amirantes to the west and south-west of Mahé is a linear archipelago of atolls and coral islands similar to arc systems found elsewhere in the tropics. It lies on a shallow bank, which varies in depth from 11 m over parts of the rim, to 70 m in the centre. No island rises more than 3 m above sea level. Desroches is the largest (394 ha) and the closest to Mahé (220 km south-west). The Alphonse group (Alphonse Atoll and St François Atoll) is included as part of the Amirantes, although separated from the other islands by a deep trench. About 350 km further south-west again is the Farquhar group, consisting of Farquhar Atoll (ten islands), Providence Atoll (two islands) and St Pierre (a single raised coral island).

At the far south-western corner of Seychelles is the Aldabra group, a complex of raised atolls and islands. Aldabra atoll comprises 46 islands, the largest of which, Grande Terre, covers most of its 15,380 ha. The rest of the group comprises Assumption (1,171 ha), Astove (661 ha) and Cosmoledo (19 islands, 460 ha). Aldabra is the world's largest raised coral atoll, with a highest point of 10 m, and among many other features is famous for its large population of endemic giant tortoises *Aldabrachelys gigantea*. Of the other coralline islands, only St Pierre reaches more than 3 m above the sea. Finally, two other coralline islands, Platte (54 ha) and Coëtivy (931 ha), are isolated from each other and from all other islands; they lie $c.$150 km and 300 km, respectively, south of the main granitics.

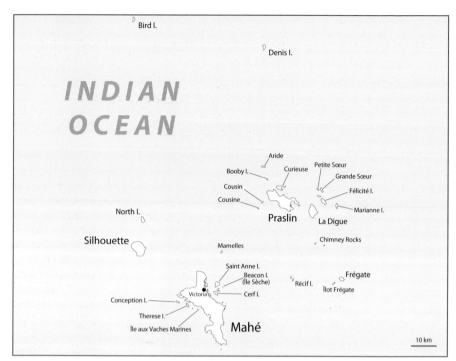

Map 4. The granitic Seychelles, including the coralline islands of Bird and Denis.

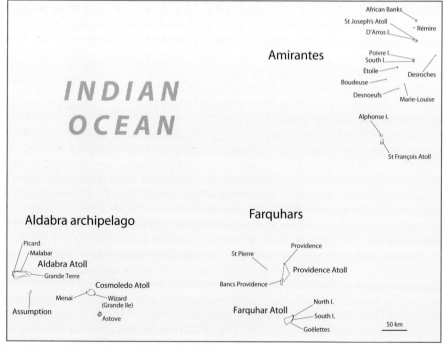

Map 5. The coralline Seychelles, excluding Platte and Coëtivy.

Climate

Seychelles has the least variable climate of the region, with low annual and diurnal variation. The south-east monsoon, in May–October in the granitics and a month or so later in Aldabra and Farquhar, brings relatively dry weather, especially in the south. Temperatures of 25–30°C are typical at this time of year, with humidity around 80%. Between November and April (or somewhat later in Aldabra and Farquhar), winds are variable, although usually from the north-west, and temperatures on average about 3°C higher, along with frequent rainstorms. Annual rainfall is about 2,400 mm in the granitics, 1,500 mm in the Amirantes and just 1,000 mm at Aldabra. The period between the two monsoons may be calm with little rainfall. Seychelles lies mostly outside the cyclone belt, but exceptionally Aldabra and Farquhar may be affected.

Vegetation

Seychelles possesses elements of a much more ancient flora than is found on any other remote oceanic islands. The granitic islands have c.850 species of flowering plants and ferns, with a native flora of 222 species, including 69 endemic species and one endemic family, the Medusagynaceae. Very little primary forest remains, but at the same time practically all the original assemblage of flora survives, making it possible to construct a reasonably good picture of the primeval vegetation. The affinities of the native flora of the granitics are mainly Asian and Mascarene, with a much smaller African component. By contrast, the flora of the low coralline islands are widespread pantropical or Indo-Pacific species, while in the Aldabra group there are strong affinities to Madagascar alongside the typical coralline assemblage of plants. The flora of the coralline islands is around 250 species with no endemics, other than in the Aldabra group which has 26 endemic taxa including 22 species showing strong affinities to Madagascar.

Habitat type and best sites	Characteristic birds (not present in this habitat type on all islands)
Moist forest, above 400 m: Morne Seychellois National Park, Silhouette National Park	Seychelles Scops Owl, Seychelles Bulbul
Dry forest 100–400 m: lower slopes of Mahé, Silhouette, plus Praslin, La Digue and other islands larger than 2 km^2	Seychelles Kestrel, Seychelles Bulbul, Seychelles Swiftlet, Seychelles Blue Pigeon, Seychelles Paradise Flycatcher (La Digue), Seychelles White-eye
Lowland and coastal vegetation	Seychelles Swiftlet, Seychelles Blue Pigeon, Yellow Bittern
Mangroves: some granitics, Aldabra group	Striated Heron, frigatebirds, Red-footed Booby
Low granite islands: islands less than 200 ha	Seychelles Warbler, Seychelles Magpie-robin, Seychelles Fody, Seychelles White-eye, breeding seabirds (up to ten species), roosting frigatebirds
Coral islands and atolls	Crab-plover, other migrant waders, Saunders's Tern, Aldabra Drongo, Aldabra Fody, endemic races, notably of White-throated Rail, Souimanga Sunbird, Madagascar White-eye, Madagascar Sacred Ibis and Madagascar Turtle Dove, seabird colonies including both species of frigatebird and all three boobies, tropicbirds, terns.

Moist forest This includes mountain moss forest and intermediate forest, the main characteristic of the habitat being abundant rainfall. Mountain moss forest is confined to the two highest islands, Mahé and Silhouette, which rise to 905 m and 740 m, respectively (by contrast the next highest islands of Praslin and La Digue reach 350 m and 333 m, respectively).

Mountain-tops are flanked by steep slopes and sheltered ravines where peat soil has accumulated, hosting a range of endemic plants. On moderate slopes and valley sides, deeper soils hosted higher forest, now mostly replaced by cinnamon and other introduced trees; however, amongst these exotics, the native species are still present. This habitat is characterised by a surprising paucity of birds and a remarkable silence save for the occasional sound of endemic tree frogs (family Sooglossidae) and the cackle of Seychelles Bulbul.

The forests of the granitic Seychelles, seen here at Vallée de Mai, Praslin, contain many of the remaining endemic species of plants, including several striking palm species, and some of the birds (Roger Safford).

Dry forest This habitat is characterised by less rainfall than moist forest habitat and includes the lower slopes of Mahé and Silhouette together with most of the landmass of other larger granitic islands, including Praslin, La Digue, Curieuse, Marianne, Félicité and North. On summits the rocky *Glacis* habitat of drier areas has less peat and a distinct vegetation. Some slopes retain native vegetation, including Vallée de Mai and Fond Ferdinand on Praslin, dominated by six endemic palms and several endemic *Pandanus*. Palms include the legendary Coco de Mer *Lodoicea maldivica*, restricted to Praslin and Curieuse. There are many invasive species but at least at Vallée de Mai and to a lesser extent elsewhere these are gradually being removed or controlled. La Réserve on Mahé and parts of Silhouette are similar, harbouring the same endemic palms other than Coco de Mer. In heavily degraded areas, endemic species may survive, including some which would normally be regarded as forest understorey plants, but Coco de Mer can survive in remarkably impoverished conditions.

Lowland and coastal vegetation Well-drained, easily worked soils and flat land meant these areas were the first to be cleared for cultivation. In some places there is regrowth of secondary forest, but for the most part tremendous human pressure remains. Today, the original lowland forest has largely gone, although vestiges remain on Silhouette, Félicité and parts of southern Mahé.

Cultivation Agriculture for arable crops commenced soon after human settlement in 1770 and caused a decline in soil fertility, opening up the large-scale development of coconut plantations. In addition, pineapples, cinnamon, bamboo and other crops replaced native vegetation. Coconuts dominated Seychelles agriculture until the 1970s, interspersed by periods of prominence for vanilla and cinnamon cultivation. Vanilla was economically important in the late 19th century and, as its cultivation requires shade, some tree canopy was encouraged. Cinnamon has the opposite requirement, with the consequence that slow-growing shade-bearing trees suffered widespread clearance.

Today, tourism has almost totally replaced agriculture as a foreign exchange earner and the great majority of what little cultivation survives in the 21st century serves the needs of the local population and tourism industry. This is now largely restricted to the coastal plateau, except for tea cultivation within Morne Seychellois National Park.

Mangroves In the granitic Seychelles, extensive land reclamation has destroyed most mangrove habitat, but pockets survive, mainly on the north-west coast of Mahé and on parts of Curieuse and Silhouette, and there are also significant areas on Aldabra, Cosmoledo and Astove.

Littoral zone Except the endemic Balfour's Pandanus *Pandanus balfouri*, all coastal natives are widespread Indo-Pacific or pantropical species, characteristically adapted for long-distance dispersal. The vegetation has been re-shaped by humans and the introduction of exotics mainly via Mauritius has caused the number of coastal plant species to double to more than 100, but native plant species have proved remarkably resilient.

Low granite islands On smaller granitic islands such as Aride, Cousine, Cousin, Frégate and Récif the native vegetation is dominated by widespread salt-resistant species. Endemics are fewer than at high altitudes. Badamier

The granitic islands of Seychelles include small islands such as Aride, seen here; all are fragments left over from breakup of Gondwana. They are typically steep and hilly, and covered in dense vegetation, rich in endemic species (C. Bell).

and Takamaka woodland is characteristic and there is a high density of *Pisonia* on rat-free seabird islands. Most of the low granitic islands are rat free, some having never hosted rats, others following successful eradication programmes. As a result, seabirds are common, breeding on Aride, Cousin and Cousine in numbers large enough to justify the status of these islands as Important Bird Areas; rare endemic birds and other land fauna also survive.

Coralline islands and atolls Native plants on the low islands of the Amirantes and Farquhar group are almost entirely pantropical or Indo-Pacific. A large proportion of species are introduced, and abandoned cultivated plantations, especially coconuts but also cotton and sisal, dominate large areas. There are no landbirds other than introduced species but some six sites qualify as Important Bird Areas as they host important seabird colonies.

The older, raised islands of the Aldabra group and St Pierre (Farquhar group) have a much richer flora, Aldabra remarkably so for an atoll. Aldabra has the most extensive intact mangrove stands in Seychelles and these are also important components of the flora of Cosmoledo and Astove. The flora of all but Aldabra is highly modified; St Pierre is the most disturbed, the native vegetation having been almost completely removed to permit guano extraction.

In contrast to the granitic islands, the coralline Seychelles – here St François – are mostly very low, just a few metres above sea level, and with pantropical island species dominating the vegetation (A. Duhec).

Comoros

Geography

The Comoros (Map 6) form an archipelago of volcanic origin in the Mozambique Channel, between Africa and Madagascar and *c.*300 km from either, at 11°23'–13°S and 43°13'–45°18'E. There are four main islands, from west to east: Grande Comore (1,024 km²), Mohéli (211 km²), Anjouan (424 km²) and Mayotte (374 km²). Nearest-neighbour distances, excluding islets, are between 39 and 67 km. The islands have never been connected to each other or to any other landmass. The archipelago has been permanently inhabited since the around the ninth century, and now supports an extremely dense human population, especially on Anjouan.

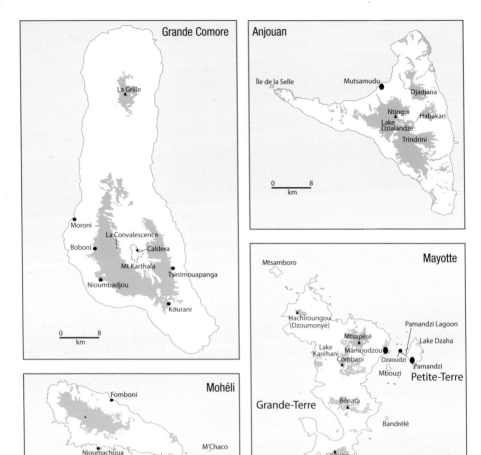

Map 6. The Comoros, showing forest and main localities mentioned in the text. Forest extent after Institut Géographique National (1993, 1995); this source appears to include secondary, degraded and exotic forest as well as primary forest (see page 34). Montane heath and bushland occurs between the caldera and forest on Mt Karthala, Grande Comore, but is not shown as its extent, especially northern limit, has not been mapped.

Grande Comore includes the archipelago's highest point and only active volcano, Mt Karthala (2,361 m); this erupted many times in the 20th century and has continued to do so since 2000. Mt Karthala is linked to a lower (1,087 m) extinct volcano, La Grille, by a saddle at c.500 m. Apart from Mt Karthala crater, relief on Grande Comore is generally smooth, with no valleys or (apart from one crater lake) surface water, as rainfall percolates into the porous rock. The older islands of Mohéli, Anjouan and Mayotte are highly eroded, with sharp ridges, precipitous slopes and numerous valleys, especially on Anjouan. The central ridge of Mohéli reaches 790 m; Anjouan and Mayotte have more complex topography, reaching 1,595 m and 660 m, respectively. Mohéli, Anjouan and Mayotte have only very short watercourses or small lakes, mostly craters, notably Dzialandze on Anjouan and Boundouni on Mohéli; Mayotte has Dzaha on Petite-Terre, with Karéhani on Grande-Terre, the latter including an important marshy area. There are very small mudflats around Grande Comore, and larger ones around the other islands, particularly Mayotte, which includes the remarkable and often bird-rich tidal wetland referred to here as Pamandzi lagoon (sometimes also Vasière des Badamiers) on Petite-Terre. To the south and east of Mohéli lie 11 islets, while Mayotte has about 20 (excluding Petite-Terre) on or inside its large fringing reef.

Climate

The climate of the Comoros is moist tropical, with mean temperatures at sea level of 24–28°C. November to April is the hottest period, with maxima of *c*.35°C, and this is also when much of the mean average rainfall of around 2,000 mm (1,000–5,600 mm) falls. The northern and eastern flanks of Mt Karthala, the volcano on Grande Comore, receive the highest rainfall in the archipelago; the driest parts are the northern lowlands of Grand Comore and Mayotte. The Comoros frequently experience tropical depressions, and occasionally more severe storms or cyclones.

Vegetation

Forest is believed to have covered most of the islands before the arrival of man. However, most has been lost (Map 6), and what remains is under considerable pressure for extraction and cultivation, even at high altitude on Mt Karthala. Originally, four main terrestrial types existed; coastal vegetation, poorly known but apparently with mixed trees and grass; mid-altitude humid forest; high-altitude forest (cloud forest); and montane tree-heath (ericaceous forest). Native forest cover (other than mangrove; see wetlands below) is now limited to the highest parts of all four islands; areas below 500 m have been almost entirely converted, and above this limit the remaining forest has been considerably but variably modified by agriculture. In most lowland areas, Coconut Palms *Cocos nucifera*, Tamarind *Tamarindus indicus*, Eucalyptus, Teak *Tectona grandis* and Ylang-ylang *Cananga odorata* form the dominant tree cover, with Strawberry Guava *Psidium cattleianum* abundant on the lower slopes, accompanied by a wide selection of other common pantropical or exotic tree and shrub species such as Lantana *Lantana camara*, Rose-apple *Syzygium jambos* and *Clidemia hirta*. Drier areas in the rainshadows of mountains may support African baobabs *Adansonia digitata* and other pantropical dryland species, especially grasses.

Habitat type and best-known or most visited sites	Characteristic birds (species in bold type are largely restricted to this habitat; *Comoro-endemic race; #island-endemic race)
Grande Comore high-altitude forest (cloud forest) and montane heath and bushland (Mt Karthala)	Madagascar Harrier, Greater Vasa Parrot*, Lesser Vasa Parrot*, **Karthala Scops Owl**, Madagascar Spinetail#, **Grande Comore Bulbul**, Madagascar Cuckoo-shrike*, Grande Comore Brush Warbler, **Humblot's Flycatcher**, **Karthala White-eye**, African Stonechat*, Comoro Thrush#, Madagascar Green Sunbird#, Humblot's Sunbird#
Grande Comore mid-altitude (montane) humid forest (Mt Karthala)	Frances's Sparrowhawk#, Madagascar Harrier, **Grande Comore Drongo**, Greater Vasa Parrot*, Lesser Vasa Parrot*, Cuckoo-roller#, Madagascar Cuckoo-shrike#, Grande Comore Brush Warbler, Comoro Thrush#, Madagascar Green Sunbird#, Humblot's Sunbird#
Grande Comore lowlands	Frances's Sparrowhawk#, Madagascar Bulbul, Madagascar Green Sunbird#, Humblot's Sunbird#
Mohéli humid forest and lowlands	Madagascar Harrier, Madagascar Green Pigeon#, Greater Vasa Parrot, **Moheli Scops Owl**, Cuckoo-roller, **Moheli Bulbul**, **Moheli Brush Warbler**, Madagascar Brush Warbler#, Blue Vanga#, Comoro Thrush#, Madagascar Green Sunbird#, Humblot's Sunbird#
Anjouan humid forest and lowlands	Frances's Sparrowhawk#, Madagascar Harrier, Greater Vasa Parrot*, Lesser Vasa Parrot*, **Anjouan Scops Owl**, Cuckoo-roller#, Madagascar Brush Warbler#, Comoro Thrush#, Crested Drongo#, **Anjouan Sunbird**
Mayotte humid forest and lowlands	Madagascar Pond Heron, Frances's Sparrowhawk#, **Mayotte Scops Owl**, Cuckoo-roller, **Mayotte Sunbird**, **Mayotte Drongo**
Most forest types, all islands on Comoros	Comoro Olive Pigeon, Madagascar Turtle Dove*, Comoro Blue Pigeon, African Palm Swift*, African Black Swift*, Madagascar Paradise Flycatcher# (four island-endemic races), Madagascar White-eye# (four island-endemic races), Comoro Fody# (four island-endemic races, rare on Mayotte)
Coastal wetlands, all islands on Comoros	Crab-plover, Madagascar Malachite Kingfisher*, waders, terns

The summit of Mt Karthala, on Grande Comore, is a patchwork of mainly heath-type shrubs and trees, low herbaceous vegetation, areas of recent lava flow and bare ground. The summit zones of Reunion are often similar (Frank Hawkins).

On Mt Karthala, the most extensive forested area, four major habitats have been identified. Plantations at 400–1,000 m are of bananas, coffee, mango *Mangifera indica* and guava *Psidium* sp., with a few native species dominating the canopy; open areas are dominated by introduced grasses. Montane forest, often with underplanted bananas, at 800–1,200 m, is characterised by a rich array of native tree species, with tree-ferns under the canopy. Cloud-forest at 1,200–1,800 m has many of the same tree species but more mosses and ferns. Around the crater, at 1,800–2,300 m, montane heath and bushland is dominated by the striking tree-heath *Philippia comorensis*. No up-to-date forest coverage survey is available, and the areas mapped (Map 6) are based on long outdated information and so are undoubtedly over-optimistic. The plantations and montane forests are particularly patchy, with many gaps caused by cultivation and wood extraction. Some areas of cloud-forest are more intact; in 2006, this type was estimated to cover about 6,300 ha.

On Grande Comore, the majority of the endemic birds occur in the intact and underplanted cloud-forest and tree-heath of Mt Karthala; a few, such as Grande Comore Brush Warbler, also occur on La Grille. Several other endemic birds on Grand Comore are found mainly at lower altitudes, some in degraded habitats.

The uppermost 100 m of the central ridge of Mohéli is covered by a narrow band of primary humid forest (around 11.5 km^2 in 1996) mixed with underplanted and secondary forest. This is home to all of the endemic bird species and races, although some also occur outside the forest, down to sea level. A similar pattern of vegetation is found on Anjouan, with some of the more invasive introduced species (e.g. Strawberry Guava) absent from the lowlands, and relict patches of rainforest especially in the south and east (much underplanted and cleared in all but the steepest areas) mainly at 1,200–1,500 m; the summit of Ntingui, Anjouan, is covered in very low vegetation. In the 1990s, probably less than 20 km^2 of primary forest survived, and forest loss has continued since.

Most of the forest of Mayotte was destroyed before 1900, leaving an area of *c.*11.5 km^2 (in 1999) covered in primary forest, mostly humid, with small areas of drier forest especially in the south. A further 146 km^2 of 'secondary' forest surrounds the primary forest, mostly on hilltops above 300 m. The rest of the island is covered in plantations and urban areas, the former consisting mostly of Ylang-ylang, Coconut, Mango and Sugarcane. Most of the endemic bird species occur in secondary formations as well as primary.

Most freshwater lakes on the Comoros are of open water with fringing herbaceous vegetation, although Lake Karéhani, Mayotte, has water lilies *Nymphaea* sp. Small areas of mangroves occur on all islands, mostly Mayotte (with 750 ha).

Mascarenes (Mauritius, Reunion, Rodrigues)

Geography

The Mascarenes consist of three strikingly different islands – from west to east, Reunion, Mauritius and Rodrigues – which owe their origins to volcanic activity of varying ages but apparently independent causes. Mauritius (Map 7) has mostly mild relief, with low plains in the north and east rising to a central plateau reaching $c.$700 m in the south-west. The terrain is broken by small (no higher than 824 m) but spectacular mountains, and by the Black River Gorge system in the south-west. The 200 km of coastline is almost completely encircled by a fringing reef. On the same submarine platform lie ten low, coralline or raised volcanic islets, together with many smaller rocks. Mauritian satellites include the extraordinary northern islets of Round I and Serpent I, which harbour some of the rarest species of animal and plant in the world. Some other, more degraded, islands have been partially ecologically restored, notably Ile aux Aigrettes off the south-east coast.

Geologically recent volcanic activity on Reunion (Map 8) has given the island the highest peak in the Malagasy region, Piton des Neiges (now extinct as a volcano) at 3,069 m. Piton de la Fournaise (2,631 m) is still active. Eruptions and lava-flows from these volcanoes have given the island a cone-like shape, with 60% of it above 1,000 m. The seabed around Reunion has a very steep drop-off close to the coast and the island therefore lacks significant coral reefs, mangroves or satellites, except the tiny Petite Ile off the south coast. Rodrigues (Map 9), while superficially composed of relatively recent lava-flows, is the heavily eroded and ancient basaltic remnant of a volcano. This is reflected in its low profile, with a highest point just 393 m. Some of the valleys that radiate from this summit are steep-sided, but mostly the island undulates gently down to an extensive coastal plain that is surrounded by a wide lagoon with a large fringing reef, and around 20 sandy, coralline or raised volcanic islets, several of outstanding actual or potential ecological interest.

Wetlands are few. Mauritius and Reunion both hold numerous rivers and streams, and Mauritius also has scattered small ponds, crater lakes and reservoirs, some with limited fringing vegetation, but no major marshes; a drained marsh, the Mare aux Songes on Mauritius, has proven a rich source of subfossil material but is not a significant wildlife habitat today. The Mascarenes hold various coastal wetlands, notably the Étangs de St-Paul and du Gol on Reunion, and Terre Rouge estuary on Mauritius. Small areas of mangroves are found on Mauritius and Rodrigues.

Climate

The Mascarenes have a tropical seasonal climate, at sea level with a relatively cool (22–24°C) season from May to September, becoming somewhat drier and warmer in October and November, and hot (23–28°C) and wet from December to April, during which time tropical depressions and cyclones may bring intense wind and rainfall. Rodrigues is the driest island (annual rainfall $c.$1,000–1,700 mm) but is severely influenced by cyclones. Mauritius and Reunion both receive around 1,000–1,800 mm in the lowlands but 3,000–5,000 mm in the uplands, although Reunion locally receives very high totals of up to 10,000 mm. Temperatures in the uplands of Mauritius and mid-altitudes of Reunion are typically 16–20°C, but regularly fall to below zero on the summits of Reunion.

Vegetation

The current native vegetation in the Mascarenes is a sad relict of the past. Only on the uplands of Reunion are conditions remotely similar to those found before the arrival of European sailors, the first of whom (the Portuguese navigator Tristão da Cunha) arrived in 1507. On all islands, lowland areas have been subject to a much higher impact than high-altitude areas; even on Reunion, no more than 1% of lowland forest remains.

Serious colonisation of the islands did not start until the Dutch established a staging post on Mauritius in 1606, but from then until 1638 there was no stable, resident population. The initial Dutch attempts at a colony were abandoned in 1710, by which time the rapacious consumption of tortoises, dodos and several other species of terrestrial vertebrate had resulted in their extinction. Forest clearance was limited but accelerated rapidly under French (1722–1810) and then British (1810–1968) rule; agriculture, especially sugarcane and for a time also tea, was expanded and intensified. Further clearance took place in the 1970s and even 1980s, such that by the 1990s less than 5% of native vegetation remained, with around 1–2% in a reasonable condition, none of it pristine (Map 7).

On Reunion, colonisation started somewhat later, by the French in 1665, and their efforts were sustained (albeit with different political orientations and leadership) until the present day. As on Mauritius, colonial activity focused on agriculture, primarily for export, and the supply of meat and produce to ships. This meant the establishment of plantations, mostly at the expense of native vegetation, yet 25% of the estimated original extent of Reunion's habitats are still in a good state (Map 8).

On Rodrigues, the first 17th-century visitors described the native ecosystems and many wondrous species, but settlement, with attendant habitat destruction, from the 18th century accounted for all of the remaining native forest by the mid-20th century (Map 9); working from the scattered remaining native plants, restoration is beginning to show results.

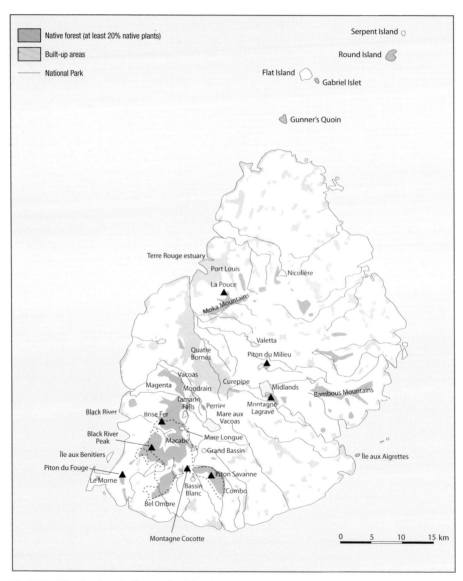

Map 7. Mauritius, showing native forest and main locations mentioned in the text including the approximate boundary of the Black River Gorges National Park, after Cheke and Hume (2008).

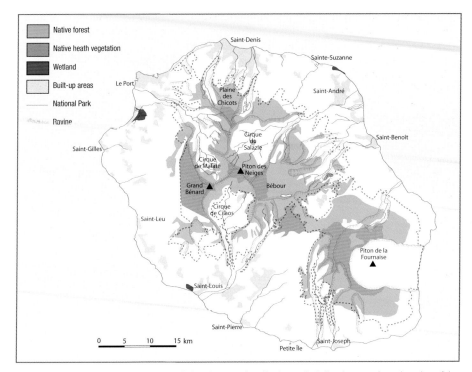

Map 8. Reunion, showing native forest and main locations mentioned in the text including the approximate boundary of the Reunion National Park, after Cheke and Hume (2008).

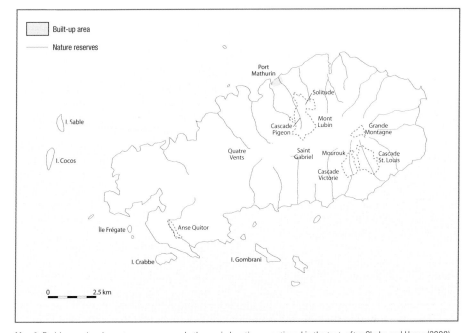

Map 9. Rodrigues, showing nature reserves and other main locations mentioned in the text, after Cheke and Hume (2008).

Montagne Lion and its surroundings in SE Mauritius, showing sugarcane cultivation on lowlands and lower slopes, with forest, often heavily infested with exotics, confined to mountain slopes. In the background is the fringing reef; the oval-shaped island in the lagoon is the restored Ile aux Aigrettes, home to many rare native species including birds (Roger Safford).

Five natural plant formations have been identified on the Mascarenes, arranged across broad moisture and altitudinal zones. Dry lowland forests dominated by native palms, screw-pines *Pandanus* spp. and other trees were present from sea level to 200 m elevation in areas with less than 1,000 mm annual rainfall; this habitat is virtually extinct, along with some of the region's most spectacular endemic animals that lived in it; relicts occur on Round I (Mauritius) and in a few places on Reunion. Semi-dry sclerophyllous forests occurred between sea level and 360 m on all sides of Mauritius and Rodrigues, but were restricted to 200–750 m on the western slopes of Reunion, where remnants survive. This ecosystem has an average annual rainfall of 1,000–1,500 mm and is characterised by a mixture of species including many ebonies. On Rodrigues, this forest seems to have had a rather open understorey, in the past apparently maintained by the bewildering densities of tortoises, all now extinct. The remnants (on all islands) of these two forest types are now rather depauperate in native bird species, although particularly important for Mauritius Kestrel; only the two Rodrigues endemics are confined to them. However, many of the extinct species throughout the Mascarenes would have occurred here.

Lowland rainforests occur on Mauritius above 360 m (where they have also been referred to as 'upland forests', as distinct from the semi-dry sclerophyllous forests which are often called 'lowland forests') and on Reunion throughout the eastern lowlands from the coast to 800–900 m and, on the western side, from 750 to 1,100 m, where average annual rainfall is 1,500–6,000 mm. These forests have a canopy up to 30 m high and contain the richest plant communities of the Mascarenes, with a particular abundance of canopy species in the family Sapotaceae, and numerous orchids and ferns. This is generally the habitat richest in native birds on Mauritius; all the endemic landbirds are present. A good range of endemic landbirds occurs in this habitat on Reunion, although none is confined to it.

The remaining montane forests are naturally virtually confined to the higher island of Reunion, with a tiny area on Mauritius. Dense cloud-forests on Reunion are still extensive where rainfall exceeds (often considerably) 2,000 mm, on eastern slopes at 800–1,900 m and on western slopes at 1,100–2,000 m; an area exists on Mauritius, above 750 m on Montagne Cocotte where rainfall is 4,500–5,500 mm. These forests have a 6–10 m canopy and are rich in epiphytes (orchids, ferns, mosses and lichens). Reunion Cuckoo-shrike is confined to a small area of this habitat in the north of the island, but most other native Reunion birds occur more widely; the relict patch on Mauritius is also rich in native birds, most of which can be seen there. Reunion's montane forests also include large areas of three plant communities each dominated by single species: Tamarin des Hauts *Acacia heterophylla* forest (remarkably like *Acacia koa* forests in Hawaii), heath with *Erica reunionensis* (reminiscent of Mt Karthala, Grande Comore), or hyperhumid screw-pine forest with *Pandanus montanus*.

Above the treeline on Reunion alone (at 1,800–2,000 m), where winter frosts are regular, is a unique subalpine scrub, with some notable endemic species. The summits of the volcanoes are covered by large mineral areas with endemic grasses and orchids, and heath-like thickets. A few areas on Mauritius also support a somewhat similar heath-like vegetation. All of these areas are frequented by white-eyes, with two species (grey and olive) on each of Mauritius and Reunion.

A general feature of the surviving vegetation of the Mascarenes is infestation by invasive alien plant species. The degradation takes the form of a gradual shift in floristic composition towards exotic species, but the mechanisms responsible are not fully understood; clearly, however, this presents a chronic problem for native, forest-dependent wildlife. The most important species include Privet *Ligustrum robustum*, Strawberry Guava *Psidium cattleianum* (known locally by the misleading name 'Chinese Guava' or *Goyave de Chine*), Rose-apple (or *Jamrosa*) *Syzygium jambos*, Traveller's Tree *Ravenala madagascariensis* and many others.

Native birds are generally associated with native forests, although the Mauritius and Reunion Grey White-eyes are more or less ubiquitous on both islands, while the Mascarene Swiftlet and Martin are also widespread and apparently not dependent on native vegetation. Despite their limited extent, the wetlands of the Mascarenes held a surprising diversity of endemic waterbirds, but these are now almost entirely extinct, leaving only the non-endemic Common Moorhen and Striated Heron, along with migratory waders and exotic ducks.

Habitat type and best-known or most visited sites	Characteristic birds (species in bold type are largely restricted to this habitat; *island-endemic race)
Mauritius: rainforests of south-west (Macchabée-Brise Fer, Plaine Champagne, Bel Ombre, Savanne Mountains, Bassin Blanc)	White-tailed Tropicbird, **Pink Pigeon**, **Echo Parakeet**, Mauritius Bulbul, **Mauritius Cuckoo-shrike**, Mascarene Paradise Flycatcher*, **Mauritius Olive White-eye**, Mauritius Grey White-eye, **Mauritius Fody**
Mauritius semi-dry sclerophyllous ('lowland') forests (Black River Gorges)	**Mauritius Kestrel**, Mauritius Bulbul, Mauritius Cuckoo-shrike, Mauritius Grey White-eye
Mauritius northern islets (Round I, Serpent I; landing usually not permitted)	Wedge-tailed Shearwater, **Trindade, Kermadec and Herald Petrels** (only colony in the Indian Ocean), vagrant petrels, Masked Booby, White-tailed Tropicbird, Red-tailed Tropicbird, Sooty Tern, Lesser Noddy, Brown Noddy
Mauritius: Ile aux Aigrettes	Pink Pigeon, Mauritius Olive White-eye, Mauritius Fody
Mauritius: Plaine des Roches (Bras d'Eau)	Mascarene Paradise Flycatcher*
Mauritius: Terre Rouge Estuary	migrant waders and terns
Reunion: upland forests and subalpine scrub (especially Plaine des Chicots)	White-tailed Tropicbird, **Reunion Harrier**, **Reunion Cuckoo-shrike**, **Reunion Bulbul**, **Reunion Stonechat**, Mascarene Paradise Flycatcher*, **Reunion Olive White-eye**, Reunion Grey White-eye
Reunion lowlands	Barau's Petrel (crossing north and west coasts in evening), Reunion Grey White-eye
Mauritius and Reunion (other habitats)	All habitats: Mascarene Swiftlet, Mascarene Martin; wetlands: Striated Heron, Common Moorhen
Rodrigues mainland	White-tailed Tropicbird, Red-tailed Tropicbird, **Rodrigues Warbler**, **Rodrigues Fody**
Rodrigues islets	Roseate Tern, Sooty Tern, Lesser Noddy, Brown Noddy, Fairy Tern, White-tailed Tropicbird
At sea around Mauritius and Reunion	As for Mauritius northern islets; also Subantarctic Skua, Tropical Shearwater, Flesh-footed Shearwater, Barau's Petrel, Mascarene Petrel (Reunion), other seabirds (petrels, storm-petrels etc.)

Outer Islands

The Malagasy Region is rich in small, remote islands, many with great importance for wildlife, especially breeding seabirds. Many are included in Seychelles, but six others, in many ways similar but some far away, complete the picture. In the Mozambique Channel are, from north to south, the Iles Glorieuses, Juan de Nova and Europa, and to the east of Madagascar are Agalega, Tromelin and St Brandon (also known as the Cargados Carajos Shoals). The eastern Outer Islands form part of an otherwise submerged bank extending over at least 50,000 km^2 between Mauritius and Seychelles. Few visitors, perhaps apart from those on cruise ships, are likely to visit these remote outposts, and they represent a particularly poorly known part of the region, sometimes ignored entirely in regional works, despite their often spectacular wildlife and geography. The relatively ecologically intact Europa is magnificent, while the potential for ecological restoration of some others is immense.

On the Glorieuses (two islands: Grande Glorieuse and Ile du Lys), seabird colonies include Sooty Terns and Common Noddies. On Juan de Nova, a long history of human impact has impoverished the fauna and flora but a colony of Sooty Terns has been estimated to number *c.*2 million pairs, along with a small colony of Greater Crested Terns. On Agalega, small numbers of seabirds now occur but more remarkable is the only population of Glossy Ibises in the region outside Madagascar. Tromelin has breeding Masked and Red-footed Boobies. St Brandon – 16 small coral islands and hundreds of tiny sandy islets extending over 60 km or 1,000 km^2 of ocean – has both species of frigatebirds and of noddies, with boobies as on Tromelin, along with terns, tropicbirds and Wedge-tailed Shearwaters. Landbirds on the Outer Islands are few, and mostly exotic; various vagrants have also been recorded and more frequent visits will surely reveal many others.

Europa, the largest of the Outer Islands (2,223 ha) has several vegetation types and has suffered relatively limited human impacts over the last 40 years at least. Accordingly, seabird populations are impressive, including a large colony of Sooty Terns, an endemic race of White-tailed Tropicbird (*Phaethon lepturus europae*), the region's largest population of Red-tailed Tropicbirds, and its second-largest populations (after Aldabra) of Red-footed Booby and Great and Lesser Frigatebirds; also noteworthy are a small population of Madagascar Pond Herons and the only landbird taxon endemic to the Outer Islands: Madagascar White-eye *Zosterops maderaspatanus voeltzkowi*.

These six islands or island groups are whenever possible referred to by their individual names, but collectively as the Outer Islands, although this term may, in strictly Seychelles and Mauritian contexts, be used to refer to the coralline Seychelles, or to Agalega and St Brandon, respectively.

BIOGEOGRAPHY: COLONISATION, DIFFERENTIATION AND SURVIVAL*

A species-poor and endemic-rich avifauna

The Malagasy region avifauna is striking in that, when compared to other island regions of similar size, it is poor in species (382 regularly occurring) but rich in endemics (41% of those species; of course both figures are liable to increase when species are 'split'). Madagascar itself is the main cause of both of these anomalies; richness and endemism for the islands surrounding Madagascar are more typical of those of oceanic islands worldwide. The following comparisons refer to totals including seabirds, vagrants and known extinct species, with figures outside the Malagasy region taken from the Avibase website (Bird Studies Canada 2012). Madagascar (587,000 km^2) has lower species diversity (290 species including seabirds and vagrants) than several smaller tropical islands, including Sumatra (710 species; 473,000 km^2), Java (510 species; 128,000 km^2) and Sulawesi (473 species; 175,000 km^2). However, the level of endemism (32% when seabirds and vagrants are included; 53% when considering only breeding birds) is much higher than for other such islands, both smaller (Sumatra 5%, Java 6%, Sulawesi 22%) and larger (Borneo 8%).

Biogeographers have long tried to explain Madagascar's avian species-poverty in terms of the long duration of its isolation compared to islands such as Borneo and Sumatra, which were connected to continents much more recently. However, while ancient separation seems to explain high endemism in Madagascar, does it really explain low species richness?

*This section is a summarised version of Warren *et al.* (2013).

Islands and their ages

Islands of the Malagasy region can be broadly divided into three groups, distinguished by both age and geology. Madagascar, the granitic Seychelles and the submerged Saya de Malha and Nazareth banks, are all relicts of the break-up of Gondwana, along with India, Antarctica and Africa. Madagascar–India–Seychelles broke off from the rest between 160 and 115 million years ago, India–Seychelles split from Madagascar around 88 million years ago, and India separated from Seychelles around 64 million years ago.

The Comoro and Mascarene archipelagos are of volcanic origin and of intermediate age. Mayotte is estimated at 10–15 million years old, Anjouan at 11.5, Mohéli at five and Grande Comore 0.5 million years. The Mascarenes, unlike the Comoros, are separated from each other by fractures in the Earth's crust, and each has developed independently. Rodrigues sits on a seamount that is probably around 16 million years old, Mauritius is probably around 8–9 million years and Reunion around 2–5 million years.

The raised coralline islands of the Aldabra group along with other small low-lying islands represent the youngest archipelagos in the region, emerging 15,000–125,000 years ago. However, they sit on much older seamounts, which may have been submerged and emerged many times over.

Given the wide range of ages of islands in this region (15,000 to 64 million years), and knowledge from the fossil record that the average species' lifespan is 1–10 million years, it is evident that many bird species in the region are probably much younger than the islands they occupy, while some species are probably older than their islands (their presence due to immigration from elsewhere), and still others may be of similar age to their islands and have arisen through the isolation of colonising populations following island formation (or most recent emergence).

Changes since island formation

The presence, size, shape and isolation of western Indian Ocean islands have been affected greatly over the last 30 million years by sea level fluctuations. Just in the last 500,000 years, sea levels dropped as far as 145 m below that at present, for periods as long as 50,000 years. A drop in sea level of 80 m or more would reveal a number of 'new' islands in addition to current islands in the chain between India and Madagascar, including the Saya de Malha and Nazareth banks. The present islands in the chain (the Laccadives, Maldives, Chagos and St Brandon, as well as Seychelles) would also have been much larger in extent than they currently are, and would therefore have been available as stepping stones for birds dispersing between India and the Madagascar region. The Seychelles, for instance, once covered 55,000 km^2 compared to its current surface area of 235 km^2; Saya de Malha, almost halfway from Madagascar to the Maldives and now totally submerged, once covered around 46,000 km^2. To the west of Madagascar and the Seychelles, it is likely that the distance to Africa was reduced in the period between 45 and 26 million years ago. For instance, two further islands would have existed between Mayotte and Madagascar, facilitating dispersal of species across this chain to Madagascar and beyond.

Isolation and missing families

Madagascar is large and appears to have had time to reach an equilibrium between immigration, speciation and extinction, as predicted by theories of species-area relationships. However the Malagasy avifauna is impoverished not only in absolute numbers of species, but also in the absence of numerous tropical bird lineages, including ones present in Africa (Box 1). The absence of numerous bird families from the Mascarenes, Comoros and Seychelles is typical of such oceanic islands worldwide; absences are accounted for both by geographic isolation, reducing the

---

BOX 1. LANDBIRD AND WATERBIRD FAMILIES THAT BREED IN SUB-SAHARAN AFRICA BUT ARE ABSENT FROM THE MALAGASY REGION
(seabirds, vagrants and exotics in the Malagasy region are excluded)

Struthionidae, Balaenicipitidae, Gruidae, Heliornithidae, Otididae, Burhinidae, Rhynchopidae, Musophagidae, Coliidae, Trogonidae, Phoeniculidae, Bucerotidae, Capitonidae, Indicatoridae, Picidae, Eurylaimidae, Pittidae, Phylloscopidae, Platysteiridae, Picathartidae, Timaliidae, Paridae, Remizidae, Certhiidae, Promeropidae, Oriolidae, Laniidae, Malaconotidae, Prionopidae, Buphagidae, Passeridae, Viduidae, Fringillidae, Emberizidae

likelihood that potential immigrants arrive, and by small size, reducing the likelihood of arrival, establishment and persistence. Unknown extinctions aside, the long duration of Madagascar's isolation does indeed seem the most plausible explanation for such absences; lineages originating in the last 80 million years that reached other large islands at times of former land connections might not have made it to Madagascar. Many of the species concerned are forest birds, and their absence could be related to the general low dispersal ability of such species.

Of the taxa that did reach the region post-isolation, many are waterbirds that are disposed to wander, raptors, and a diverse set of landbirds (Box 2). Some are migratory (cuckoos, nightjars, swifts, bee-eaters, swallows). Many others have high dispersal ability, being found on many remote islands (pigeons, owls, white-eyes, bulbuls, sunbirds). Still others tend to irrupt following climatic events (sandgrouse, larks and starlings).

> **BOX 2. LANDBIRD AND WATERBIRD FAMILIES NATIVE TO THE MALAGASY REGION**
> (seabirds and vagrants are excluded)
>
> Podicipedidae, Phalacrocoracidae, Anhingidae, Ardeidae, Scopidae, Ciconiidae, Threskiornithidae, Phoenicopteridae, Anatidae, Accipitridae, Falconidae, Phasianidae, Turnicidae, Rallidae, Jacanidae, Rostratulidae, Dromadidae (non-breeding), Recurvirostridae, Glareolidae, Charadriidae, Scolopacidae, Laridae, Sternidae, Pteroclidae, Columbidae, Psittacidae, Cuculidae, Strigidae, Tytonidae, Caprimulgidae, Apodidae, Alcedinidae, Meropidae, Coraciidae, Upupidae, Alaudidae, Hirundinidae, Motacillidae, Campephagidae, Pycnonotidae, Turdidae, Muscicapidae, Monarchidae, Cisticolidae, Locustellidae, Acrocephalidae, Nectariniidae, Zosteropidae, Dicruridae, Corvidae, Sturnidae, Ploceidae, Estrildidae
>
> Also the following endemic families: Mesitornithidae, Brachypteraciidae, Leptosomidae, Philepittidae, Bernieridae, Vangidae
>
> Extinct in the region: Aepyornithidae, Pelecanidae

Origins of Malagasy birds

The geographical affinities of Malagasy birds can often be inferred based on either molecular analyses or other taxonomic sources. These reveal predominantly African and Asian affinities, with just a few Australasian cases. Madagascar is vastly more isolated from Asia (India is 3,796 km away) than it is from Africa (413 km), yet many of the bird lineages colonising the Malagasy region have not done so from the nearest continental source (Africa). This observation holds even when the effect of the islands surrounding Madagascar as potential stepping stones for colonisation are taken into account: the granitic Seychelles are 2,718 km from India, and the longest single crossing in the Mozambique Channel is a mere 302 km from Mayotte to the west coast of Madagascar.

For many bird lineages, we now have molecular clock-based datings of arrival time in the region. Could any of Madagascar's birds have ancient ancestors that occupied the supercontinent Gondwana, rather than having arrived after break-up? This process (called Gondwanan vicariance) has been suggested to explain the presence of some amphibian and reptile taxa of Asian affinity. However, this does not seem to be the case for birds, as all Malagasy lineages originated well after the last land connection (Seychelles to India) was broken. Even the elephant birds, or rather their ancestors, probably flew to Madagascar (from a Gondwanan fragment, probably Australasia–Antarctica), where they became flightless. Flighted dispersal also explains the arrival of other endemic (and ancient) orders: the cuckoo-roller and mesites.

Arrival across the sea

Remarkably, for a few species, evidence points to some birds arriving across the full breadth of the Indian Ocean from Australasia, examples being the ancestors of the Madagascar Cuckoo-shrike, blue pigeons, vasa parrots and Madagascar Teal. However, most appear to have arrived from Africa or Asia and could have been aided by stepping stones or winds. At various times, additional stepping stone islands have existed in the Mozambique Channel and in the Indian Ocean between Asia and the Malagasy region. Further, the now-prevailing east–west winds (assisting colonisation from Asia) were occasionally reversed (likewise from Africa). These opportunities were probably presented to the ancestors of a number of living taxa, including endemic families such as asities, ground-rollers and tetrakas, but also the less distinctive (i.e. divergent from neighbouring continental forms) birds, many of which arrived within the last ten million years. It might be expected that endemic genera diverged during an intermediate period, and in several cases, such as Madagascar Starling and the vasa parrots, this seems to be the case. However, several species not currently treated in endemic genera are actually 'old', such as Grey-headed Lovebird, while some endemic genera such as fodies, blue pigeons, and even Dodo and Solitaire, appear to be younger.

A few lineages have radiated across the region, and most of these have fairly recent origins; in other words, although the oldest lineages have probably stayed put on Madagascar (with notable exceptions including the cuckoo-roller, vasa parrots and Blue Vanga), some younger ones have spread. Why this should be is not entirely clear. Certainly many of the islands colonised are themselves young, often being less (and most much less) than 15 million years old. However this does not explain why lineages that colonised Madagascar relatively long ago (> 15 million years) so rarely went on to colonise the younger islands after they were formed. All are competent fliers, but perhaps they have become specialised to particular conditions on Madagascar, and/or are not well equipped to become established on remote oceanic islands.

Expansion, differentiation and survival on islands

For those groups with several species on the different islands, such as white-eyes and *Hypsipetes* bulbuls, research suggests rapid expansion across numerous islands, followed by differentiation of separate populations to produce the many single-island or single-archipelago endemic species that we now find. Some of these island populations might then go extinct, leaving odd gaps or isolated species; perhaps this is why there is no sparrowhawk on Mohéli, and the Karthala White-eye is found only on one mountain-top on Grande Comore. Some related findings from DNA-based studies are intriguing: the endemic Mauritius and Comoro Fodies may be each other's closest relatives; likewise the four Mascarene white-eyes and Karthala White-eye. However, the fodies and white-eye of Madagascar are not part of either of these lineages. It is difficult to imagine how fodies or white-eyes might have dispersed between the Comoros and Mascarenes without colonising Madagascar en route. Did a fody and a white-eye perhaps go extinct on Madagascar?

In contrast to sudden, rapid expansion across existing (old) islands, some islands are simply colonised as they emerge, (e.g. the Madagasar White-eye on Europa in the Mozambique Channel). What determines the presence of a species on a particular island? Some species are simply good dispersers, yet even these may fail if an earlier colonist already occupies its niche. Then again, if the earlier colonist has specialised sufficiently, this might permit a second, more generalist species to become established: on Mauritius and Reunion we find a rare case of two white-eyes occurring together, and this is explained on each island by the specialisation of the olive species, probably allowing subsequent establishment of the grey (so-called double colonisation). What appears never to have happened is speciation within any island surrounding Madagascar. Likewise, Malagasy island bird species hardly ever vary across their islands. A notable exception is Reunion Grey-White-eye, which shows conspicuous plumage colour variation at a microgeographic scale on the topographically complex island of Reunion; but so far only one species has been recognised.

Diversification and niches

Presented with an abundance of unoccupied ecological niches on arrival in the Malagasy region, birds evolved in two main ways. Some, on Madagascar alone, underwent classic 'adaptive radiations' in which a single species arriving on the island diversified, giving rise to many species inhabiting a variety of environments or using a variety of resources, and differing in traits required to exploit these; the two most striking and undisputable examples are vangas (Vangidae; of which one relatively recently made it to the Comoros) and the tetrakas (Bernieridae).

Others, especially on surrounding islands, evolved without such diversification, with lineages arriving in these archipelagos yielding a maximum of two species per island. Most of the avian diversity in the Comoros, Seychelles and Mascarenes is the product of multiple immigration events and independent evolution resulting from the geographical isolation of populations on islands, rather than the ecological isolation (between environments or niches) observed within Madagascar (a likely exception being Reunion Grey White-eye).

Nonetheless, vacant avian ecological niches have for the large part been filled, even without such great diversification. For example, pigeons included large flightless, ground-dwelling birds, possibly eating crabs and shellfish as well as fruit (the Dodo on Mauritius and Solitaire on Rodrigues), white-eyes evolved curved bills and took on a nectar-feeding 'sunbird' niche (Mascarene olive white-eyes), and several weavers lost their gregarious habits and developed long pointed bills associated with increasing insect and nectar feeding (fodies); but once established on a particular island, none of these lineages underwent further diversification at species level.

Variation within Madagascar

Some species, such as Crested Drongo and Madagascar White-eye, hardly vary at all across the big island; this could be the result of expansion after earlier contractions caused by prehistoric climate changes. Others, such as Long-billed Tetraka and the wood rails (genus *Mentocrex*) show genetic variation by geography, perhaps reflecting contraction into small natural 'refugia', again as a result of climatic shifts. Within-island variation in plumage is fairly frequent: several species are paler in the west or south and darker in the east, the different forms being traditionally treated as races; examples include Greater and Lesser Vasa Parrots, Grey-headed Lovebird, Madagascar Green Pigeon, Madagascar Cuckoo-shrike, Souimanga Sunbird and Long-billed Tetraka.

The disappearance of 'Benson's Rock Thrush *Monticola bensoni*' from recent lists is a good illustration of a taxonomic dilemma: repeated genetic analyses have rejected its separation, even at racial (subspecies) level, from Forest Rock Thush *M. sharpei*. However, the same studies on Madagascar's rock thrushes showed that striking changes in appearance can occur without genetic divergence being detected: the rufous-backed population (*erythronotus*) on Amber Mountain is obviously distinct in colour from other Forest Rock Thrush populations yet, like Benson's, is indistinct genetically, this result being based on a limited number of genetic markers; we therefore continue to recognise it as a race. However, for Madagascar Scops Owls in the west and east, levels of genetic differentiation, and the existence of intermediates on the central plateau, suggest that race is the right level of distinction. Other variable species, with several very distinct races, worth examining via new sequencing methods that can assess many thousands of genetic polymorphisms, include Crested Coua, Common Jery and Madagascar Magpie-robin.

Prehistoric extinction?

The grim toll of unambiguously human-caused extinctions is reviewed on p. 54 (Extinct species). But could pre-human extinction in Madagascar account not only for some of gaps in distribution, but also for the overall low bird species richness and recent population expansions suggested by weak geographic variation? If the Malagasy bird species most susceptible to environmental change became extinct prior to human arrival on the island, it could explain why the latter has apparently resulted in comparatively few bird extinctions on Madagascar compared to, say, the Mascarenes. We may never know, as the fossil record is missing for most of the period in question, but palaeontology and phylogenetic analyses of speciation and extinction rates continue, and anyone working in this region has learned to expect the unexpected!

ENDEMIC FAMILIES AND GENERA

The exceptional number, and proportion, of species in the Malagasy region that are found nowhere else is well known. Equally astonishing are the rates of endemism at higher taxonomic levels, of which those of greatest interest to visiting birdwatchers are likely to be genera and families. These are briefly described in this section.

We begin with the six endemic families: mesites, ground-rollers, cuckoo-roller, asities, tetrakas and vangas. The validity of the first three as families has rarely if ever been questioned. The tetrakas were only recognised as a separate family in the last few years, while for a time the asities were treated as a 'mere' subfamily of the broadbills, a treatment now abandoned in the light of DNA evidence showing the true age and evolutionary history of these birds. However, similar studies on the vangas and their relatives suggest that the vanga family should include some Asian and African species; if so, it would no longer be endemic to the Malagasy region, although the Malagasy members do form a discrete group with their own evolutionary history. These families include 45 species (excluding extinctions) in 32 genera, of which all but two species are restricted to Madagascar: the exceptions are Cuckoo-roller and Blue Vanga, which have also reached the Comoros. Of these six families, two – the mesites and cuckoo-roller – are so extraordinary, ancient and lacking in any close relatives elsewhere that they have recently been recognised as endemic orders, Mesitornithiformes and Leptosomiformes, respectively.

Ten more genera, containing 21 species, which are not in endemic families are restricted to Madagascar; another eight (22 additional species) are endemic to the wider Malagasy region. Many of these have been shown conclusively to be among the most ancient lineages in their respective families, and are often referred to as 'basal taxa'.

We also draw attention to what might be called new genera-in-waiting: another seven endemic species, not yet studied (or not conclusively so) by molecular techniques, which are treated in non-endemic genera but are so distinctive that they may in future be shown to belong in endemic genera. In any case, they are of great interest for visiting birdwatchers or scientists.

Endemic families and their genera

MESITORNITHIDAE (mesites) Family of uncertain affinities, with three species in two genera, *Mesitornis* and *Monias*. Traditionally placed in Gruiformes (cranes, rails and allies) but DNA analysis has shown this to be incorrect and suggests a closer link to sandgrouse and pigeons. Ground-dwelling birds, limited to native forest, with elongated bodies, full, wide tails, shortish legs and short, straight (*Mesitornis*) or longer and decurved (*Monias*) bills. Wings rather weak; capable of flying only short distances. Calls are quiet and hissing or sneezing; songs are loud, medium-low whistles, in duets or choruses. All feed on terrestrial invertebrates, small vertebrates and seeds and fruit, located by turning over leaves or digging in the substrate. All three species are threatened.

Mesitornis Two species (White-breasted Mesite and Brown Mesite) with short legs and bills and reddish-brown upperparts, broad-based tails and thick plumage. Apparently monogamous, with sexes similar; sing in closely coordinated duets.

Monias One species, Subdesert Mesite, extraordinary on (at least) morphological, vocal and behavioural grounds; threatened and restricted to south-west Madagascar. A terrestrial rail-like bird of spiny forest with medium-long legs and bill, the latter decurved; sexually dichromatic, living in groups of up to ten led by dominant ♀; reproduction variably polygynous or polyandrous; sings in complex choruses.

BRACHYPTERACIIDAE (ground-rollers) Related to rollers (Coraciidae) but sufficiently divergent to be treated as a family apart, containing five species in four genera. Shorter wings and longer legs than rollers, associated with their ability to uncover, pursue and capture prey on ground, whereas rollers typically swoop down and pick it up. Medium-sized, short-winged birds; tail full to very long, bill stout, legs strong and in most species long; eyes large. Mainly ground-dwelling, although one species, Short-legged Ground-roller, calls and locates prey (usually on ground) from tree perches. Plumage cryptic in some, bright in others, but striking in all. Often secretive and difficult to observe, although can become relatively conspicuous and tame when calling in breeding season. Nest in cavities. Birds of primary rainforest and spiny forest, rightly among the most prized target species for the visiting birder.

Brachypteracias One species, Short-legged Ground-roller. Large ground-roller, notable for short legs and long hind-toe, reflecting arboreal habits in primary rainforest.

Geobiastes One species, Scaly Ground-roller, formerly placed in *Brachypteracias* but DNA and morphology confirm distinctness. Large, strictly terrestrial ground-roller with extraordinarily cryptic yet crisply marked and subtly colourful plumage, unexpectedly relieved by sky-blue tail tip and pink legs, restricted to primary lowland rainforest.

Atelornis Two species, Pitta-like Ground-roller and Rufous-headed Ground-roller. Small to medium-sized ground-rollers, almost entirely terrestrial, relatively colourful although the two species are strikingly distinct from each other, restricted to rainforest where often separated by altitude.

Uratelornis One species, Long-tailed Ground-roller, restricted to spiny forest of south-west Madagascar. Structurally, differs from *Atelornis* by much longer tail and tarsi, and more exposed nostrils, seemingly trivial distinctions relative to unique appearance, in plumage as well as structure, in equally remarkable spiny forest habitat. Mechanical 'wing-cracking' sound possibly unique in the family.

LEPTOSOMIDAE (Cuckoo-roller) Single genus with one species on Madagascar and Comoros; 1–2 races on Comoros may deserve separate species status.

Leptosomus One species, Cuckoo-roller. Described as a 'living fossil', and one of the world's most evolutionarily distinct birds, with no close relatives; not closely related to either rollers or cuckoos. Locally common and easily observed. Superficially resembles a large *Coracias* roller: a medium-large arboreal bird, with huge-looking head, heavy body, large wings and small, short feet. Feet used zygodactylly (two toes pointing forward, two back) but not as completely so as cuckoos or woodpeckers, as fourth toe extends laterally rather than clearly backwards. Loral feathers point forward then curve up and back, covering base of bill; this, together with blunt crest on rear of crown adds to large-headed appearance. Loud wailing calls, often given during swooping display-flights high above canopy, are highly evocative of forests of Madagascar and the Comoros. Strongly sexually dichromatic, distinctions showing in young before fledging age, as soon as they have lost their white down. Hole-nester, chicks with vigorous threat display recalling an owl's. Carnivorous, taking arthropods and small vertebrates. Biological knowledge remarkably scant for a species of such exceptional interest.

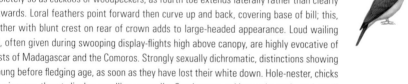

PHILEPITTIDAE (asities) Passerines (order Passeriformes) are usually divided into the 'suboscines' (suborder Tyranni) and 'oscines' (suborder Passeri); the asities are the only suboscines in the Malagasy region. Two genera, *Philepitta* and *Neodrepanis*, with two species in each, quite different in form and habits yet similar in many behaviours, and in fundamental structural characters of syrinx, tarsal scutellation (scales) and feathers. Very long isolation on Madagascar clearly indicates family status, sister to a clade composed of Asian broadbills plus African Green Broadbill *Pseudocalyptomena graueri*. Strongly sexually dichromatic, with distinct breeding and non-breeding ♂ plumages, the former showing strong metallic and iridescent components with bright yellow on underparts in three out of four species, and velvet black in the forth. Breeding ♂♂ of all species have caruncles (bare skin, sometimes loosely termed wattles) around eyes, with intense blue and green structural colours; caruncles grow and atrophy (or are reabsorbed) annually, unusually among birds, and are the subject of pioneering studies on colours of avian skin. Nests are suspended globes. All species restricted to primary forest formations, three in rainforest, and one in western deciduous forest.

Philepitta asities Two species, Velvet Asity in the east, Schlegel's Asity in the west. Plump, short-tailed, mostly frugivorous suboscines restricted to forest. Short, slightly decurved bill, ♂♂ with bright yellow underparts and spectacular blue and green caruncles. Caruncles covered in tiny cone-shaped papillae (small protuberances) in Velvet Asity, or broadly rounded undulations or 'bubbles' in Schlegel's; extensible by muscle contraction at least in Velvet. Tongue forked and brush-tipped, indicating adaptation for nectarivory. Velvet Asity has a dispersed lek display system and incubation and care of young is by ♀ only, the only known instance of lek polygyny in the Malagasy region, a characteristic well known in some other frugivorous, tropical families such as birds-of-paradise (Paradisaeidae) and manakins (Pipridae).

Neodrepanis sunbird asities Two species, Common Sunbird Asity and Yellow-bellied Sunbird Asity, largely separated by altitude. Tiny, sunbird-like suboscines, formerly considered to be sunbirds (Nectariniidae), and scientific name reflecting convergence with Hawaiian honeycreepers *Drepanis*, as true relationship to *Philepitta* revealed long after first description. Feed predominantly on nectar using long, markedly decurved bills; short tails, and in ♂♂ bright yellow underparts, iridescent blue upperparts and spectacular blue and green caruncles, which are non-extensible unlike in Velvet Asity. Calls high-pitched and stuttering; in both species, wings may make highly audible whirring sound in flight, suspected to be produced or amplified by strongly emarginated first primary in ♂♂; sound is under bird's control, produced in specific contexts around nests and in courtship.

BERNIERIDAE (tetrakas) One of the biggest recent surprises in Malagasy ornithology – a new family, and adaptive radiation, revealed by DNA analysis, bringing together 11 species, currently treated in eight genera (based on morphology; to be further tested). The name 'tetrakas' based on Malagasy name for some species; 'Malagasy warblers', sometimes used, refers to their membership of the large 'sylvioid' assemblage, and consistent with general proportions and sombre coloration, but confusing as there are other (non-bernierid) sylvioid birds in Madagascar. Until recently, its constituents were variously placed in bulbuls Pycnonotidae (current *Xanthomixis* and *Bernieria*), babblers Timaliidae (*Crossleyia*, *Oxylabes*, *Thamnornis*, *Hartertula*; latter also sometimes placed in jeries *Neomixis*) and the large Old World warbler family Sylviidae (*Thamnornis* again, *Randia*, *Cryptosylvicola*). Some variety of bill shapes and overall proportions but the radiation is not comparable to that of the vangas; slender bills typical of insectivorous birds, although very long in one species, Long-billed Tetraka. In most species, ♂ larger than ♀, this being exceptionally marked in Long-billed Tetraka in which difference in wing lengths is around 16%, and in bill an extraordinary 25%. Coloration mostly green above and yellow or whitish below, with few striking plumage patterns such as yellow supercilium in Madagascar Yellowbrow and white on head and throat in White-throated Oxylabes. All species mainly associated with primary forest, all but three restricted to humid forest. Most are understorey- or ground-dwelling. Breeding and general habits unremarkable as far as understood, but poorly known. Several species gregarious in mixed and single-species groups.

Crossleyia One species, Madagascar Yellowbrow, of montane rainforest. A small, largely terrestrial warbler-like bird, often in dense mossy forest. Dark green above and yellow on throat and on (very striking) supercilium; calls high-pitched and penetrating, and song, noticed rather rarely, is high and rambling.

Xanthomixis Four species of medium-sized understorey passerines. Evidence for sister relationship between Madagascar Yellowbrow and rare Dusky Tetraka suggests genera should be reconsidered. Forage on slim branches, mossy trunks or on ground, often in mixed-species flocks; three species restricted to rainforest, one to broad-leaved dry forest of south-west (Appert's Tetraka *X. apperti*).

Hartertula One species, Wedge-tailed Tetraka. Frequents the rainforest understorey with other species such as tetrakas *Bernieria* and *Xanthomixis*. Dull green above and yellowish below, with grey ear-coverts and an odd, triangular and pointed bill. Song long, complex and high-pitched.

Randia One species, Rand's Warbler. A small warbler-like bird of rainforest canopy, with slightly decurved bill. Greyer than other members of the family, lacking green or yellow, some individuals showing a pale orange base to the lower mandible. Sings for long periods from tall treetops.

Thamnornis One species, Thamnornis. Restricted to sub-arid, mainly spiny, forest. Dull grey-green above and whitish-yellow below, with greenish edges to primary feathers, a strong, slightly decurved bill and whitish tips to outer tail feathers. Forages on ground or in low shrubs, ascending to canopy rarely and mostly to sing. Reminiscent of brush warblers *Nesillas*, but shows strong similarities, particularly in bill and nostril shape, to Wedge-tailed Tetraka.

Cryptosylvicola One species, Cryptic Warbler. Recalls a leaf-warbler *Phylloscopus* (absent in Malagasy region except as vagrant) or jery *Neomixis*, green above and whitish below with some yellow on the throat, a 'loose' longish tail, often slightly spread, and dark legs. Very poorly known: completely overlooked, even though common in its montane forest habitat, until its discovery in 1992, singing a loud, harsh and distinctive song for long periods from elevated perches.

Oxylabes One species, White-throated Oxylabes. Unusually colourful for the family, with rich rufous, white and brown coloration, juveniles surprisingly plain and, in parts, greenish. Usually in dense, low rainforest vegetation or on ground, in company with other species.

Bernieria One species, Long-billed Tetraka. A robust passerine of rainforest and dry forest understorey, often in mixed-species groups. ♂ larger than ♀ and has disproportionately much longer bill; plumage dull green above and pale yellow below.

VANGIDAE (vangas) Fifteen genera and 21 species restricted to Madagascar and (one species) the Comoros. One genus with four species (*Newtonia*), one with three (*Xenopirostris*), one with two (*Calicalicus*), the remainder monospecific. Closest relatives are six shrike-like genera found in Africa (helmet-shrikes *Prionops*, and shrike-flycatchers *Megabyas* and *Bias*) and Asia (woodshrikes *Tephrodornis*, flycatcher-shrikes *Hemipus* and philentomas *Philentoma*), and arguably the family should include these species; however, the Malagasy species are descended from a single ancestor and so form a valid taxonomic group. Primarily insectivorous, but several also take small vertebrates; a few, also fruit. Sexes similar in most species, slight to strong sexual dichromatism in others. Largely arboreal, apart from Crossley's Vanga. Few other features in common: considering single origin, an unbelievably diverse group of arboreal passerines, and perhaps the world's most spectacular adaptive radiation. Majority are somewhat shrike-like, robust birds with a stout, hooked bill; but others include thin and warbler-like (*Newtonia*), broad and flycatcher-like (*Pseudobias*), and bizarre forms such as the laterally compressed bill of *Xenopirostris*, the sickle-shaped bill of *Falculea*, and the colourful, swollen casque of *Euryceros*. All but one genus restricted to Madagascar: *Cyanolanius* has spread to the Comoros. Several recent additions to the family confirmed by DNA analysis: *Pseudobias*, *Mystacornis*, *Tylas* and *Newtonia*.

Newtonia Four species of newtonias, formerly treated as flycatchers (Muscicapidae) or warblers (Sylviidae'), habits more recalling the latter; soft plumage texture, unspotted young and partial fusion of toes unlike true flycatchers (but like other vangas). Bill slender and pointed, with long, strong rictal bristles. Plumage very plain and dull. Insectivorous, arboreal birds of forest and scrub, often joing mixed-species flocks associated with other small birds.

Calicalicus Two allopatric species, Red-tailed Vanga (widespread) and Red-shouldered Vanga (restricted to south-west Madagascar). Small, arboreal, forest insectivores reminiscent of large tits *Parus*, but occasionally taking small vertebrates, frequently in mixed-species flocks. ♂ colourful, but ♀ plainer.

Tylas One species, Tylas Vanga. A medium-sized vanga, sexes similar. Plumage pattern like Pollen's Vanga or Madagascar Cuckoo-shrike and plumage variations intriguingly mirror those of Pollen's. Widespread in humid forests; rare and local in dry forests and mangroves.

Hypositta One species, Nuthatch Vanga. A small, short-billed vanga of humid forest that gleans prey from surface as it ascends tree trunks like a nuthatch (Sittidae). Striking blue-grey plumage with red bill. Recently described 'Blutschli's Vanga *Hypositta perdita*' is not a valid species: type specimens are juvenile White-throated Oxylabes.

Cyanolanius One species, Blue Vanga. The only vanga to occur outside Madagascar, having reached the Comoros; Comoro populations may be treated as a separate species. A small vanga, foraging acrobatically for invertebrates in forest canopy, spectacularly coloured in brilliant blue and white; ♀ duller.

Mystacornis One species, Crossley's Vanga. Perhaps the most aberrant vanga, being terrestrial in rainforest, with plumage predominantly of browns, greys and black, like the babblers with which it was long placed. Bill long and pointed, plumage soft and lax. Sexes strikingly different in plumage, but both elegantly marked.

Leptopterus One species, Chabert Vanga. A small, pied vanga with a bright blue, unfeathered eye-ring. Often in flocks, pure or mixed-species. Widespread in all wooded habitats, including plantations and forest fragments. More aerial than most vangas. Genus formerly often included White-headed and Blue Vangas, but these have been shown not to be its closest relatives.

Pseudobias One species, Ward's Vanga. With black-and-white plumage, black breast-band, flycatching habits and bright blue, unfeathered eye-ring, formerly assumed to be a monarch (Monarchidae) or wattle-eye (Platysteiridae). A fairly common and conspicuous bird of primary forests, often joining mixed-species flocks.

Vanga One species, Hook-billed Vanga. A large, powerful vanga with striking, black-and-white plumage, and long, heavy bill with hooked tip. Widespread in all wooded habitats on Madagascar, solitary and arboreal. Diet includes invertebrates and small vertebrates.

Euryceros One species, the incredible Helmet Vanga. A large, black-and-rufous vanga with a large, bright blue bill, which is long, very deep, and laterally compressed. Restricted to lowland rainforest in north-east Madagascar. A sit-and-wait predator, sallying or sally-gleaning from a perch for large invertebrates and small vertebrates.

Schetba One species, Rufous Vanga. Closely related to Helmet Vanga, another sit-and-wait predator of invertebrates and small vertebrates, but bill unspectacular. Widespread in primary rainforest and dry forest.

Oriolia One species, Bernier's Vanga. A medium-large vanga, ♂ entirely black; ♀ brown above and buffy-brown below, barred black. Uncommon and local in lowland rainforest in north-east Madagascar, primarily insectivorous.

Xenopirostris Three allopatric species of large vanga, patterned in black, white and (sometimes greenish-) grey, some showing orange on underparts. ♂ greyer with more extensive black on head than ♀. Most striking feature is bill: deep, laterally compressed, with concave cutting edge to lower mandible, in two species so much so as to leave a narrow gap between mandibles. Found in forest, feeding on invertebrates and small vertebrates.

Artamella One species, White-headed Vanga. A simply marked, black-and-white (♀ duller) vanga, plumage recalling *Falculea*, its closest relative despite structural similarity to Blue and Chabert Vangas. Widespread in most forest habitats, and primarily insectivorous.

Falculea One species, Sickle-billed Vanga. A large, black-and-white vanga, with very long, decurved bill, unique among the region's birds; in this respect and in foraging and flocking behaviour (but not in plumage), recalls African wood-hoopoes. Found in dry forest and mangroves.

Genera endemic to Madagascar, not in endemic families

Lophotibis One species, Madagascar Crested Ibis. An old lineage of ibises, with plumage unique among them, showing brown body and white wings. Unlike other ibises in the region, lower mandible is shorter than upper; has long nostrils, large bare skin around eyes, strong legs and short, strong toes with big claws. Forest.

Eutriorchis One species, Madagascar Serpent Eagle. The only raptor genus endemic to Madagascar, adult plumage resembling some *Accipiter* species including sympatric Henst's Goshawk *A. henstii* and structurally also recalls snake eagles, especially Asian serpent eagles *Spilornis* (with similar crest or 'cape') or Congo Serpent Eagle *Dryotriorchis spectabilis*. On closer observation many distinctions obvious, DNA evidence revealing membership of a group of raptors unrelated to any of these, including honey-buzzards *Pernis*, cuckoo-hawks *Aviceda* and some Old World vultures (*Gypohierax*, *Gypaetus* and *Neophron*). Rainforest; secretive and seldom observed.

Margaroperdix One species, Madagascar Partridge. The only galliform endemic to Madagascar, although introduced to Reunion. Male beautifully marked, ♀ much more cryptic, and slightly smaller. Lacks spurs on tarsi. Superficially resembles a partridge or francolin, but true affinities with quails, despite latter's smaller size. Open country and bushland.

Mentocrex Two species of wood rails, Madagascar Wood Rail and recently described Tsingy Wood Rail. Medium-sized, reddish-brown, forest-dwelling rails, formerly sometimes treated in genus *Canirallus*, with the West African Grey-throated Rail *C. oculeus*, but DNA confirms remoteness from this group, instead suggesting affinity with flufftails *Sarothrura*. Furthermore, flufftails (inhabiting Africa and Madagascar) appear so divergent from remaining rails as to merit their own family Sarothruridae, in which *Mentocrex* would presumably belong.

Coua Nine species of couas extant; large, colourful Snail-eating Coua *C. delalandei* presumed extinct but survived at least into 19th century. Sometimes treated as an endemic subfamily of the cuckoos, or in a subfamily together with closest relatives, the coucals *Centropus*. Large, long-legged and sometimes colourful cuckoos; plumage soft and lax with dove-like colours of pastel pinks, purples and peach, but entirely dark blue in Blue Coua. Tail long, feet large. Colourful bare skin around eye, and long eye-lashes. Terrestrial or arboreal; when in trees, some species walk or run along branches like African turacos (Musophagidae). Nestlings lack natal down, and palate and tongue have patterns of raised white or blue spots that contrast with the red mouth-lining; begging nestlings reveal these spectacular patterns while, in at least some species, emitting equally remarkable rattling hiss. Most in primary forest, some also in secondary habitats.

Gactornis One species, Collared Nightjar. Formerly placed in *Caprimulgus* but revealed by DNA to be highly divergent from all other genera and species of nightjars studied: apparently sister (and basal) to all caprimulgids apart from the eared nightjars *Eurostopodus*. Unusually among nightjars, eggs white, laid on low shrub rather than ground; also has pronounced facial discs, unique plumage including narrow collar on side and rear of neck, and apparently unusual sallying behaviour. Inexplicably, voice unknown. Found mainly in rainforest.

Neomixis Three species of jeries, small, largely arboreal passerines. Formerly placed in babblers (Timaliidae, or occasionally warblers 'Sylviidae'), but DNA reveals jeries to be an old lineage in the African warblers (Cisticolidae), consistent with their domed nests. Bill finely pointed. Plumage mainly green above, variably yellow on underparts. One species, Common Jery, found in wooded habitats throughout, including highly degraded areas, and is one of Madagascar's commonest birds; others more restricted to forest.

Amphilais One species, Grey Emu-tail. A secretive, streaked, marshland 'warbler'. Previously placed in genus *Dromaeocerus* along with Brown Emu-tail '*D.*' *brunneus*, because of similar long, 'decomposed' (emu-like) tail and restriction to Madagascar; however, morphology suggests few other true similarities and *seebohmi* is therefore placed in its own genus, *Amphilais*, while Brown Emu-tail has been found to belong with African (not Asian) species of *Bradypterus* warbler. However, DNA analysis only available for Brown Emu-tail, not Grey, so this genus' affinities are unconfirmed.

Hartlaubius One species, Madagascar Starling, an unremarkable-looking brown, black and white starling, formerly often placed in genus *Saroglossa* along with Spot-winged Starling *S. spiloptera* of Asia. DNA reveals completely different ancestry, in a genetically distinctive clade together with (but well separated from) Amethyst Starling *Cinnyricinclus leucogaster* of Africa; both represent old lineages of starlings. May be mainly frugivorous but takes insects and other animal prey.

Lepidopygia One species, Madagascar Mannikin, in the waxbill family (Estrildidae) but well removed from Asian *Lonchura* and African *Spermestes* mannikins or munias, which it resembles in gross structural features. Closer to *Lonchura*, along with the silverbills *Euodice* but this is another Malagasy example of an old lineage with primitive features deserving its own genus. A tiny, rather plain-plumaged estrildid with a small bill and unique black bib. Display recalls African waxbills *Estrilda* more than *Lonchura*.

Genera endemic to the Malagasy region, not in endemic families

Dryolimnas One species (or two if birds of Aldabra and Madagascar are split from one another), related to Asian and Australasian rails including genus *Lewinia*, Slaty-breasted Rail *Gallirallus striatus* and Snoring Rail *Aramidopsis plateni* (revision of rail genera is needed); not close to *Mentocrex*. Medium-sized reddish-brown rails of wetland and dry habitats. Aldabran race is the last surviving flightless bird in the Indian Ocean.

Nesoenas Two species of Malagasy pigeons, Pink Pigeon (Mauritius) and Madagascar

Turtle Dove (Madagascar, Seychelles and Comoros; four distinctive races of the latter, one introduced elsewhere). Closely related to *Streptopelia* and especially *Spilopelia*. Medium-large pigeons, with deep voices and warm colour tones, the pink body and rufous tail of the Pink Pigeon being particularly distinctive. Forest and woodland.

Alectroenas Three allopatric species of fruit-eating, blue pigeons of wooded habitats on Madagascar, Comoros and Seychelles. Like *Foudia* fodies and *Nesoenas* doves, biogeographically unusual in having both spread and diversified into several species collectively occurring throughout the region, but not having reached beyond it. Short wings and tails, the feathers around the head and neck forming distinct hackles that can be raised and lowered at will; also show variably developed, bare red orbital skin. Strictly arboreal: tightly clenched feet noted on Seychelles species makes walking on flat surface almost impossible. Plumage largely deep blue; tail purplish-red in some species, head blue or silvery-white. Distant affinities with Indo-Pacific fruit pigeons, in particular Cloven-feathered Dove *Drepanoptila holosericea* of New Caledonia, with *Ptilinopus* fruit doves representing the ancestral group.

Coracopsis Three species of medium to large vasa parrots on Madagascar, Seychelles, Comoros and perhaps formerly Reunion (as the extinct Mascarene Parrot *Mascarinus mascarinus* may belong in *Coracopsis*). Characterised most obviously by unusual dull blackish plumage. Bill broad and heavy, tail long and slightly rounded. Studies of Greater Vasa Parrot *C. vasa* have revealed polygynandrous breeding system unique among birds; other taxa less well studied, but breeding biology appears less outlandish. Highly distinct from all other parrots, although closest relative may be Pesquet's Parrot *Psittrichas fulgidus* of New Guinea. Seychelles Black Parrot once treated as a race of Lesser Vasa Parrot, but DNA shows the two are not each other's closest relatives; Comoro-endemic form *sibilans* of Lesser Vasa is distinctive and may also be treated as a species.

Phedina One species, Mascarene Martin, with races on Madagascar and Mascarenes (Mauritius and Reunion); Madagascar birds partially migratory to south-east Africa in non-breeding season. Stocky, brown hirundines with streaked underparts, building open nests on flat surfaces in cavities. Shares a common ancestor with tunnel-nesting Brazza's Martin *Phedinopsis brazzae* of Congo Basin and Banded Martin *Riparia cincta* also of Africa, but not particularly close to either.

Humblotia One species, Humblot's Flycatcher, the only genus endemic to the Comoros (only in forest on Mt Karthala), and the region's only representative of the 'Old World flycatchers' (Muscicapidae), with typical broad-based, flattened (and, unusually, yellow) bill: close to the widespread *Muscicapa*, with streaked plumage and short, broad bill as in that genus but differs in extremely soft and fluffy plumage, the forecrown feathers forming the beginnings of a crest.

Nesillas Four (possibly up to seven) species of brush warblers on Madagascar and Comoros; another recently (1980s) extinct on Aldabra. Medium-sized, olive-, greenish and grey-brown warblers with rather long, strong bills, short rounded wings and long graduated tails. Inhabit mainly bushes and undergrowth, sometimes (or in some taxa) lower and middle levels of forest. Voice mainly harsh chacking and rattles, but some taxa have more varied songs. Most taxa allopatric, but two sympatric on Mohéli, Comoros. Membership of the reed warblers Acrocephalidae confirmed, unlike two emu-tails which appear to belong among the grass and bush warblers (Locustellidae). Up to three forms here treated as races of Madagascar Brush Warbler (*obscura* of karst in western Madagascar; *moheliensis* and *longicaudata* of Mohéli and Anjouan, Comoros) may be best treated as distinct species; DNA not yet studied.

Foudia Seven species (ten taxa) of fodies on Madagascar, Seychelles, Comoros, Mauritius and Rodrigues: highly characteristic weavers (Ploceidae) of the entire Malagasy region. Sexes strongly dimorphic in all species except Seychelles Fody, breeding-plumaged ♂♂ with brilliant red coloration on, at least, head and often rump, largely replaced by yellow in Rodrigues Fody *F. flavicans*. Bill hard, conical, robust or slender, former typical of seed-eating species, latter associated with nectar-feeding and insectivory. Recall African queleas *Quelea* and especially bishops *Euplectes* but Asian *Ploceus* species may be closest relatives. Most inhabit forest, but some, especially Madagascar Fody, occupy open country. Build simple domed nests, relatively crude-looking compared to many weavers but with some unique features such as an external roofing layer.

New 'genera-in-waiting'

'*Falco*' *zoniventris* Banded Kestrel A strong-billed, grey falcon from Madagascar, believed to be closest to Grey Kestrel *F. ardosiaceus* and Dickinson's Kestrel *F. dickinsoni* of Africa but with pale irides and unusually streaked and barred plumage. Large *Falco* genus may deserve splitting into more genera.

'*Sarothrura*' *watersi* Slender-billed 'Flufftail' A small rail from Madagascar, placed with the flufftails based on plumage and structure, but differs in fine bill, nostril shape, pointed wings, short tarsi, apparent absence of spotting on adults, and distinctive calls. Rarely seen, but behaviour said to recall small crakes more than flufftails; a separate genus *Lemurolimnas* has been proposed.

'*Ninox*' *superciliaris* White-browed Owl A medium-sized, dark-eyed owl from Madagascar lacking ear-tufts, long placed in Asian genus *Ninox*, but definitely does not belong there, DNA revealing a much closer relationship to a very different owl genus, *Athene*. Being isolated and morphologically distinct from (other) *Athene* and related species, White-browed, and some other *Athene*-like owls, may prove to deserve their own genera, of which none has ever been proposed for this species.

'*Zoonavena*' *grandidieri* Madagascar Spinetail A small, spine-tailed swift found in and near forests on Madagascar and the Comoros. Presumed congeners occur in South Asia and on São Tomé, giving the genus a rather unlikely distribution; research should test for true relationship or convergence, and resulting possibility of different genera, among the three species.

'*Mirafra*' *hova* Madagascar Lark A small, unremarkable, streaked lark from Madagascar, long placed (like other such larks) in *Mirafra*, but recently shown to belong with the unstreaked sparrow-larks *Eremopterix*. Given extreme plumage distinctness from other *Eremopterix*, might merit its own genus.

'*Ploceus*' *nelicourvi* Nelicourvi Weaver A rainforest-living weaver from Madagascar with plumage pattern highly unusual among *Ploceus* species. This large genus probably requires subdivision; an endemic genus *Nelicurvius* was originally established for this species and may prove justifiable.

'*Ploceus*' *sakalava* Sakalava Weaver Another distinctive weaver from Madagascar, with pinkish-red bare skin around eye and yellow head (in breeding ♂). Similar comments apply as for Nelicourvi: if not in *Ploceus*, both could belong in endemic *Nelicurvius*, but another genus, *Saka*, has been proposed for Sakalava Weaver alone if it proves not to be close to Nelicourvi.

Endemic species not in endemic families or genera

Around 97 species treated as endemic to the region are or seem to be readily recognisable as local members of more widespread genera. In most cases, there is only one representative in the region, or per archipelago, with common names accordingly, such as Madagascar Kestrel, Seychelles Magpie-robin, Comoro Thrush and Mascarene Paradise Flycatcher. However, Madagascar has two endemic *Anas* ducks, three *Accipiter* hawks (one of which also occurs on the Comoros), two *Corythornis* kingfishers (one also on the Comoros) and two *Monticola* rock thrushes, while Grande Comore and Mohéli (Comoros) have two *Hypsipetes* bulbuls, and Grande Comore, Mauritius and Reunion each have two white-eyes *Zosterops*.

Endemic races

Not surprisingly, many species that occur in, but are not endemic to, the Malagasy region have endemic races. Equally unsurprisingly, in these days of splitters and lumpers, their classification is variable and unstable. Most of these species have large ranges in Africa – a few in Asia or more widely – and a single race (variably distinct) only in Madagascar; the more distinctive and so of most interest to visiting birdwatchers are listed in Box 3, but others, less striking, are Reed Cormorant, Purple Heron, Grey Heron, Hamerkop, African Openbill and Marsh Owl. Three of these non-endemic species have evolved separate endemic races on Madagascar and the Comoros (also in Box 3), while three others – Peregrine Falcon, Common Moorhen and White-fronted Plover – occur in the same regionally endemic race in Madagascar and elsewhere in the region. Finally, the peripatetic Striated Heron has endemic races on Madagascar, the Comoros, Coralline Seychelles and Granitic Seychelles, and populations of uncertain affinity on the Mascarenes.

Some of these races are easily distinguishable from their African and Asian counterparts in the field, and a case may be, and sometimes has been, made for splitting them at species level, adding yet more to the region's haul of endemics; examples proposed are 'Madagascar' Black Swift and Three-banded Plover, but there could be others. Rather few have been investigated at the molecular level; one that has been is Madagascar Stonechat, which is morphologically not very distinct from African forms, yet has DNA suggesting long isolation.

Among resident (and presumed native) landbirds, remarkably few are considered identical to Asian and/or African populations, i.e. *not* endemic at any taxonomic level: three are presumably recent arrivals from Africa on the Comoros: Cape Turtle Dove, Tambourine Dove and Bronze Mannikin, together with others that have reached Madagascar: Barn Owl, Little Swift, Olive Bee-eater, Pied Crow, Bat Hawk, Black ('Yellow-billed') Kite, Common Quail, Harlequin Quail and Helmeted Guineafowl. However, DNA analysis including Malagasy populations of some of these may yet throw up surprises.

Extinct species

Around the world, higher proportions of birds on islands are threatened with extinction or have become extinct than in neighbouring continental areas, and the Malagasy region is no different. At least 58 endemic species, and numerous genera, are known to have become extinct, meaning that the birds of the region were once even more amazing than they are now. The great majority of these have disappeared since people arrived, undoubtedly through human agency. The elephant birds of Madagascar were perhaps the most ancient group of all the region's birds, surely forming a seventh endemic family (Aepyornithidae) and probably a third endemic order. All now extinct, these flightless giants, with tiny, apparently functionless wings, seem to have been slow-moving vegetarians, not the fierce predators of legend. They are believed to have been the heaviest birds that ever lived, the largest perhaps around 350 kg (Ostrich *Struthio camelus* weighs 90–130 kg). They lived only in Madagascar, and their eggshell remains are still frequently found in the south-west; these were the largest eggs of any vertebrate including dinosaurs, vastly larger than Ostrich eggs. At least five (perhaps 7–8) species (and two genera) are recognised. The last populations died out before 1650, after a long period of overlap with human habitation of the island. Their extinction was probably caused by a combination of hunting, habitat destruction and climate change. Madagascar has lost at least 18 other species over the last 20,000 years; most extinctions happened long ago but two (Snail-eating Coua and Alaotra Grebe) disappeared in the last 200 years, the grebe as recently as the 1980s.

BOX 3. ISOLATED RACES IDENTIFIABLE IN THE FIELD

The taxonomy used in this book is largely 'conservative' (see **Names and classification**) in its formal recognition of species. However, many, indeed most, birders are interested in races, particularly where they are identifiable as such in the field and exist in populations that are known or likely to be largely isolated from one another, based on geographical separation of presumably sedentary populations, a lack of intermediates, and/or DNA analysis. Here, we draw attention to such races, but note that some less visually distinctive races not listed here might prove to be more so genetically. Whether or not these are tomorrow's splits (several have already been recognised as species by some authors) is for others to determine, but, whatever we choose to call them, they deserve visitors' attention.

Endemic races of non-endemic species

Tropical Shearwater (*bailloni, temptator*; also non-endemic *nicolae*), Striated Heron (*rutenbergi, degens, crawfordi, rhizophorae*), African Darter (*vulsini*), Little Bittern (*podiceps*), White-backed Duck (*insularis*), Peregrine Falcon (*radama*), Common Moorhen (*pyrrhorrhoa*), Three-banded Plover (*bifrontatus*), Kelp Gull (*melisandae*), Namaqua Dove (*aliena*), Thick-billed Cuckoo (*audeberti*), African Palm Swift (*gracilis, griveaudi*), African Black Swift (*balstoni, mayottensis*), Alpine Swift (*willsi*), Broad-billed Roller (*glaucurus*), Brown-throated Martin (*cowani*), African Stonechat (*sibilla, tsaratananae, voeltzkowi*).

Races of regional endemic species

Madagascar, distinctive and (possibly or definitely) geographically isolated races:
Red-capped Coua (*olivaceiceps*), Crested Coua (*pyropyga*), Mascarene Martin (*madagascariensis*; nominate on Mauritius and Reunion), Forest Rock Thrush (*erythronotus*), Madagascar Magpie-robin (*pica*), Stripe-throated Jery (*pallidior*), Common Jery (*debilis, decaryi*), Spectacled Tetraka (*fulvescens*), Madagasar Brush Warbler (*obscura*), Tylas Vanga (*albigularis*), Chabert Vanga (*schistocercus*).

Granitic Seychelles:
Madagascar Turtle Dove (*rostrata*).

Coralline Seychelles (Aldabra Group), differing clearly from Madagascar race (and, for sunbird, each other):
White-throated Rail (*aldabranus*), Madagascar Turtle Dove (*coppingeri*), Madagascar Bulbul (*rostratus*), Souimanga Sunbird (*buchenorum, aldabrensis, abbotti*), Madagascar White-eye (*aldabrensis*).

Comoros, single race on all occupied islands differing clearly from races elsewhere:
Madagascar Turtle Dove (*comorensis*), Madagascar Green Pigeon (*griveaudi*), Greater Vasa Parrot (*comorensis*), Lesser Vasa Parrot (*sibilans*), Madagascar Spinetail (*mariae*), Madagascar Malachite Kingfisher (*johannae*), Crested Drongo (*potior*).

Comoros, different races per island, differing clearly from Madagascar race and each other:
Frances's Sparrowhawk (*griveaudi, pusillus, brutus*), Cuckoo-roller (*gracilis, intermedius*), Madagascar Cuckoo-shrike (*cucullata, moheliensis*), Madagascar Paradise Flycatcher (*comorensis, voeltzkowiana, vulpina, pretiosa*), Madagasar Brush Warbler (*moheliensis, longicaudata*), Madagascar Green Sunbird (*moebii, voeltzkowi*), Madagascar White-eye (*kirki, comorensis, anjuanensis, mayottensis*), Blue Vanga (*comorensis, bensoni*).

Comoros, endemics with clearly different races per island:
Comoro Thrush (*bewsheri, comorensis, moheliensis*), Humblot's Sunbird (*humbloti, mohelica*), Comoro Fody (*eminentissima, consobrina, anjuanensis, algondae*).

Mauritius and Reunion, endemic with clearly different races per island:
Mascarene Paradise Flycatcher (*desolata* and *bourbonnensis*).

The Mascarene Islands have been hardest hit, with about 21 of 34 or so extinctions during the last 400 years, effectively since human habitation. These happened mainly as a result of habitat loss, hunting and the introduction of pernicious nest predators such as rats and (on Mauritius) Long-tailed Macaques. The most dramatic and tragic of these extinctions are the Dodo and Solitaire of Mauritius and Rodrigues, respectively, and an amazing array of parrots and starlings including the wonderful Hoopoe Starling of Reunion. Reunion also had an ibis, and there were endemic night herons, shelducks, and more rails, blue, pink and other pigeons.

On the granitic Seychelles, only two, probably three, bird species extinctions are known to have occurred, and none on the Comoros, partly perhaps because so few vertebrate fossil remains have been found. However, another species became extinct as recently as the 1980s on the coralline Seychelles: Aldabra Brush Warbler. In the remainder of this book, we exclude species known to be long extinct, but do include this warbler and Alaotra Grebe, as these form part of the truly recent avifauna of the region. Sadly, we do not hold out realistic hopes of their rediscovery, although stranger things have happened.

Distribution of species across the region

	Madagascar	Comoros	Granitic Seychelles	Coralline Seychelles	Mauritius	Rodrigues	Reunion	Outer Islands	Whole region
Total species	292	139	225	190	108	53	104	78	488
Endemic species	108	18	13	7	8	2	8	0	185
Endemic genera	40	1	0	0	0	0	0	0	50
Resident	200	61	50	47	38	20	40	30	284
Breeding migrant	5	0	3	3	5	1	2	1	10
Non-breeding migrant	47	39	38	36	30	16	28	31	77
Vagrant	41	37	138	108	31	15	31	16	197
Introduced	6	8	9	9	23	10	24	11	33

LIST OF BIRD ORGANISATIONS

General

African Bird Club (ABC) A UK registered charity with a focus on African ornithology including the islands of the western Indian Ocean. It publishes a twice-yearly colour Bulletin. The latest sightings from the region appear in the Recent Reports section of the Bulletin.
Website: www.africanbirdclub.org   Email: info@africanbirdclub.org

Madagascar

Asity Madagascar A national, membership-based NGO, Asity Madagascar works to improve understanding of the biodiversity of Madagascar and its conservation in natural ecosystems, and the promotion of its scientific, social, economic, cultural and ecological importance in Madagascar and worldwide. It has large projects in the western wetlands and eastern rainforest. National Partner of BirdLife International.
Website: www.asitymadagascar.org

Vahatra A national NGO, whose name means 'grassroots' in Malagasy, Vahatra has a dual mission, first to improve information on the biodiversity of Madagascar, particularly the terrestrial fauna, in support of conservation; and second to build the capacity in ecology and conservation biology of young Malagasy naturalists and graduate students. Vahatra publishes the journal *Malagasy Nature* (downloadable from its website), and has produced numerous books on Malagasy biodiversity.
Website: www.vahatra.mg

Seychelles

Island Conservation Society (ICS) A national NGO with a special focus on Aride Island Nature Reserve, which is owned by its associated UK registered charity, Island Conservation Society UK. It is active in the outer islands as conservation advisors and managers for Islands Development Company and operating programmes in collaboration with private tourism operators at Alphonse, Desroches and Silhouette. ICS also runs conservation, rehabilitation and educational programmes throughout the islands and has focused especially on Cosmoledo and Farquhar.
Website: www.islandconservationsociety.com

Nature Seychelles A national NGO with a mission to improve the conservation of biodiversity through scientific, management, educational and training programmes, Nature Seychelles manages Cousin Island Nature Reserve. It is also active in island restoration in association with other partners, including the translocation of endemic birds between islands. It coordinates the Seychelles Seabird Group and Seychelles Magpie-robin Recovery Team, and publishes the newsletter *Zwazo*, which can be downloaded from their website. National Partner of BirdLife International.
Website: www.natureseychelles.org

Seychelles Bird Records Committee (SBRC) Collects information and records relating to the status of all birds observed in Seychelles and maintains a Seychelles list of accepted species. In particular, SBRC assesses reports of all vagrants. Summaries of accepted records are published on the website. Latest news and other information on the birds of Seychelles also appear there. Record forms can be downloaded or obtained by email.
Website: www.seychellesbirdrecordscommittee.com

Seychelles Islands Foundation (SIF) A public trust established by the Seychelles government to manage and protect the World Heritage Sites of Aldabra and Vallée de Mai. The board of trustees, appointed by the President, has 14 members, including five representative organisations concerned with the conservation of wildlife and natural history or national academies of science.
Website: www.sif.sc

Union of the Comoros
Dahari A Comorian NGO created in 2013 following a successful programme of work, its mission is to forge partnerships with local communities for rural development and sustainable national resource management, with special reference to biodiversity and with strong expertise on the birds of the Comoros; based on Anjouan but covering also Mohéli and Grande Comore.
Website: www.daharicomores.org

Mayotte
Groupe d'Études et de Protection des Oiseaux de Mayotte (GEPOMAY) This group has been studying the birds of Mayotte and their habitats since 2010, with a large database of bird records and special knowledge of rare or endemic species. It works with numerous partners including the French Committee for IUCN, the Natural History Museum (Paris), Protected Areas authorities and the Regional Council.
Website: www.gepomay.fr

Mauritius
Mauritian Wildlife Foundation (MWF) The only Mauritian NGO exclusively concerned with the preservation of the nation's endangered plant and animal species. Hands-on conservation projects are carried out in Mauritius including the offshore islets and Rodrigues, working closely with local and international partners, while also raising national awareness of conservation issues through an education programme. National Partner of BirdLife International.
Website: www.mauritian-wildlife.org

Reunion
Société d'Études Ornithologiques de La Réunion (SEOR) A membership-based NGO created in 1997, SEOR aims to promote the conservation of birds of Reunion and its environment, be the main contact for planners and managers of the natural environment in relation to ornithological studies, and promote knowledge and awareness among children and adults of environmental protection, of which birds are an essential component.
Website: www.seor.fr

International conservation NGOs and research institutions
Madagascar and the other islands of the Malagasy region are, not surprisingly, a magnet for organisations devoted to research and conservation of fauna and flora, including birds. **The Peregrine Fund** (http://www.peregrinefund.org) specialises in raptors and owls on Madagascar. **Durrell Wildlife Conservation Trust** (www.durrell.org) works on conservation of a range of species, including endemic ducks. Larger conservation NGOs active in Madagascar and benefiting birds include **WWF** (www.panda.org), the **Wildlife Conservation Society** (www.wcs.org) and **Conservation International** (www.conservation.org). **BirdLife International** is represented by its national Partner NGOs in Madagascar, Seychelles, Mauritius and France. BirdLife is also the official **IUCN Red List** Authority for birds (www.iucnredlist.org); factsheets on all the world's bird species can be found on the BirdLife website (www.birdlife.org/datazone), along with many other information resources on the region's birds. It is not possible to mention all of the **universities, museums and other academic institutions**, mainly in Europe and USA but also South Africa and Japan, among others, with research programmes on the region's fauna, although at the time of writing surprisingly few deal directly with birds.

PLATE SECTION

PART 1 Larger birds (non-passerines):
Malagasy Region   Plates 1–36

PART 2 Landbirds ('near-passerines' and passerines)
Divided into seven subregions:

Madagascar   Plates 37–61

Granitic Seychelles   Plates 62–65

Coralline Seychelles   Plates 66–69

Comoros   Plates 70–79

Mauritius and Rodrigues   Plates 80–84

Reunion   Plates 85–89

Outer Islands   Plates 90–92

PART 3 Vagrants: Malagasy Region   Plates 93–124

PLATE 1: MALAGASY REGION: ALBATROSSES AND PETRELS I

Shy Albatross *Thalassarche cauta*   90 cm
Larger than all other mollymawks (*Thalassarche* species) with paler grey (not black) back, mainly white head and *very narrow dark edge to whitish underwing; dark 'thumb-mark' at base of leading edge of underwing* and inconspicuous pale grey shawl over cheeks and neck. Bill base always pale; imm has dark tip, also pale grey head and collar. Race *steadi* ('White-capped Albatross') confirmed and nominate probably occurs; ads dubiously distinguishable at sea; *steadi* has paler face and less yellow (more green) on culmen; imms inseparable. See vagrant Black-browed and Wandering Albatrosses (Plate 93). **Voice** Not recorded in region. **SHH** Breeds off Australia and on New Zealand's subantarctic islands, and ranges extensively across the Southern Ocean. Scarce and irregular off S & E Madagascar and rare around Mascarenes, all so far Jul–Nov, mostly imms; usually well offshore. Mostly alone. Very stately flight, with occasional slow, deep wingbeats. Races often treated as species. **NT** (both races/species)

Indian Yellow-nosed Albatross *Thalassarche carteri*   76 cm
Slighter and smaller than Shy, with *wider dark borders to whitish underwings*, darker back and whitish head with *mostly black, long, slender bill*; adult has yellow stripe on culmen; imm bill all black and has pale grey shawl over neck and cheeks, extending to breast-sides. Adult from potential vagrant Atlantic Yellow-nosed by whitish head (just cheeks with greyish wash; Atlantic has darker greyish wash on neck), and pointed upper tip to yellow stripe on bill (rounded on Atlantic). From adult Kelp Gull by large size, black bill, greyish tail, dark borders to underwings (not just tip) and stiff-winged, shearing flight. See also vagrant Black-browed and Wandering Albatrosses, Plate 93. **Voice** Not recorded in region. **SHH** Breeds S Indian Ocean islands, dispersing across 30°–50°S. Irregular, sometimes common off S Madagascar, rare in Mascarenes; Apr–Dec, most Jun–Aug. Usually well offshore. Mostly alone; slow, stately flight with occasional wingbeats. **EN**

Southern Giant Petrel *Macronectes giganteus*   87 cm
From albatrosses by shorter, narrower wings, shorter and more massive bill, and more laboured flight. Most plumages very similar to Northern Giant Petrel but white-morph ad unique to Southern. Ad dark-morph Southern has whitish head and throat, little contrast between upper- and underparts, and grey-brown underwing with pale leading edge. Northern has darker head than breast, distinct contrast between pale underparts and darker upperparts and uniform dark underwing. At all ages, Southern has *bill uniform yellow-orange with greenish tip; bill of Northern has dark red tip contrasting with base.* Imm plumage of both species usually all dark; bill colour is best distinction. **Voice** Not recorded in region. **SHH** Circumpolar in Southern Ocean. Ads stay close to breeding grounds but imms move further N, regularly reaching tropics. Scarce visitor (mostly imm) Mar–Sep (especially Jul) to Madagascar and Mascarenes, vagrant to granitic Seychelles and St Brandon; not recorded in Mozambique Channel. Usually alone, well offshore, but sometimes close to coast. Flight more hurried and less stately than albatrosses.

Northern Giant Petrel *Macronectes halli*   87 cm
See Southern Giant Petrel for identification and behaviour. **Voice** Not recorded in region. **SHH** Circumpolar in Southern Ocean. Ads remain close to colonies but imms more dispersive. Apparently rarer than Southern, but many records of giant petrels not specifically identified; known with certainty from two records, S Madagascar, Oct, and Reunion, Sep.

Cape Petrel *Daption capense*   39 cm
From other black-and-white seabirds by *pale patches on upperwing and back.* **Voice** Not recorded in region. **SHH** Breeds on islands in Southern Ocean, dispersing widely; follows cool currents into tropics. Scarce visitor, mostly May–Sep, to S of region, especially S Madagascar, S Mozambique Channel; rare in Mascarenes and vagrant to Aldabra group. Usually alone or in small groups; rapid shearing flight, wheeling with stiff wingbeats.

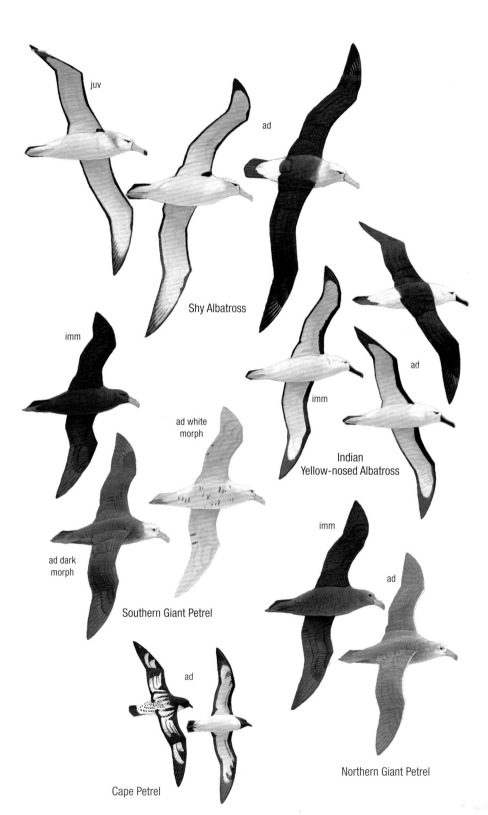

PLATE 2: MALAGASY REGION: PETRELS II

Mascarene Petrel *Pseudobulweria atterima*   35 cm
Mid-sized all-dark brown petrel, somewhat paler on belly and central underwing; head usually appears slightly darker. From *very similar Great-winged with difficulty by slightly longer neck, squarer head with thicker bill* (more than half the depth of the head at the bill base) lacking conspicuous hook, *slimmer build, longer, more tapered and pointed rear end*, parallel-sided wings with somewhat blunter tip, and shorter hand. *Flight lower and less energetic*. Also similar to Jouanin's Petrel, from which separated by straighter wings, more massive head, less active and zigzagging flight. Bill much thicker than otherwise similar Wedge-tailed shearwater. **Voice** At colonies, a high, unearthly, drawn-out scream *crrreeeeeee-awwwww* with variable whinnying quality. **SHH** Endemic to Reunion, breeding in high mountains; formerly bred Rodrigues, vagrant Mauritius, probably dispersing Sep–Mar to S of Mascarenes. Effortless, relaxed flight, usually low above water. **CR**

Great-winged Petrel *Pterodroma macroptera*   39 cm
Most similar to Mascarene Petrel, which see. From dark-morph Trindade and Kermadec Petrels by dark underwing. *Larger with thicker bill and fuller belly than Jouanin's and Bulwer's Petrels; tail shorter*. From Wedge-tailed and Flesh-footed Shearwaters by thick, blunt bill, easier, more wheeling flight. See also vagrant shearwaters and petrels, Plates 94 and 95. **Voice** Not recorded in region. **SHH** Breeds in S Atlantic and Indian Oceans; also around Australia and New Zealand. Scarce in S & SW of region, mostly in S Mozambique Channel and off S Madagascar, also S of Mascarenes; records mostly Oct–Mar. Strong wheeling flight, languid flapping.

Bulwer's Petrel *Bulweria bulweria*   27 cm
Smaller than other dark petrels, with slimmer bill, *long, wedge-shaped (when spread) or pointed tail, pale band on upperwing-coverts and erratic flight*, mostly low above water; see similar Jouanin's Petrel. Matsudaira's and Swinhoe's Storm-petrels are smaller, with shorter, forked tails and variable pale shafts to primaries. **Voice** At breeding sites, a long series of *woof* notes; also a duet of two high notes alternating with a longer, single, lower note. **SHH** Breeds N Atlantic and Pacific, ranging across entire tropics. Migrant, mostly S of Reunion and in S Mozambique Channel, Dec–Feb; one Jun vagrant to Seychelles; isolated records ashore, including egg-laying, on Round I, Mauritius. Usually alone, flying low over water, twisting and turning.

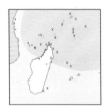

Jouanin's Petrel *Bulweria fallax*   31 cm
From Bulwer's Petrel by *somewhat larger size, less erratic flight, lack of brown carpal bar (except worn birds) and thicker bill*. Lacks pale under-primary patches of dark-morph Trindade and Kermadec Petrels; larger than storm-petrels and lacks forked tail. Slimmer with more vigorous, twisting flight and more angled wings than very similar Mascarene Petrel; head shorter and tail projects further beyond short uppertail-coverts. Great-winged Petrel larger, with shorter wings and tail, and less heavy bill. **Voice** Not recorded in region. **SHH** Breeds Socotra and probably other localities in NW Indian Ocean. Regular migrant in N of region, at least Dec–Feb, Seychelles, N Madagascar, E to Mascarenes; also SE Madagascar.

Barau's Petrel *Pterodroma baraui*   38 cm
From other petrels in region by *pale underwing, with dark bar from carpal joint rearwards across coverts; black mask and cap and narrow white forehead*, clear white underparts and grey upperparts with dark M on upperwing. Tropical Shearwater is blackish above. See also vagrant Black-winged Petrel (Plate 95). **Voice** Around colonies, muffled, raucous *oaoou* followed by a rapid *kekekeke*, often repeated twice; also *gor-wick*. **SHH** Endemic breeder (Sep–Mar) in highlands of Reunion; disperses E to C & N Indian Ocean, rarely E to Australia. Breeding-season feeding movements to seas SE, S & SW of Madagascar; rarely seen from land in S Madagascar. Flight strong, with deep swoops in stronger winds; otherwise appears 'aimless', low above water; in flocks near coasts of Reunion (W & SW) in evening before return to colonies, crossing coast before dusk. **EN**

Soft-plumaged Petrel *Pterodroma mollis*   34 cm
From other petrels by combination of *grey upperwing with dark M and dark underwings*; whitish underparts with dark breast-band; lacks black cap of Barau's. Very rare dark morph greyer than other dark petrels, with black M on upperwing (hard to see) and pale inner underwing-coverts. **Voice** Not recorded in region. **SHH** Breeds around New Zealand and islands of S Indian and Atlantic Oceans; fairly common May–Sep, S Africa and waters to S of Malagasy region; few records in region but probably regular S & SE of Madagascar and S Mozambique Channel. Very powerful swooping flight, quick wingbeats.

PLATE 3: MALAGASY REGION: PETRELS III

Trindade Petrel *Pterodroma arminjoniana*   37 cm

Exists as pale (*c.*30%), intermediate (*c.*15%) and dark (*c.*55%) morphs. *From Kermadec with difficulty by lack of whitish primary shafts on upperwing*, and slightly longer tail; pale morph slightly larger than Herald, with *longer bill and darker face (especially lores)*. Often lacks clear breast-band, but this is shown to a varying extent by some intermediate birds, which can also have dark underwings with only a very indistinct whitish flash. Some birds on Round I show mixed characters (hybridisation occurs) and impossible to identify confidently to species on current knowledge. These three species *from other petrels by conspicuous whitish flash in under-primaries*, underwing otherwise mostly dark; in addition, pale morph from Barau's by paler cap and mask. Larger than Soft-plumaged or potential vagrant Black-winged Petrel. Also similar to small skuas/jaegers; wings longer and slimmer, and held straighter, with relaxed flight, fewer wingbeats, bill thicker. See also potential or actual vagrant Sooty or Short-tailed Shearwaters (Plate 95), in relation to dark-morph Trindade and Kermadec. **Voice** Rapid *keh keh keh...* repeated up to 20 times, mellowing to a lower *klou klou klou...* in pursuit flight over breeding grounds. **SHH** Atlantic (islands off Brazil); otherwise breeds only on Round I (Mauritius), dispersing to C & S Indian Ocean, and regular around Seychelles Bank mainly Apr–Sep; not recorded in Mozambique Channel. Active at colonies by day. Flight easy and relaxed with deep wingbeats and buoyant towering in strong winds. Usually alone or in small groups. [Alt: Round Island Petrel; name is sometimes used collectively for all species on this plate, recognising difficulty of identification]

Kermadec Petrel *Pterodroma neglecta*   39 cm

Pale, intermediate and dark morphs in region. Stockier than Trindade and Herald Petrels with shorter tail, and *wing-flashes on upperwing*; see Trindade Petrel for further distinctions from that species and Herald Petrel, and other similar seabirds. Kermadec Petrel easily confused (more so than congeners) with small skuas/jaegers owing to upperwing flash, stocky build and sometimes aggressive habits, but has longer, slimmer wings and flight is more buoyant and 'easier'. **Voice** A distinctive phrase: *kek or kyek keraaooouw*, followed by a *wuk-oo-wuk-oo-wuk*, repeated up to four times. **SHH** Subtropical S Pacific, dispersing widely across S, C & N Pacific; also Indian Ocean, only from Round I (Mauritius) (*c.*10% of breeding petrels) and vagrant (bred once), Cousin I, Seychelles; dispersal unknown. Active at colonies by day. Flight similar to Trindade Petrel; relaxed and buoyant, deep wingbeats; attacks other seabirds, in manner of skuas/jaegers, unlike Trindade or Herald.

Herald Petrel *Pterodroma heraldica*   35 cm

Similar to pale-morph Trindade and Kermadec Petrels, but *slimmer and greyer; white lores and peppering around eyes*. See Trindade Petrel for further distinctions. **Voice** Similar to Trindade: repeated *kyek-kyek-kyek*; also high-pitched, sharp, single note repeated rapidly for up to 10 s, longer than Trindade. **SHH** Subtropical S Pacific, some dispersing to N Pacific; also Indian Ocean, where breeds only on Round I (Mauritius) (*c.*2% of petrels) and vagrant to Cousin I (Seychelles); dispersal from there unknown. Active at colonies by day. All-dark birds in Pacific, formerly included in this species, appear to be a different species; no such birds recorded in Indian Ocean.

PLATE 4: MALAGASY REGION: SHEARWATERS

Cory's Shearwater *Calonectris diomedea* 46 cm
From rare pale-morph Wedge-tailed Shearwater by *larger size, paler uppertail-coverts, yellow bill with dark tip, and less wedge-shaped, shorter tail*. Much larger and browner above than other dark-and-light petrels of the region. Race *borealis* confirmed in region; from potential vagrant nominate ('Scopoli's Shearwater') by solidly dark under primaries, lacking whitish bases of nominate, and slightly larger size. See also vagrant Streaked Shearwater (Plate 95). **Voice** Not recorded in region. **SHH** Breeds Mediterranean and Macaronesian islands, migrating S and abundant off South Africa Nov–May. Regular in waters SW of region, seen occasionally near S Madagascar, once from land, mostly Dec–Feb. Flight languid and low to the water, with deep beats, wings flexed and bowed; in calm winds often resting on water. Races sometimes treated as species.

Wedge-tailed Shearwater *Puffinus pacificus* 45 cm
One of many all-dark seabirds of the region; most similar to Flesh-footed Shearwater, from which separated with difficulty by *longer, wedge-shaped tail (held pointed in level flight), narrower tail-base, relatively wide base and narrow tip to wings*, dark legs (very difficult to see) and (usually) dark bill; some Wedge-tailed have pale base to bill. Also Flesh-footed has distinct jizz, appearing heavier and shorter-tailed, wings held almost straight (less angled at carpal joint) and heavier bill. *From all-dark petrels by slimmer neck and bill, smaller head, longer narrower tail.* See Cory's Shearwater for distinctions from rare pale-morph Wedge-tailed; also vagrant Streaked, Sooty and Short-tailed Shearwaters (Plate 95). **Voice** Low repeated moaning or sobbing *ohwoo-oooo waaaah*, or *oohwwowoow-hoaarrrr*, very like a human. **SHH** Subtropical and tropical Pacific and Indian Oceans; breeds in region in Oct–Mar in Seychelles, Mascarenes and St Brandon; formerly (perhaps still) off SW Madagascar. Appears to disperse mainly E to C Indian Ocean. Pale morph not confirmed in the region but claimed from Seychelles. Unhurried flight, usually low above water with wings pushed forward; often around tuna feeding concentrations.

Flesh-footed Shearwater *Puffinus carneipes* 43 cm
See Wedge-tailed Shearwater for distinctions. **Voice** Not recorded from region. **SHH** Breeds mostly in Australasia; small population on St Paul I, S of region, moving to NW and SW Indian Ocean as far as South Africa. Large shearwater with slow, deep, languid wingbeats, shearing flight in stronger winds; sometimes around tuna feeding concentrations.

Tropical Shearwater *Puffinus bailloni* 31 cm
Smaller than other dark-and-light seabirds of the region, with simpler pattern, blackish above and whitish below, and more flapping and hurried flight. Race *nicolae* slightly smaller and browner above than others, undertail-coverts dark with white tips (whitish with dark central stripe in nominate). Race *temptator* has underwing 'cloudy', slightly larger than nominate, bill bluish-grey and possibly longer than other races. See also vagrant Little Shearwater (Plate 95). **Voice** At night on breeding grounds, loud, medium-pitched crooning *korrooo-herr korrooo-herr*, or *kerrreee-hoorr kerreee-horrr*, repeated endlessly. **SHH** N Indian Ocean (race *persicus*), S Pacific (race *dichrous*); nominate breeds on Reunion and Europa (formerly Mauritius), *nicolae* on Seychelles and Maldives (probably Chagos), *temptator* on Mohéli (Comoros). Fast, agile, jinking flight, with rapid flapping and short glides. Mostly pelagic, often in feeding flocks with terns and noddies over tuna. Comoro race *temptator* apparently closer to *persicus* than to other Malagasy races, and these two are sometimes treated as a separate species, Persian Shearwater.

PLATE 5: MALAGASY REGION: STORM-PETRELS

Wilson's Storm-petrel *Oceanites oceanicus* 17 cm

Small storm-petrel, the only one known from region with combination of *all-dark belly* (sometimes hard to see) and white rump; smaller than Black-bellied and White-bellied, *with distinct pale band on upper greater coverts* and more erratic, swallow-like flight. **Voice** Not recorded in region. **SHH** Breeds Antarctic islands, migrating N of equator in all oceans; in region scarce, most regular Apr–Jun and Nov, usually offshore. Usually flies low over water, in swooping, erratic flight. Patters over feeding areas with wings in V over back; follows boats.

White-faced Storm-petrel *Pelagodroma marina* 20 cm

From all other storm-petrels by *white on face*; slightly larger than Wilson's. *Very distinctive feeding style, jumping forwards across water on tip-toe.* **Voice** Not recorded in region. **SHH** Breeds N & S Atlantic and Australasia; some of latter birds migrate across Indian Ocean. Vagrant or scarce passage migrant, recorded May, Sep, Nov, mostly around Seychelles; also Mozambique Channel and Mascarenes. Erratic weaving flight and distinctive feeding action; usually singly or in loose flocks.

Black-bellied Storm-petrel *Fregetta tropica* 20 cm

From White-bellied by *dark ventral line on white belly* (often hard to see, and rare individuals lack this), longer legs (feet often visible beyond tail), and darker upperwing with fewer pale fringes. Wilson's lacks white belly-sides and is smaller, with more swooping, erratic flight. **Voice** Not recorded in region. **SHH** Breeds subantarctic islands in S Indian Ocean and S of New Zealand, moving to N Indian Ocean in Aug–Sep; in Malagasy region, mostly recorded in S; probably scarce, but perhaps widespread, passage migrant, Apr–Jun and Sep–Nov. Singly or in small groups, flying low over water with *rocking, sweeping flight, often dragging one foot in water*.

White-bellied Storm-petrel *Fregetta grallaria* 20 cm

See Black-bellied for distinctions. **Voice** Not recorded in region. **SHH** Breeds in subtropical and temperate regions throughout Southern Ocean but rare in S Indian Ocean, dispersing N. Several reports in Malagasy region (Mozambique Channel) but none fully documented; may be regular, probably mostly Apr–Jun and Sep–Nov. Behaviour like Black-bellied.

Swinhoe's Storm-petrel *Oceanodroma monorhis* 20 cm

Large size and dark rump separate this species from Wilson's, Black-bellied and White-bellied. Smaller and shorter-tailed than Bulwer's and Jouanin's Petrels; tail fork and pale shafts of upper primaries offer clear distinctions from these species but both hard to see; flight less forceful and slower. From Matsudaira's by smaller size, usually less obvious pale patch on primaries, narrower and shorter wings, proportionately wider and shorter tail, and larger head, narrower more sharply defined ulnar bar, uniform sooty-brown lacking slight contrast of Matsudaira's between blacker head and browner body, and more rapid, erratic and nightjar-like flight; Matsudaira's is languid with shallower wingbeats. **Voice** Not recorded in region. **SHH** Breeds NW Pacific, moving to Indian Ocean, mostly N of equator. Scarce in region, mostly in N waters and S Mozambique Channel, Sep–Jun (mostly Oct–Feb). Singly or in very small groups, flying low above water; also pattering over surface.

Matsudaira's Storm-petrel *Oceanodroma matsudairae* 25 cm

See Swinhoe's for distinctions; much larger than other storm-petrels, with dark rump. From Bulwer's and Jouanin's Petrels by *shorter, forked tail and pale patch at base of upper primaries*, both sometimes hard to see, and slower, more languid flight. **Voice** Not recorded in region. **SHH** Breeds NW Pacific, moving to tropical Indian Ocean; some present all year. In region, very few records, mainly in N: confirmed in Seychelles, probably also in Mozambique Channel (near Comoros) and around Mascarenes. Flies low above water with languid, shallow wingbeats, even in strong winds; usually alone.

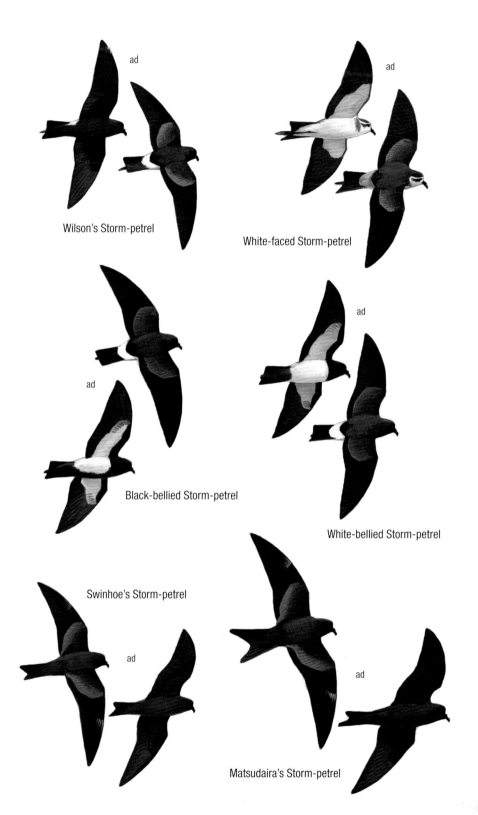

PLATE 6: MALAGASY REGION: TROPICBIRDS AND BOOBIES

White-tailed Tropicbird *Phaethon lepturus*   37 cm + 40 cm tail streamers

Ad from other tropicbirds by *yellow bill and very long white tail*, usually obvious; imm from larger Red-tailed by *extensive black on primaries*. Race *europae* endemic to Europa, with golden wash to head, underparts and (especially) central tail-feathers. See also vagrant Red-billed Tropicbird (Plate 96). **Voice** Quiet guttural clicks and chucks, e.g. *karakarakarakara... kikikikiki-kik... clik... clik... clik-clik kikikik* and variants. **SHH** Breeds in all tropical oceans. Widespread in region, but in Madagascar only in N, and absent from C Mozambique Channel. In Seychelles, common in granitics and Aldabra group, rare in Amirantes and Farquhar group. Scarce on Comoros, most occurring on Mayotte islets and Petite Terre. Common on Mascarenes, often far inland over forest in mountains. Breeds year-round, some dispersing to C Indian Ocean. At sea, flies high with steady wingbeats; feeds far out to sea by plunge-diving, and rests on sea with tail cocked.

Red-tailed Tropicbird *Phaethon rubricauda*   46 cm + 40 cm tail streamers

Ad from White-tailed by *larger size and stockier build, red bill, all-white wings* and red tail-streamers (hard to see at sea); imm has (like ad) *remiges mostly white, appearing much whiter-winged than other tropicbirds*. See also vagrant Red-billed Tropicbird (Plate 96). **Voice** Harsh disyllabic *kelek*, and a loud goose-like *aahnk-aahnk* during courtship flight. **SHH** Tropical Indian and Pacific Oceans, although absent N of Equator in Indian Ocean; in region, very local, breeds mainly Oct–Mar on Nosy Ve (SW Madagascar), Aldabra, Cosmoledo, Aride (rare), several islands around Mauritius, Rodrigues, Europa and possibly St Brandon; pelagic, feeding far out to sea and dispersing mainly to C Indian Ocean. Flies high, with direct flight; feeds by plunge-diving; often investigates boats at sea and rests on sea with tail cocked.

Masked Booby *Sula dactylatra*   86 cm

Ad from white-morph Red-footed Booby by *black tail and tertials; innermost secondaries black*, thus lacking white gap of Red-footed; also *black mask and orange-yellow bill*; latter also distinguishes it from vagrant ad Cape and Australasian Gannets (see Plate 96). Imm from Brown Booby by paler upperparts and mantle contrasting with dark primaries and secondaries, and white hind-collar; from imm gannets by dark markings on otherwise pale underwing-coverts. Local race is *melanops*, also in NW Indian Ocean. **Voice** High-pitched double honk, sometimes given at sea. **SHH** Pantropical; in region, colonies on Seychelles (Cosmoledo and Boudeuse), Comoros (M'Chaco I, off Mohéli), Mauritius (Serpent I), Tromelin and St Brandon. Vagrant to Madagascar and Reunion. Usually pelagic when not breeding; mostly solitary or in small groups. Steady, powerful flight. Feeds by plunge-diving, often over feeding concentrations.

Red-footed Booby *Sula sula*   71 cm

White-morph ad from ad Masked Booby and vagrant ad gannets *by white tail, tertials and inner secondaries; most show unique dark patch on primary coverts of underwing*. Ads of other morphs distinctive. Imm from other boobies by *dark underwing* and lack of clear dark/light demarcations or clear white patches anywhere on body. Imm white morph similar to ad brown morph, but bill and facial skin darker. **Voice** Rapid call, rising and falling in volume, *rah-rah-rah-RAH-rah-rah*, given mostly near colonies. **SHH** Pantropical, with scattered colonies of race *rubripes* in region, mostly in Seychelles (Cosmoledo, Aldabra, Farquhar and Marie-Louise, entirely white morph) and on Outer Islands (Tromelin, St Brandon, Europa); otherwise regular throughout Seychelles, scarce off Madagascar and Comoros and vagrant to Mascarenes. Usually pelagic but within general area of colony; flies quite high with steady wingbeats, and plunge-dives or chases flying-fish. Usually alone or in small groups.

Brown Booby *Sula leucogaster*   69 cm

Ad lacks white collar of imm Masked, and is more uniform brown above, *lacking contrast especially between primaries and mantle* plus coverts; lower border of brown breast-band straight, lacking inverted V of Masked; imm Red-footed lacks whitish mottling on belly of imm Brown. **Voice** ♂ gives loud, high-pitched wheezing whistle, ♀ a harsh quack or honk. **SHH** Pantropical between 30°N and S; race *plotus* breeds C Pacific to Indian Ocean. Sparse in region, only breeding in small numbers off NW Madagascar (Quatre Frères, in Nosy Mitsio group) and Seychelles (Cosmoledo and Boudeuse; has also bred Aldabra; non-breeders regular to granitic Seychelles. Flight strong, direct; mostly pelagic, feeding fairly close to colony, plunge-diving from low angle and catching flying-fish in flight.

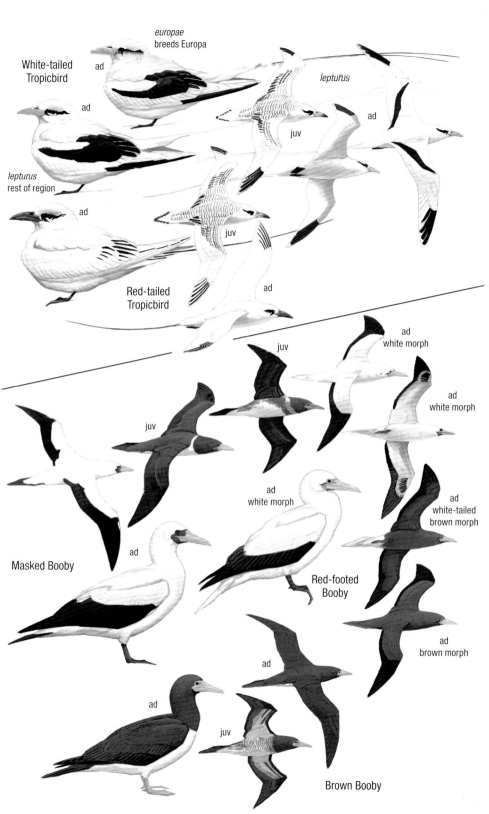

PLATE 7: MALAGASY REGION: CORMORANT, DARTER AND FRIGATEBIRDS

Reed Cormorant *Phalacrocorax africanus* — 55 cm

Smaller than African Darter with *shorter neck, tail and bill*; does not soar. Much smaller than potential vagrant Great Cormorant (see Plate 97); imm lacks clear white underparts, and bill and neck much shorter. Madagascar endemic race *pictilis* larger than nominate; ad has larger and less rounded black spots at tips of scapulars, tertials and upperwing-coverts than nominate; juv less white below. **Voice** Usually silent; bleats and hisses reported on breeding grounds. **SHH** Sub-Saharan Africa. Resident in Madagascar, where widespread to 1,500 m but never numerous; scarcer in SW, only on rivers; rare on C plateau. Vagrant (race not determined) to Seychelles (Mahé and Aldabra) and Comoros. Usually alone or in small groups, on fresh water (rivers, lakes) or in mangroves, usually with floating vegetation; often perches on trees to dry wings. [Alt: Long-tailed Cormorant]

African Darter *Anhinga rufa* — 91 cm

From Reed Cormorant by long, sharp-tipped bill, *extraordinarily long neck, and long full tail; soars regularly*. Madagascar endemic race *vulsini* has head browner than nominate, black borders to washed-out cheek-stripe of ♂ less obvious; greater coverts paler and greyer. **Voice** Usually silent; croaks and quacks given at breeding sites. **SHH** Sub-Saharan Africa except Horn. Resident in Madagascar, where most common in W & NW; rare or absent E, C plateau, SE, SW & S. Possibly vagrant to Aldabra. Usually on calm, deep freshwater rivers and lakes; sometimes in mangroves. Perches on nearby trees or sandbanks; soars high, often over dry places (towns, savanna). Usually singly or in small groups.

Lesser Frigatebird *Fregata ariel* — 76 cm

Slim, small frigatebird; *white flash on underwing distinctive at all ages*, and ad/subad ♀ and imm ♂ show combination of black throat and white breast. See Great Frigatebird for further distinctions. W Indian Ocean race *iredalei* distinguished mainly by measurements but possibly invalid. **Voice** Whistles and chucking calls on breeding grounds; silent at sea. **SHH** W Pacific and Indian Oceans; also Atlantic, off Brazil. Breeds on Aldabra, Europa and St Brandon; possibly also Quatre Frères, Mitsio Is, Madagascar, and islet off Mohéli, Comoros. Roosts at many places in coralline Seychelles and fewer in granitics, also Glorieuses and Tromelin; widespread off N & NW Madagascar and Comoros. Occasional around Mascarenes, often after cyclones. Behaviour like Great Frigatebird but far less kleptoparasitic.

Great Frigatebird *Fregata minor* — 95 cm

From Lesser at all ages by larger size (especially ♀) and *lack of white spur on flanks extending to axillaries*; most evident on ad ♂, juv may have some white on axillaries but further back, not at level of dark breast-band. *Ad and third/fourth-year C lack dark on breast or throat*, contrasting with white breast. W Indian Ocean race *aldabrensis* distinguished mainly by measurements but possibly invalid. **Voice** Falsetto warbling in display; also creaking and whistling. Silent at sea. **SHH** Pacific, Indian and locally Atlantic Oceans. Breeds on Aldabra, Cosmoledo, Europa and St Brandon, and possibly Quatre Frères in Mitsio Is, Madagascar. Roosts at many places in Seychelles, Comoros (islet off Mohéli; may breed), N & NW Madagascar, Glorieuses; widespread at sea and along coasts in this area, vagrant to far S & SE Madagascar and Mascarenes, the latter after cyclones. Generally high-flying, slow-moving birds, soaring to great heights; will chase terns and boobies, especially near breeding grounds; most food taken from or near sea surface.

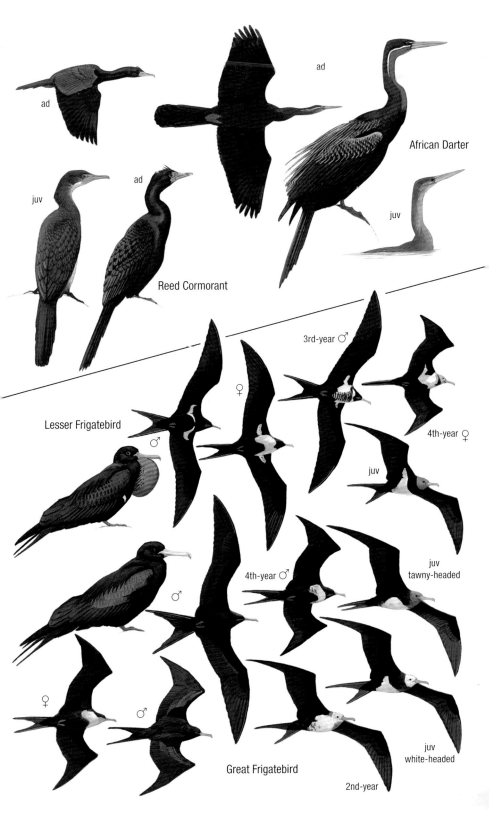

PLATE 8: MALAGASY REGION: HERONS I

Grey Heron *Ardea cinerea* 94 cm
From other large resident and vagrant herons (see also Plate 98) by combination of *pale foreneck and forehead*; lacks rufous on neck or belly. Juv has greyer sides to neck, resembling young Madagascar Heron, but has *white on leading edge of wing and pale chin and belly*. Endemic race *firasa* has longer bill and legs, brighter courtship coloration than nominate. **Voice** A low, hoarse *graank* or *fraaank*, given in flight, sometimes repeated 2–5 times at 1–2 s intervals. **SHH** Eurasia and sub-Saharan Africa; in region, resident Madagascar and Comoros (race *firasa*), Seychelles (nominate race), Juan de Nova (race uncertain); probably vagrant Mascarenes. In Madagascar mostly coastal in W & N, especially around mangroves, coastal lakes and rivers; rare in highlands and in E & S.

Madagascar Heron *Ardea humbloti* 96 cm
Lack of noticeable pale or rufous areas in plumage distinguishes from other herons; chin dark, bill mostly pale. Much more massive than dark-morph Little Egret. **Voice** Low bark *karr* or *aaark* similar to Grey Heron. **SHH** Endemic to Madagascar and Comoros, but rare and not proven to breed in Comoros. Vagrant to African mainland (Tanzania). In Madagascar mostly on W estuaries, rivers and lakes as far S as Mangoky R, more patchy in N on coast and rivers; also regular at L Alaotra. Rare elsewhere on central plateau, and on coast in S & E. **EN**

Purple Heron *Ardea purpurea* 84 cm
Slimmer and much more rufous than other resident herons, *especially on neck and underwing*; back darker. Madagascar endemic race *madagascariensis* darker than nominate especially on upperparts. **Voice** Deep *craank*, like other *Ardea* species. **SHH** Eurasia and Africa. Race *madagascariensis* resident in Madagascar; vagrant (nominate race?) to granitic Seychelles (annual), Aldabra group and Comoros. In Madagascar, fairly common in wetlands throughout, in reedbeds and in or near other dense vegetation; scarce in more open areas. Commonest in W, patchy on C plateau and scarce in dry S.

Squacco Heron *Ardeola ralloides* 45 cm
Smaller size and contrast of white wings with *buff-brown breast and upperparts* distinguish from all egrets; additionally from Cattle Egret by more horizontal posture, shorter neck, blue bill and more extensive buff colour during breeding season. From Madagascar Pond Heron by more pointed wings and buffy breeding plumage; in non-breeding plumage, with difficulty by *overall mid-brown cap and mantle*, compared to darker chocolate-brown on Madagascar Pond Heron, which also has broader, darker edges to long feathers of nape and neck. *Juv lacks dark primary tips of Madagascar Pond Heron*. See also vagrant Indian Pond Heron (Plate 98). **Voice** Low croaking and grunting *graak* or *grek*, also *ough...ough...wugh* and variants; somewhat higher-pitched than Madagascar Pond Heron. Flight call a quiet *kaw* or *koor*, often prolonged into a short sequence *kaw kaw kaw kaw kow* or *kek-kek-kek*, or similar. **SHH** Sub-Saharan Africa to Iran and Europe. Resident in Madagascar, where common below c.1,600 m except in drier parts of S & SW, where scarce and limited to rivers and coast. Vagrant to Seychelles and Mauritius.

Madagascar Pond Heron *Ardeola idae* 47 cm
In breeding plumage, from Cattle Egret by absence of buffy colouring on head and breast (sometimes a tinge), shorter, more retracted neck, shorter legs projecting only slightly beyond tail, and *bright blue bill-base*. Slightly larger and stockier than Squacco Heron with, in flight, blunt, square-cut wingtips; in non-breeding plumage, with difficulty by *dark chocolate-brown cap and back*, almost blackish, compared to mid-brown of Squacco Heron. Also broader, darker edges to long feathers of nape and neck. Juv similar, greyish primary tips further distinguishing it from Squacco Heron. **Voice** Flight call, often given when disturbed, a somewhat low-pitched *graak*; also longer *garrr-ga garrrkk garrrkkk garrkkk* especially when agitated; louder, more raucous than equivalent of Squacco Heron. **SHH** Endemic breeder in region, in Madagascar, Aldabra, Mayotte and Europa; migrant elsewhere in Comoros, and vagrant to Reunion and Seychelles E of Aldabra. Migrates to E & C Africa May–Sep. In Madagascar, patchy and scarce, commonest in W & N, also in scattered colonies on C plateau and rare in S. **EN**

PLATE 9: MALAGASY REGION: HERONS II

Striated Heron *Butorides striata*   41 cm
At all ages, identified by *small size*, in flight with neck slightly stretched, yellow legs extending beyond tail, and *dark appearance*. Streaky juv distinguished from scarcer, more secretive Yellow Bittern or Little Bittern by larger size, darker upperparts and *lack of pale buff patches on wing-coverts*; from juv Black-crowned Night Heron by much smaller size, thinner bill, shorter legs and lack of pale spots on upperparts including wings, scapulars and mantle. In Madagascar and granitic Seychelles, resembles African races; elsewhere more like Asian birds. **Voice** Harsh, slightly hollow, 'affronted' *quiaak*, given singly or in sequence on take-off. Song a croaking boom *croooor...croooor*, or *whooor...whooor*. **SHH** Old World tropics, South America and many oceanic islands. In Madagascar up to 1,500 m; also on most islands in region. Endemic races on Madagascar (*rutenbergi*), Comoros (*rhizophorae*), and granitic (*degens*) and coralline Seychelles (*crawfordi*); on Mauritius and Rodrigues, race *javanica* as in SE Asia; on Agalega and Reunion, race uncertain. Catholic habitat choice in shallow fresh, brackish and/or saltwater habitats, including open coasts. Opportunistic and versatile feeder, often in the open, frequently first noticed when flushed, calling. [Alt: Green-backed Heron]

Black-crowned Night Heron *Nycticorax nycticorax*   61 cm
A distinctive, medium-sized, stocky heron, *often seen at dawn and dusk*. Flight silhouette characteristic, with *short neck, stocky body*, toes that barely extend beyond tail, and broad, rounded wings. Ad plumage distinctive. Juv from juv Madagascar Pond Heron and juv Squacco Heron by larger size and dark, not white, wings; see also Striated Heron. **Voice** Flight call a hoarse, crow-like *goak*, sometimes higher *gwik* given on take-off, as when leaving roost in evening or in response to disturbance. **SHH** Temperate and tropical regions worldwide; widespread in Madagascar, and recent (since 1995) colonist of Seychelles. Breeds and forages in varied wetland habitats with fresh, brackish or salt water. Primarily crepuscular and nocturnal; during day usually hunched up and sleeping in secluded, shady trees, and often seen in flight at dusk.

Yellow Bittern *Ixobrychus sinensis*   35 cm
Three *Ixobrychus* species occur in the region and the two breeders do not overlap, but with vagrancy all three recorded on Seychelles and could occur elsewhere. Closest similarity is between Yellow Bittern and vagrant nominate Little Bittern (see Plate 98). Yellow Bittern *paler and more yellowish-brown* at all ages than Little Bittern, *especially on mantle and scapulars*, which are glossy black on ad ♂ Little Bittern and tawny-brown on Yellow Bittern; upperwing-coverts much darker, more rufous-brown on Little Bittern, especially in Malagasy race. Throat pale rufous-buff in Little Bittern, whitish shading to light ochraceous-buff in Yellow Bittern. Juv Yellow Bittern shows broader, buff edging on upperparts (narrower, more rufous on Little), and lacks blackish centres to brown foreneck streaks of Little. Vagrant Cinnamon Bittern is distinctive; see Plate 98. **Voice** A soft repetitive *oo-oo-oo* in breeding season, recalling Zebra Dove but more monotonous and continuous; also a harsh *kak-kak* in flight. **SHH** S & E Asia, with isolated population resident on granitic Seychelles, especially Mahé, Praslin and La Digue, on small freshwater marshes behind coastal plains; vagrant Amirantes. Solitary or in pairs, usually hidden in dense vegetation and most often seen in flight in early morning and late afternoon. Flight rather erratic, low and rapid with occasional glides, occasionally perching briefly in view before dropping out of sight.

Little Bittern *Ixobrychus minutus*   30 cm (*podiceps*)
Smallest heron in Madagascar, closest to Striated Heron; easily distinguished from this and other regularly occurring species in flight by *black rounded wings with contrasting pale secondary-coverts*, bulging neck, toes extending beyond tail, and alternating glides and quick wingbeats. See also Yellow Bittern and (vagrant) Cinnamon Bittern (see Plate 98). Madagascar resident endemic race *podiceps* much smaller, darker and more rufous than other races. **Voice** ♂ advertising call a low croaking, grunting or barking *whump*, repeated at 3–5 s intervals. Flight call a sharp rasping *eerk...eerk*. **SHH** Europe, Africa and Australasia. Race *podiceps* resident in Madagascar up to 1,600 m in wetlands with dense reeds, grasses or other aquatic vegetation, occasionally mangroves; infrequently reported but probably overlooked. Vagrant to Seychelles, probably migratory nominate race (see Plate 98), which may also occur in Madagascar.

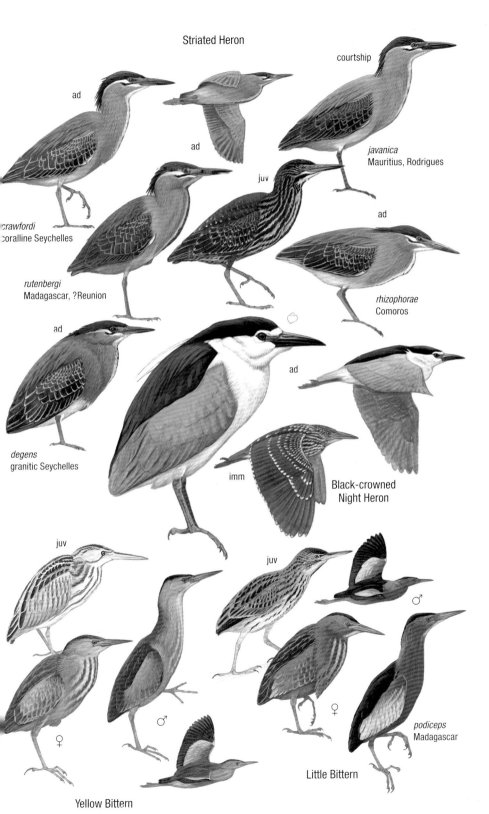

PLATE 10: MALAGASY REGION: HERONS III

Great Egret *Ardea alba* 95 cm
The largest white egret in the region, of comparable size to Grey and Madagascar Herons; *plumes on back (but not head)* and displays similar to *Egretta* species; from other white herons by *larger size, slower flight, long legs extending well past tail*, deeper neck-bend, *dark feet* and black gape-line reaching behind eye. **Voice** In flight a hoarse rattling *carrrrrk* or shorter *cakkk*, often when disturbed; ads at nest make growling and croaking noises, with bill-claps. **SHH** Widespread globally. Breeding resident in Madagascar up to 1,600 m, locally abundant in W & N but limited habitat in E & S, and on Comoros (rarer, but on all islands). Vagrant to granitic and coralline Seychelles (Intermediate Egret also vagrant to Seychelles – see Plate 98), and Mauritius. Uses wide range of wetland areas, both fresh and salt water; breeds colonially in trees including in C Antananarivo.

Cattle Egret *Ardea ibis* 51 cm
From white-morph Little Egret by smaller size, *thick yellow-orange bill*, shorter yellow-green legs and yellow feet, shorter, thicker neck, stockier build, and buff elements to breeding plumage. In flight, has shorter, rounded wings and faster wingbeats. See also Madagascar Pond Heron and Squacco Heron. **Voice** Not very vocal except at colonies where gives a constant chorus of low *craak* notes. In flight, a short harsh *aark* or disyllabic *aark-aark*. **SHH** Widespread globally. Often abundant in Madagascar up to 1,800 m and on Seychelles; more local on Comoros. Sporadic records, often long-staying but sometimes disappearing, on Mauritius and Reunion; vagrant to W Outer Islands. Inhabits all kinds of wetlands, but also grasslands; often in areas of intensive human activity, on Seychelles including rubbish dumps and Victoria fish market. Nests in colonies in trees, including in Antananarivo. Granitic Seychelles birds sometimes recognised as endemic *seychellarum* but may represent a mixture resulting from influxes of race *coromandus* (Asia) and *ibis* (all other populations).

Black Heron *Egretta ardesiaca* 51 cm
Combination of *all-black bill, yellow eyes and black legs with yellow feet* is diagnostic; during courtship, feet more reddish. Foraging 'umbrella posture', with wings spread over head, unique. Distinguished from dark-morph Little Egret by smaller size, lack of white anywhere in plumage and shorter legs, which are especially noticeable in flight. **Voice** Rather silent but gives harsh, rasping cries in alarm. **SHH** Sub-Saharan Africa. In Malagasy region, restricted to Madagascar and vagrancy unknown. Found up to 1,500 m, breeding in Antananarivo and elsewhere. Restricted to shallow freshwater and brackish wetlands, less often saline mudflats or mangroves. Nesting very poorly known, probably in marshes more often than in trees. [Alt: Black Egret]

Little Egret *Egretta garzetta* 60 cm
White morph from other white herons by combination of *dark bill and legs, yellow feet, medium size* and slim build; dark morph from Black Heron by (usually) white primary coverts, longer bill and neck, and pale lores. Race *dimorpha* ('Dimorphic Egret') found in E Africa and Malagasy region; dark birds sometimes outnumber white, and some intermediates bluish-grey speckled with white; feet yellow or olive, this colour sometimes extending up tarsi, which may be diagnostic, and *lores generally bright yellow turning deep pink when breeding*, as do feet. See also vagrant nominate race (Plate 98). **Voice** Deep growling or grunting at breeding sites; also bill-clapping. Low *karrk* rarely given in flight. **SHH** Widespread globally. Race *dimorpha* breeds E Africa (mainly coastal), Madagascar, Aldabra group and Europa; probably this race also rare visitor to Comoros, Glorieuses and Juan de Nova. Formerly bred on Mauritius and Reunion; no recent records. Nominate race vagrant to granitic Seychelles. 'Dimorphic Egret' sometimes treated as separate species or included in Western Reef Egret *E. gularis*.

PLATE 11: MALAGASY REGION: HAMERKOP, STORKS AND FLAMINGOS

Hamerkop *Scopus umbretta* — 56 cm
At rest, recalls a small stork but distinctive outline produced by *large crest protruding straight back from head* in line with heavy, curved bill; hence name. In flight, recognised by *broad wings, energetic, strong wingbeats*, and short tail with legs not protruding (reminiscent of a raptor; also soars), and often also by calls. Madagascar endemic race *tenuirostris* has bill less deep than African races, with subtle differences in tail barring. **Voice** Vocal when with others or in display, and in flight, when may continually call *taket-taket...*. **SHH** Sub-Saharan Africa and SW Arabia. Occurs patchily or locally at low density up to 1,700 m in Madagascar; well distributed on C plateau including Antananarivo but scarce in W and rare in S. Often noticed by gigantic spherical nests in trees. Associated with a wide range of wetlands, mostly freshwater but sometimes saline wetlands and coastal plains.

Yellow-billed Stork *Mycteria ibis* — 105 cm
Black-and-white pattern unique among resident large waterbirds, but confusable with vagrant Pink-backed Pelican and White Stork; see Plate 97. **Voice** Generally silent, possibly odd whining or hissing calls as in Africa. **SHH** Sub-Saharan Africa, where common, and Madagascar, where a rare and highly threatened resident breeder (but not an endemic taxon). Occasionally wanders outside heartland in lowlands of CW Madagascar. Found in shallow wetlands, such as large marshes, coastal lagoons, on mudflats, rivers with sandy banks, rice fields, and lakes, usually below 150 m, rarely to 1,500 m.

African Openbill *Anastomus lamelligerus* — 87 cm
Small, dark stork, easily distinguished from other large waterbirds by *unique bill* (looking thick when gap in bill not discernible) and *very dark plumage*. Madagascar Heron has dark grey rather than glossy blackish-brown plumage and pointed, often pale bill, and flies with neck retracted. Madagascar endemic race *madagascariensis* differs from African birds by bill structure. **Voice** Uncertain in Madagascar, but probably as in Africa, where gives loud, raucous croaks or honks. **SHH** Sub-Saharan Africa, where common, and Madagascar, where a rare and highly threatened resident breeder. Mainly found in large freshwater wetlands, more rarely seashores, estuaries and mangroves, usually below 150 m, rarely to 1,500 m.

Greater Flamingo *Phoenicopterus roseus* — 132 cm
Unmistakable as flamingos; tall, long-legged, fragile-looking pink, red and black waterbirds, wading in shallows or swimming. Ad distinguished from Lesser Flamingo by *bright scarlet upper- and underwing-coverts*, generally paler pink plumage, *large pink bill tipped black* (Lesser has bill dark at all ages); also much larger, especially taller. Youngest imms much larger, taller, browner than Lesser Flamingo; subad mainly white with black primaries, grey legs and bill **Voice** *A low, clear honking arrra-arrra or urruk-urrk*; in flight, a more melodious *kuwwuk-kuwwuk*, much like a goose; both usually in chorus. **SHH** Breeds locally in Africa, S Europe and S & SW Asia; non-breeders wander widely. In Malagasy region, breeds only on Seychelles (Aldabra: world's only flamingo colony on a coral atoll), formerly Madagascar and historically elsewhere. Non-breeders present year-round, most in W Madagascar, typically a few thousand birds. Vagrant to granitic Seychelles, Comoros and Mozambique Channel islands. Found especially on large lakes, lagoons and estuaries of W Madagascar, but also in small groups on other shallow-water wetlands. On Aldabra, uses open, shallow, inland pools.

Lesser Flamingo *Phoeniconaias minor* — 85 cm
Ad distinguished from Greater Flamingo by darker pink plumage, *upper- and underwing-coverts mottled and blotched deep dark red* (not bright scarlet) and *bill dark red* (not pink). Much smaller; in mixed groups in shallow waterbodies, Greater Flamingos stand above backs of Lesser Flamingos. Imm dark-billed and usually greyer than Greater. **Voice** A high-pitched *chissick*, or *kwirrick*; also, a low bleating murmur *murrrh-errh, murrr-errh* especially when walking or resting, very different to Greater. **SHH** Breeds mainly in Rift Valley lakes of E Africa, a few in S Africa and elsewhere; non-breeders wander widely. In Malagasy region, restricted to W & SW Madagascar, especially Lake Ihotry; rarely >30 km inland. Vagrants on Mozambique Channel islands, with old or unconfirmed reports on Mauritius, Reunion and Rodrigues; no breeding ever proven in region. Inhabits similar sites to Greater Flamingo but avoids riverbeds or smaller lakes. **NT**

PLATE 12: MALAGASY REGION: IBISES AND SPOONBILL

Madagascar Sacred Ibis *Threskiornis bernieri* 76 cm
The *only largely white ibis* in its range. See also vagrant African Sacred Ibis (Plate 97). *Eyes pale grey or whitish* in Madagascar (nominate race), bluer on Aldabra (*abbotti*). **Voice** Generally silent away from nesting sites, but in flight sometimes gives raucous croak. **SHH** Endemic to Malagasy region, with small and declining population. Confined to W, S & extreme NE Madagascar (<2,000 birds), almost all close to coast, and Aldabra (200–500 birds), where small numbers found throughout but chiefly on Grande Terre (breeding), Picard, Polymnie and Malabar Is (all non-breeding). Vagrant Comoros (Mayotte). On Madagascar, frequents muddy estuaries, estuarine sandbanks, tidal mudflats, muddy bays and bars, sandbanks and mudbanks along rivers, mangroves and brackish lakes; much less often in freshwater wetlands. On Aldabra, in woodland and open mangrove, foraging in leaf-litter inland, on seashores, lagoon margins and especially brackish inland pools. Formerly treated as conspecific with African Sacred Ibis but display and morphological differences support species split. **EN**

Glossy Ibis *Plegadis falcinellus* 60 cm
The *only small, dark ibis* in the region. Long, downcurved bill and long, slender legs and feet shown otherwise only by unrelated Eurasian Curlew and Whimbrel, which are much paler overall with complex brownish markings. Can appear rather duck-like in flight, but flattened outline with bill and legs outstretched, slightly drooped wings, and rapid wingbeats interspaced with short glides, distinctive. **Voice** Soft croaks and grunts in flight or at nest but otherwise silent. **SHH** Cosmopolitan. In Madagascar, numbers much reduced but still locally common in W; isolated breeding population (probably <20 birds) on Agalega. Vagrant to granitic Seychelles; pre-1940 records also Mauritius and Reunion. Occupies a variety of wetland habitats, including cultivated areas, from sea level to 1,500 m.

Madagascar Crested Ibis *Lophotibis cristata* 70 cm
A forest ibis. *White wings*, revealed when flushed, with *rufous plumage, green-tipped crest* and decurved bill make unique combination of features. **Voice** Fairly vocal, and distinctive. Advertising call usually at night, at dusk or early in morning: *sequence (5–20 s) of pumping, wheezy notes* or growling monosyllables *slightly up and down scale* werg-weerg-waug-wugh-weeerg-weergh-werg-wugh..., sometimes with a second bird contributing single *werg-wugh* and similar notes (recalling Madagascar Scops Owl, but more grating), perched or less often in flight. Also gives quieter foraging or courtship calls. **SHH** Endemic to Madagascar, where present up to 2,000 m. Races (nominate in E & N, *urschi* in W) dubiously distinct. Found in primary forest; in E also other dense shaded areas such as vanilla and oil-palm plantations, dense secondary forest, often but not always in damp places. Often common but shy as heavily hunted; not threatened but locally vulnerable. **NT**

African Spoonbill *Platalea alba* 91 cm
Easily recognised by *spoon-shaped, grey-and-red bill, and white plumage*. Other large, mainly white waterbirds in Madagascar are Yellow-billed Stork (with black remiges) and Great Egret (cleaner white, flies with neck retracted and has completely different feeding style). **Voice** Usually silent, occasionally giving *ouark ouark* in flight. **SHH** Sub-Saharan Africa and Madagascar, where characteristic of W lowland wetlands, occasionally wandering to C plateau. Rare and probably declining and threatened in Madagascar, but not an endemic taxon. Usually found on shallow inland lakes, fresh or brackish; also rivers, deltas, estuaries, coastal lagoons, flooded rice paddies. Gregarious, in groups of up to 10–15 birds, sometimes more, often with other large waterbirds such as flamingos or storks.

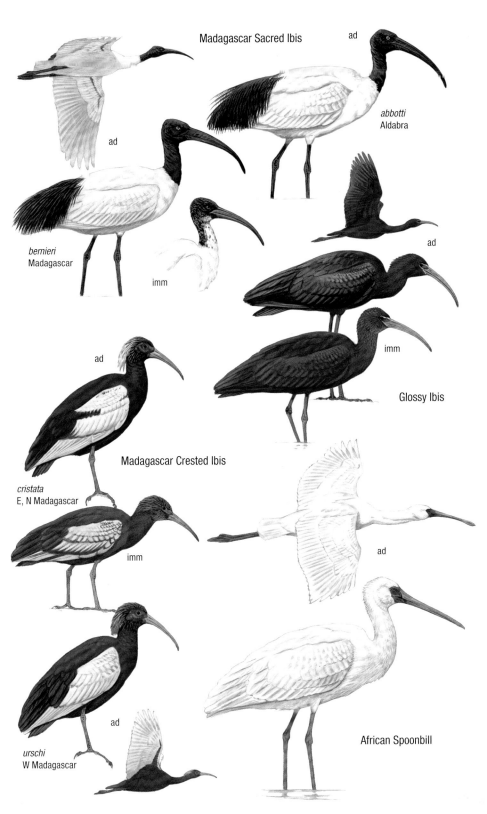

PLATE 13: MALAGASY REGION: GREBES AND PYGMY GOOSE

Little Grebe *Tachybaptus ruficollis* — 25 cm
See Madagascar Grebe for distinctions. **Voice** Song *a loud, rippling trill, going up and down scale*, often in chorus. Calls high *kik* or *pit*. **SHH** Widespread in Old World. In Madagascar, occurs up to about 1,500 m, patchy but common in appropriate habitat even in dry areas of S. Breeds on Comoros; absent from Mascarenes and Seychelles. Usually singly or in small groups on large or small lakes, usually with some surface vegetation, often on reservoirs around towns; moves to breed on seasonal ponds in W & S in rainy season. Dives for small fish and larger insects. Sometimes in large groups on larger lakes, especially in W Madagascar.

Alaotra Grebe *Tachybaptus rufolavatus* — 25 cm
From Little Grebe by *pale eye, strong bill,* and in breeding plumage by pale cinnamon wash on ear-coverts and foreneck; *whitish chin and throat*. Underparts more dusky. From Madagascar Grebe by pale eye, lack of white line behind and below eye in breeding plumage; chestnut wash on foreneck. Wings shorter than other species; never seen to fly. **Voice** Unknown. **SHH** Extinct; known only from L Alaotra, CE Madagascar, but not seen since 1988. Believed to have hybridised with Little Grebe, and last birds perhaps killed by fishing nets and introduced predatory fish. Appeared to have had similar habits to other small grebes, frequenting areas with floating vegetation and diving for fish and insects. **EX**

Madagascar Grebe *Tachybaptus pelzelnii* — 25 cm
From Little Grebe in breeding plumage by *white streak below and behind eye*, separating black cap from grey-and-rufous ear-coverts; *grey neck and chin* (chestnut on head being restricted to rear ear-coverts), and lack of swollen yellow gape-spot. More difficult *outside breeding season but whitish streak below eye usually evident*. **Voice** Song a *low-pitched, 'wooden' trill*, dropping at end, lacking musical quality of Little Grebe. Calls lower than Little. **SHH** Endemic to Madagascar; widespread except in S & SW, where absent from many places holding Little Grebe; possibly more widespread on C plateau and at higher altitudes than latter. Usually singly or in family groups, rarely in larger numbers; sometimes in mixed groups with Little Grebe, often on smaller, more vegetated waterbodies than Little, sometimes in rivers or pools in forest. Dives frequently; eats more insects than Little. **VU**

African Pygmy Goose *Nettapus auritus* — 30 cm
Highly distinct from other ducks in region; dull ♀ especially is more like a grebe, but pygmy goose is larger, has *triangular orange bill and, in flight, white patch on outer secondaries*. **Voice** Vocal, both sexes uttering repeated high-pitched twittering whistles resembling a squeaky bath toy. **SHH** Sub-Saharan Africa and Madagascar, where locally common in lowlands of W and rare elsewhere. Found on open, slow-moving or still freshwater wetlands with emergent or floating aquatic vegetation; strictly associated with water lilies and rare where these are absent.

Little Grebe

Alaotra Grebe

Madagascar Grebe

African Pygmy Goose

PLATE 14: MALAGASY REGION: DUCKS I

Fulvous Whistling Duck *Dendrocygna bicolor* — 50 cm
Medium-large size, **mainly buff plumage, creamy flank-line** and dark bar down back of neck distinctive; whitish **upper- and undertail-coverts contrast with dark tail in flight**. May float low in water, and even dive, when can be mistaken for White-backed Duck. In open water in bright light, may look very dark with white undertail, recalling Madagascar Pochard. **Voice** *Two resonant notes often repeated, tsoo-ee*. Less vocal than White-faced Whistling Duck: calls less familiar in Madagascar. **SHH** Africa, Asia and Americas. In Madagascar, most occur in lowlands of W; far less common, adaptable and ubiquitous than White-faced Whistling Duck. No confirmed vagrancy, reports on Mauritius and Reunion definitely or possibly escapes, not yet self-supporting populations. Mainly found in shallow freshwater or brackish wetlands with extensive emergent vegetation; forages in rice fields if not persecuted.

White-faced Whistling Duck *Dendrocygna viduata* — 45 cm
White face and black rear end diagnostic from Fulvous Whistling Duck. Juv much duller than ad, with chestnut breast; from similar-sized Meller's Duck by lack of streaking on breast and head, and from juv Comb Duck by lack of dark eye-line. **Voice** Very vocal and distinctive, loud, *three-note whistle swee-swee-sweeoo*, often repeated and given by many birds standing, swimming or in flight. **SHH** Africa and tropical Americas; in Malagasy region, common in Madagascar, where, with Red-billed Teal, the most familiar duck: found on most kinds of wetland, including rice fields, other wet agricultural land and even highly disturbed and polluted urban wetlands as in Antananarivo. Vagrant to Comoros and Aldabra group; recent releases in Mauritius may lead to established population.

White-backed Duck *Thalassornis leuconotus* — 40 cm
Low profile in water, *large-headed and hunch-backed silhouette and mottled brown plumage* all unique; in flight shows white rump and trailing feet. May recall Fulvous Whistling Duck; see that species. Madagascar endemic race *insularis* smaller and darker than African birds. **Voice** Usually silent, although may give various whistling calls, not documented in Madagascar. **SHH** Sub-Saharan Africa. In Madagascar, race *insularis* is endangered, population probably <2,000 restricted to W lowlands with very occasional records in E and around L Alaotra. Inhabits well-vegetated pools and lakes, marshes and dams. Mostly crepuscular, remaining in cover of emergent and floating vegetation, especially water lilies, where it can hide and sleep during hottest parts of day. Feeds by diving.

Comb Duck *Sarkidiornis melanotos* — 79 cm (♂), 64 cm (♀)
Ad is the only *large, black-and-white duck* in the region; ♂ much larger than ♀. Duller juv may be brownish and plainer, causing confusion with Meller's Duck, although lacks typical fine patterning of this and other *Anas* species. At all ages, wings all dark in flight. **Voice** Usually silent; may give soft wheezing hiss in flight. **SHH** Sub-Saharan Africa, Asia and South America. Breeding resident in Madagascar up to at least 1,500 m wherever habitat allows. Vagrant Aldabra, Comoros, probably Reunion (escape possibility not excluded). Frequents a wide range of freshwater wetland habitats; much less often in saline wetlands. [Alt: Knob-billed Duck]

Madagascar Pochard *Aythya innotata* — 46 cm
The only *Aythya* species resident in the region; this and White-backed Duck are the most persistent diving ducks, but beware other species including more similar African Pygmy Goose occasionally doing so. *White iris of ♂ and white wingbar* distinguish it from other resident waterfowl in Madagascar; this and White-backed Duck are the only ducks with running take-off from water. New reports would require elimination of vagrant Ferruginous Duck and (♀ or juv) Tufted Duck (Plate 99). **Voice** Generally very quiet, calling generally associated only with display. **SHH** Endemic to Madagascar, where probably restricted to one site (cluster of small crater lakes) in NW; last record elsewhere at L Alaotra, 1991, where no longer present. One of the world's rarest birds, following degradation of wetland habitats including exotic fish introductions: wild population *c.*20 birds, but now established in captivity. Has occurred in lakes and pools with extensive emergent vegetation and water lilies, or marsh-fringed open water with no emergent vegetation. **CR**

PLATE 15: MALAGASY REGION: DUCKS II

Madagascar Teal *Anas bernieri* — 40 cm
Dull plumage at rest relieved in flight by extremely striking, *diagnostic white panel on greater upperwing-coverts*. Commoner Red-billed Teal and Hottentot Teal both show contrasting dark caps, never present on Madagascar Teal; *bill colours also differ, being pinkish* on Madagascar Teal. **Voice** Occasionally vocal, most calls having a whistling quality; typically *vit-vit-vit* or *queck queck*; in distress a slightly hoarse *criiiiiik*. **SHH** Endemic to Madagascar; small population distributed mainly around mangroves with concentrations in CW, NW (especially Betsiboka and Mahavavy estuaries) and N. Feeds on other shallow, open water over soft mud, such as shallow open lakes (can be fresh water), estuarine mudflats and mangrove fringes. **EN**

Mallard *Anas platyrhynchos* — 55 cm
In Madagascar, domestic birds can be mistaken for much shyer Meller's Duck. Normally the only duck on Mauritius, but other escapes occur, and brown ♀ or eclipse ♂ may be misidentified as introduced Meller's Duck, although latter probably now extinct there; see that species. **Voice** Familiar quiet, nasal *raeb* in ♂, and archetypal 'quack' in ♀. **SHH** Widespread in N Hemisphere and widely introduced. In Malagasy region, domesticated birds throughout (not mapped), vagrant (probably wild) to Seychelles, and feral on Mauritius, mainly on highland streams and reservoirs of C & SW.

Meller's Duck *Anas melleri* — 65 cm
The only large *Anas* in Madagascar. Darker than brown plumages of Mallard, with *longer, greyer bill and plainer wings with white confined to trailing edge of secondaries*; striking pale underwing. Juv Comb Duck and whistling ducks, similar in size and often unexpectedly brown, may cause confusion but lack typical *Anas* plumage pattern. **Voice** ♂ gives quiet nasal *nhek* or *raaet*, often three notes; ♀ a subdued quacking *wark...wark...*, higher-pitched than that of Mallard. **SHH** Endemic to Madagascar. Mainly on forested streams and rivers of E, dispersing to adjacent areas, including lakes in deforested areas, when not breeding; mostly at 800–1,500 m; occasional in W. Introduced to Mauritius but probably now extinct there. Rare and declining due to hunting, habitat loss and degradation, and disturbance. **EN**

Red-billed Teal *Anas erythrorhyncha* — 45 cm
Contrasting dark cap and pale cheeks also in Hottentot Teal, which is smaller and dumpier, showing buff cheeks with a brown smear at rear, and blue, not red, bill. In flight, has *striking creamy secondaries* (speculum), whereas Hottentot Teal has much narrower band of pure white restricted to tips of secondaries, which are otherwise metallic green. May recall Madagascar Teal, but dark cap and pale cheeks of Red-billed diagnostic (more obvious than red bill); flying Madagascar Teal shows white stripe on greater coverts, not secondaries. **Voice** Generally quiet; ♂ gives soft whistle-like *whizzt* and ♀ a loud, sharp quack. **SHH** Africa and Madagascar, where often common throughout, sometimes in large flocks, at up to 2,000 m. Vagrant to Aldabra. Found on open or vegetated shallow lakes, marshes, reservoirs, flooded rice fields and small rivers throughout range; may rest on sea.

Hottentot Teal *Anas hottentota* — 35 cm
Noticeably the *smallest dabbling duck in the region*; fast flight also distinctive. Combination of *blue bill, dark cap and plain buff cheeks unique*; ♀ slightly duller than ♂, especially on flanks. See Red-billed and Madagascar Teals for distinctions. **Voice** Series of harsh, clicking *ke* notes when disturbed or flying, or within flock; ♀ gives a harsh quack. **SHH** Sub-Saharan Africa and Madagascar, where found throughout up to 1,500 m, although very patchy in S and C plateau. Frequents shallow freshwater marshes and lakes; feeds on land and in flooded fields including rice paddies.

Garganey *Anas querquedula* — 40 cm
Breeding-plumaged ♂ distinctive (eclipse not expected in region); ♀ and juv show *dark eyestripe and pale supercilium and loral spot*. In flight has striking white borders of roughly equal width above and below speculum. Could occur in range of Madagascar Teal, differing in head markings, bill colour and wing pattern. ♀ Mallard much larger. **Voice** Generally silent in non-breeding range; ♂ has dry rasping rattle, ♀ a low quack. **SHH** Breeds in Palearctic, migrating to tropical Africa and Asia. In Malagasy region, annual migrant in very small numbers to granitic Seychelles and Amirantes; vagrant elsewhere. Mainly on low-lying wetlands with partially submerged and emergent vegetation, usually fresh water, rarely saline.

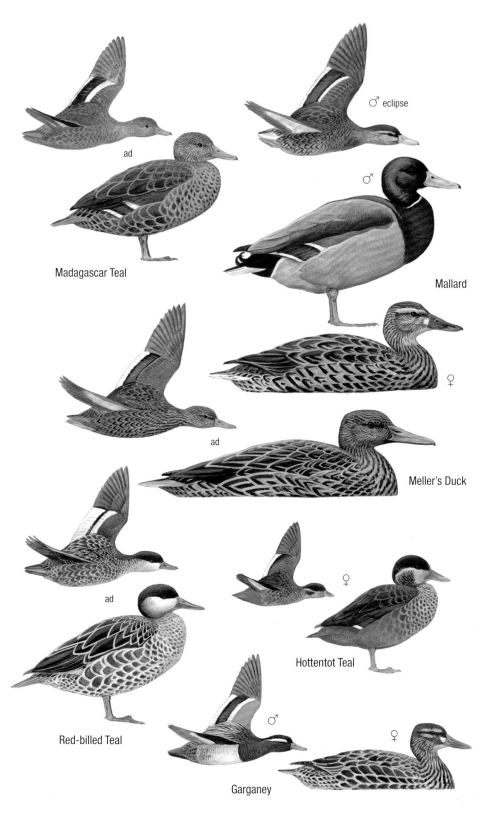

PLATE 16: MALAGASY REGION: DIURNAL RAPTORS I

Madagascar Cuckoo-hawk *Aviceda madagascariensis* 43 cm
Very similar to Madagascar Buzzard; at rest, tail often held slightly open showing slightly forked effect; *wings long, nearly to tail-tip*. Rounded head and bulging eyes; *legs short, appearing feathered to feet*; sometimes shows slight crest. Pattern of dark on breast may differ; darker on breast and paler on belly, the reverse of Madagascar Buzzard, but variable. In flight, *diagnostic pattern of dark bars on under-primaries* usually evident; tail also more strongly barred than buzzard. *Lacks dark patch on lesser underwing-coverts*. **Voice** Quiet but penetrating, medium- to high-pitched whistle: *peeew* or *peeer*. **SHH** Endemic to Madagascar, mostly below 1,500 m; scarce in forest and wooded areas throughout, unrecorded recently in far S, more regular on C plateau than buzzard. Usually singly in tops of trees, moving irregularly within canopy while hunting or resting; soars occasionally for short periods; possibly crepuscular. *Display distinctive; tilts in flight so wings point up and down, and flutters wings quickly.*

Madagascar Harrier *Circus macrosceles* 59 cm
Only harrier in Madagascar and Comoros; from Madagascar Buzzard and Black Kite by *slimmer build, hunting flight low over ground with wings in V*; female from Black Kite by white rump. **Voice** Male in display-flight gives repeated, medium-pitched, slightly hoarse whistle *whier-whier* or *wheo-wheo*. Also harsh *kekekekek* in alarm. **SHH** Endemic to Madagascar and Comoros; throughout Madagascar to *c.*1,700 m, but very patchy; now mostly around L Itasy, Ambohitantely, on Horombe plateau and Bealalana on C plateau, and in some W wetlands. Scarce or absent in far N, far S & E, but regular around L Alaotra. On Comoros, scarce and scattered on Grande Comore, Mohéli and Anjouan; vagrant to Mayotte, none recently. Singly or pairs over grassland or valley-bottom marsh; also reedbeds and lake edges; especially on Comoros, also over forest. Hunts low (2–5 m above ground), making short swoops to catch insects, birds and reptiles. Often perches on ground on termite mound or rock. Flies high calling and diving in display. **VU**

Reunion Harrier *Circus maillardi* 54 cm
The only resident raptor on Reunion. Strong sexual dimorphism and distinct juv plumage, yet unlikely to be mistaken except for very rare vagrant Black Kite, which has forked tail, numerous plumage differences and quite different flight; the only other visiting raptors on Reunion are large falcons. **Voice** ♂ calls a wailing *whier...whier*; ♀ a chattering *kukukukukuk*, pair in display calling in turn. **SHH** Endemic to Reunion, where widespread; commonest below 1,000 m. Scarce (<200 pairs) and affected by habitat degradation and poaching. Found in most habitats including the few wetlands on Reunion, but favours mosaics of open and wooded (especially native forest) habitats. ♂ performs spectacular display-flights, stooping, levelling and rising again while also rolling body from side to side. **EN**

Madagascar Buzzard *Buteo brachypterus* 50 cm
See Madagascar Cuckoo-hawk for distinctions from that species. Stockier, with *shorter tail and broader wings* than other raptors in the region; *strong dark bar on leading edge of underwing*. From potential vagrant Common Buzzard *B. buteo* (reported from Europa and Comoros) by lack of rufous in tail, whitish rump and strong dark bar on underwing. **Voice** *Loud sequence of rising notes often followed by shorter notes in more rapid sequence*: *peeer... piiiieerrr.... piiiiiiiieeer... peer..peer....peer*; or *kaarr... kiiiirrrr... kieeeeeer ... keer*. **SHH** Endemic to Madagascar; widespread to 2,500 m, but absent from treeless areas, including much of C plateau around Antananarivo. Heavy flight with strong flapping and short glides; soars frequently but rarely very high; often perches on conspicuous trees. Usually around forest edge or over wooded areas.

PLATE 17: MALAGASY REGION: DIURNAL RAPTORS II

Bat Hawk *Macheiramphus alcinus*   45 cm
From large falcons in flight (with some difficulty in twilight) by *blackish plumage with variable white patches on throat and underparts*; somewhat broader wings, more leisurely flight, *large yellow eye*; perches more horizontally. Outline recalls Broad-billed Roller, but much larger, with small, dark bill. **Voice** Usually silent; in display or in aggression, high-pitched broken whistles *kik-kik-kik-kik-keeee* or *kwieck-kwieck*. **SHH** Sub-Saharan Africa and SE Asia to New Guinea. Widespread but scarce in Madagascar; most regular in CW lowlands; also scattered sites in E (only in lowlands) and on C plateau; rare in N and far S. One unconfirmed report from Comoros (Anjouan). In degraded areas, forest, wooded cultivation, near cliffs or caves holding bat roosts; often hunts near small towns in CW Madagascar.

Black Kite *Milvus migrans*   60 cm
From other large raptors by *forked tail*, usually evident although may appear square when widely spread; uniform dark coloration with pale *upper wing-covert bar* across angled wings; distinctively agile flight with loose, easy wingbeats. In resident race *parasitus*, bill yellow in ad but black in juv. See also vagrant nominate race (Plate 101). **Voice** Whining *piiieeerrrrrr*, starts high-pitched, becoming even higher, ending in lower, slightly trilled whistle. **SHH** Africa, Eurasia to Australasia. Race *parasitus* in sub-Saharan Africa, Madagascar and Comoros, vagrant to Seychelles, Juan de Nova and Reunion. On Comoros, has declined greatly; now absent from Grande Comore and scarce, possibly migrant, on other islands. In Madagascar, widespread to c.2,000 m, commonest in W & S; scarce on C plateau and in forested parts of E. Common around wetlands and cultivation, coasts, towns and villages, often in small or larger groups, absent from dense forest. Flies low to pick up food from ground with lazy twisting flight; also soars; often follows locust swarms or smoke from bushfires. Vagrant to Aldabra group; nominate race vagrant Seychelles. Race *parasitus* of S Africa and Madagascar may belong in a separate species, resident or migrant within African and Malagasy regions, Yellow-billed Kite *M. aegyptius*.

Madagascar Fish Eagle *Haliaeetus vociferoides*   75 cm
From all other raptors by *very large size, relatively short tail and large head*. **Voice** Shrill, bright and ringing yelps reminiscent of a large gull, *ko ko koy koy koy koy*. **SHH** Endemic to Madagascar; scarce in W & NW, in mangroves, on coastal islands, lakes and large rivers. Well inland only along Tsiribihina R around Miandrivazo, and at Ankarafantsika (Ampijoroa). Commonest in CW and along NW coast. Soars or flies heavily with deep wingbeats; spends much time perched on lakeside trees or on coastal rocks, searching for fish prey; robs Black Kites for food. **CR**

Madagascar Harrier-hawk *Polyboroides radiatus*   68 cm
Largest raptor over most of Madagascar, although rare fish eagle is larger. Ad distinctive. Imm very different, but readily separated from other raptors by *large size, pale head with dark eyestripe*, bare facial skin and *pale fringes to upperparts*. **Voice** High-pitched, peevish, whining or 'feeble' *weeee-eer, qweeeer* or *huiiiiiiii*, dropping sharply at end; lacks trilled character of Black Kite call. **SHH** Endemic to Madagascar, fairly common to c.1,300 m in forest, wooded areas and cultivation; rare in open savanna and on C plateau, mostly in forest. Soars or flaps with heavy wingbeats. Often in pairs; searches crevices in trees and cliffs using long legs to extract prey such as lemurs.

PLATE 18: MALAGASY REGION: DIURNAL RAPTORS III

Madagascar Serpent Eagle *Eutriorchis astur*   66 cm

From Henst's Goshawk, with difficulty, by *slightly crested, elongated appearance to head, dark-barred upperparts* (most conspicuous on tertials and primaries, but hard to see), broader and duller breast-bars; at close range, *slightly bulging bill shape, large eye and less marked ridge over eye* give face a less severe expression than Henst's; dark bars on tail narrower than pale bars. Scales on tarsus knobbly rather than smooth, and toes short and thick. *Imm has distinctive pale fringes to upperparts, still present on some older birds, especially on nape and crown*. **Voice** Very distinctive, *rhythmic series of 3–4 quite loud low or medium- to low-pitched 'breathy' notes* at intervals of 1–2 s, usually followed by a single lower note: *wugh… wugh… wugh… wugh… ruh*. **SHH** Endemic to Madagascar; rare in rainforest in E from Tsaratanana (CN) to Ranomafana (SE), mostly below *c*.1,000 m (but higher in N), only known from primary forest. Very elusive, only below canopy, where hunts (mostly chamaeleons) by perching in sub-canopy and flying irregularly from perch to perch. Flight less powerful than Henst's and direct, with more floppy wingbeats; long tail noticeable. **EN**

Frances's Sparrowhawk *Accipiter francesiae*   28 cm (♂), 35 cm (♀)

On Comoros, no other small raptor; from possible vagrant small falcons by short wings and yellow eye. In Madagascar, ♀ and imm, with difficulty, from ad Madagascar Sparrowhawk by *barred undertail-coverts, unbarred throat with variable dark central streak*, more conspicuous in imm and sometimes vestigial in ♀, *duller and more widely spaced brown bars on breast in imm*; also shorter central toe. **Voice** Nominate gives *short, medium- to high-pitched simple pik or peek*, singly or repeated at 2–3 s; Comorian races have compound calls, possibly different between islands; short, high simple *rick-rickrick-rick-rickrickrick* (Grande Comore), high, grating, squeaking *criik-criik-criik-criik* (Anjouan) or short, high simple *rick-rick-rick-rick* (Mayotte). **SHH** Endemic to Madagascar and Comoros. In Madagascar, nominate race fairly common throughout to *c*.1,600 m, but scarce in savanna and much of C plateau. Distinctive forms on Comoros are probably a separate species (or even three species): scarce on Grande Comore (race *griveaudi*) only from sea level to 1,800 m on Mt Karthala; rare on Anjouan (*pusillus*), mostly in highlands; and very common throughout on Grande Terre, Mayotte (*brutus*); one record from Petite Terre, Mayotte, and inexplicably absent from Mohéli. Throughout range, mostly in wooded areas, including plantations and gardens, also forest; scans for (mainly) lizard prey from perch in vegetation, when often tame and approachable. Sometimes soars; quick snappy wingbeats in level flight.

Madagascar Sparrowhawk *Accipiter madagascariensis*   29 cm (♂), 40 cm (♀)

For distinctions from Frances's Sparrowhawk, see that species. Adult from Henst's Goshawk with great difficulty (especially ♀ Madagascar and ♂ Henst's), by *unbarred undertail-coverts, thinly streaked throat, not finely barred or mottled, and long central toe*. Large ♀ Henst's and small ♂ Madagascar usually distinctive by size; imms with broadly streaked underparts very similar but streaking thinner on Madagascar. **Voice** *Distinctive short, high-pitched squeaky whik, whisk or eeesk* repeated every 1–4 s, very mechanical. **SHH** Endemic to Madagascar, where scarce in primary forest throughout to about 1,000 m, rarely higher; probably commonest in S spiny forest where Henst's rare or absent. Hard to see, discreet, perches often high in canopy of primary forest; sometimes in adjacent degraded vegetation. Catches birds in determined pursuit; sometimes soars **NT**

Henst's Goshawk *Accipiter henstii*   52 cm (♂), 62 cm (♀)

See Madagascar Sparrowhawk, Frances's Sparrowhawk and Madagascar Serpent Eagle for distinctions. **Voice** Vocal, especially in breeding season; *call a clear, loud, high-pitched peer or wheer*, repeated every 2–10 s; territorial call a chanting sequence of similar notes *peer-peer-peer-peer-peer-peer…*, repeated at *c*.1 per second in a long, forceful sequence over 3–20 s; sometimes two birds together. **SHH** Endemic to Madagascar; in primary forest in N, W & E, mostly below *c*.1,000 m; scarce or absent in S & SW and on C plateau. In primary forest or forest edge; hunts by moving from perch to perch in sub-canopy; also soars. **NT**

PLATE 19: MALAGASY REGION: FALCONS I

Madagascar Kestrel *Falco newtoni*   25 cm
Polymorphic, but all morphs from Banded Kestrel by *small size, slim build, spotted or streaked rather than barred belly, dark eye and rufous upperparts*. From potential vagrant kestrels by small size and spotted or chestnut underparts. **Voice** Medium-high rising *kreee, often repeated or extended in longer series with variable tempo, kreekreeekreeeekreeeekreeeekreee...*, or *kreee... kreee...kreee...kreee...*. **SHH** Endemic to Madagascar and Aldabra. Nominate race common throughout Madagascar except in dense forest, to *c*.2,500 m, including centres of towns and cultivation. Often in pairs, on tall buildings, trees and cliffs. Scarce but widespread on Aldabra, as endemic race *aldabranus* (possibly invalid; smaller, only has pale morph); vagrant Assumption.

Mauritius Kestrel *Falco punctatus*   23 cm
The only resident Mauritian raptor. Separated from migrant falcons (usually Eleonora's Falcon and Sooty Falcons, although others could occur) by *much smaller size and distinctive profile with rounded wings*. Can be mistaken in flight for similar-sized Madagascar Turtle Dove or Spotted Dove, perhaps even Zebra Dove, but kestrel's flight is more rapid, direct and powerful, with many plumage differences including tail pattern. **Voice** A long, sharp whine, higher-pitched in ♂ than ♀; also *chip* or higher-pitched *tink-tink-tink*. **SHH** Endemic to Mauritius, formerly on brink of extinction: wild population in 1970s *c*.4 birds, but now widespread in and around Black River Gorges and Bambous Mts, population *c*.100 pairs, mainly in native forest, less often exotic forest and agricultural land; has bred in villages. Specialist predator of *Phelsuma* day-geckos. **EN**

Seychelles Kestrel *Falco araeus*   19 cm
The only resident falcon in the granitic Seychelles. World's smallest *Falco* species, much smaller, generally paler and slower-flying than vagrant falcons (Sooty, Eleonora's, Red-footed and Amur Falcons, Eurasian Hobby, Lesser Kestrel and Common Kestrel; Plates 102 and 103). Confusable in flight with Madagascar Turtle Dove or Zebra Dove, but kestrel's *flight is more rapid, direct and powerful*; many plumage differences, most obviously tail pattern. **Voice** Main call a high-pitched, forceful *shikshikshikshikshikshik*; also a shrill *shkeea-shkea-shkea-shkea*, or more rapid *kikikikiki*. **SHH** Endemic to granitic Seychelles: Mahé and its larger satellites, Praslin (reintroduced), Silhouette and North; recorded La Digue, Curieuse and Félicité. Widespread in forest, cultivation and urban areas from coast to mountains. In flight rapid wingbeats interspersed by gliding; never hovers. **VU**

Banded Kestrel *Falco zoniventris*   27 cm
From Madagascar Kestrel by *barred belly, pale eye, greyish or brownish upperparts and larger size*. Resembles Frances's Sparrowhawk more closely, from which separated by longer wings (to near tail tip) and diffusely streaked breast. **Voice** Truncated, *high-pitched chk or chik* and a short sequence of 3–5 high notes *chk-kskskss*, over 2–3 s, first note slightly lower-pitched and sequence dry and mechanical. **SHH** Endemic to Madagascar, where widespread below *c*.1,200 m, but patchy and scattered; commonest in CW to SW; apparently absent from SE and scarce in E; absent from C plateau except on margins of native forest blocks. Usually solitary, hunting from perch in large tree; rarely flies for long periods.

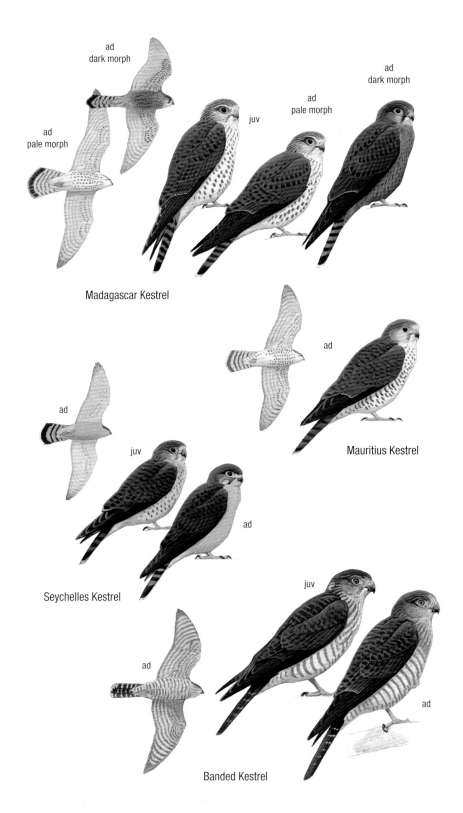

PLATE 20: MALAGASY REGION: FALCONS II

Eleonora's Falcon *Falco eleonorae*   40 cm
Most similar to Sooty Falcon: both have long, pointed, narrow wings, long, narrow tail and slender body, giving them an appearance often distinct from other falcons. Sooty typically *c*.20% smaller (no overlap in wing length) and has *slightly projecting central tail-feathers*, whereas Eleonora's has a *round-tipped tail*. Dark morph Eleonora's (<25% of population) uniquely blackish, much *darker overall than adult Sooty*, which also has contrasting darker primaries and tail tip. Pale-morph adult more similar to juv Sooty but has *very dark underwing with slightly paler primaries*, whereas Sooty has barred underwing-coverts. Juv of both morphs paler below than ad pale morph; distinguishing these from juv Sooty is challenging. Tail shape and size difference useful. *Sooty has greyer upperparts with contrastingly darker outer wings and unbarred tail above, whereas Eleonora's has uppertail barred dark grey and rufous with broad dark subterminal band and rufous tip*. Juv Sooty usually has paler underparts, streaks greyer, less brown, more diffuse or drop-shaped (and drop-streaks sometimes merging on upper breast), but streaking and background colour variable; on primaries and secondaries, pale bases contrast with darker tips and trailing edge. For distinction from Peregrine see below, and vagrant Amur Falcon, Red-footed Falcon and Eurasian Hobby, see Plate 103. **Voice** Usually silent in Malagasy region but may give *kee kee kee*. **SHH** Breeds on islands of Mediterranean Sea, Morocco and Canary Is. Migrates to SE Africa and Malagasy region, especially Madagascar where present Nov–May, preferring humid, mountainous parts of N, E & C plateau at 600–1,500 m, occasionally 2,000 m. Rare in Antananarivo; regular in tiny numbers in W Madagascar, Comoros, Mauritius and Reunion; vagrant to Seychelles (especially Aldabra; less frequent on granitics). Typically over primary and secondary humid forest, and along ecotones such as forest edge or rivers, feeding on insects on the wing. Flight elegant, powerful and active: wingbeats shallow and often slow, even languid, yet agile and powerful when chasing prey. Soars and glides above canopy and along ridges. Seen singly or in groups of up to eight.

Sooty Falcon *Falco concolor*   34 cm
Ad distinctive with *blue-grey body plumage* (not shown by any other falcon in region), black primaries and bright yellow bare parts; otherwise similar to Eleonora's, especially juvs; see that species, as well as vagrant Amur Falcon, Red-footed Falcon and Eurasian Hobby (Plate 103). **Voice** Generally silent in Malagasy region, but could give *kee kee kee* calls as in breeding areas. **SHH** Like Eleonora's Falcon, breeds mainly in passerine migration seasons (hence food supply at this time), followed by migration (mainly) to Madagascar and shift in diet to insects. Breeds in NE Africa and Middle East, and migrates to SE Africa and Malagasy region, especially Madagascar, where present Nov–May mainly in W & S and on C plateau; best known from W coastal lowlands. Small and irregular numbers elsewhere in region. In Madagascar, typically over flat, lowland forest, scrub and grassland with scattered trees; also rice fields and towns, including C Antananarivo. Preference for dry, flat areas of C, W & S Madagascar contrasts with humid, mountainous areas of N & E preferred by Eleonora's. Gregarious, occurring in flocks of up to 20, exceptionally 100.

Peregrine Falcon *Falco peregrinus*   34 cm (♂), 39 cm (♀) (*radama*)
Much larger than other resident falcons, but confusable with Sooty and Eleonora's Falcons, both of which are smaller, slimmer, more slender-winged and longer-tailed, but proportions variable and easily misjudged. Highly variable species, but small, dark endemic race *radama* distinctive in region with *black-and-white barred underparts* of ad. Juv less distinctive, but Peregrine browner (less grey) above than Sooty and Eleonora's, and paler below because of narrower, lighter brown streaks on whitish or buff background. Bat Hawk could cause confusion at dusk; see that species, and also vagrant (or migrant) 'tundra' Peregrine Falcon (race *calidus*) and vagrant Saker Falcon (both shown on Plate 102). **Voice** Usual adult call *keh-keh-keh*, mainly during breeding period, similar to other falcons. **SHH** Global, with sedentary tropical populations and migrants at higher latitudes (including *calidus*). Resident endemic race *radama* recorded throughout Madagascar and Comoros. Found in almost any habitat.

PLATE 21: MALAGASY REGION: LARGER GAMEBIRDS

Grey Francolin *Francolinus pondicerianus*   34 cm
The only francolin in the region: *rufous tail visible in flight, and dull red legs and feet unique in region's gamebirds*. Twice the size of quails and buttonquails, of which four species present on Reunion, and about half the size of very differently coloured Helmeted Guineafowl, a rare exotic on Mauritius. On Reunion, *buff-orange throat and rufous tail* distinguish it from brown ♀ or imm of somewhat smaller Madagascar Partridge; see also Red Junglefowl. Most frequently detected by loud advertising calls, dissimilar to any other species in region except, tenuously, strident calls of Common Myna. **Voice** ♂ advertising call a common sound of lowland Mauritius: a series of relatively quiet *check-check* notes, followed by much louder, rapid, high-pitched trisyllabic *ketitur* or *pateela*, repeated; ♀ may interject single, high-pitched *kila* or *pela* notes. Alarm call a high, whirring *khirr-khirr* call before taking flight or when excited. **SHH** Introduced (S Asia) to Desroches and Coëtivy (Seychelles), Mauritius (except SW uplands), Reunion (lowlands of NE, W & SW), Rodrigues, and (at least formerly) St Brandon. Prefers stony, hot, sunny areas, including sugarcane plantations and other cultivation, dry savanna and margins of roads, airstrips and paths. Most often seen walking or running on ground, where all feeding occurs.

Madagascar Partridge *Margaroperdix madagarensis*   28 cm
Beautiful *black, white and rufous* ♂ unlike any other species in region. Yellowish-brown, scaled pattern of ♀ much less distinctive but no other gamebird between size of quail and guineafowl occurs in Madagascar. On Reunion, confusable with Grey Francolin; see that species. **Voice** Contact call a low *bub...bub...*, repeated in long sequences; when flushed, soft rolling, chuckling or trilling notes in rapid succession; range of quieter calls (a cackle, also *poa-poa-poa* and a low *peet*) reported. Voice not well understood; appears to be a rather silent species, and report of loud advertising call needs confirmation. **SHH** Endemic to Madagascar, where widespread from sea level to highest mountains, avoiding closed-canopy or spiny forest, and preferring dry, open or bushy habitats, often mosaics, including scrub, grassland and weedy cultivated fields (if dry). Strictly terrestrial, secretive and quail-like; very poorly known in the wild. Introduced and fairly common on Reunion in similar habitats, throughout but particularly in uplands at 1,000–2,400 m.

Red Junglefowl *Gallus gallus*   43 cm + 28 cm tail projection (♂)
Unique in region, and familiar as ancestor of domestic chicken. Brown ♀ *much larger and darker than Grey Francolin*, with distinctive chicken gait; no populations of Helmeted Guineafowl known on Reunion. Naturalised Red Junglefowl and domestic chickens usually distinguished by the former carrying tail horizontally in both sexes, absence of a comb in ♀, dark or slate-grey leg colour, and an annual eclipse moult in ♂. **Voice** ♂ gives familiar *cock-a-doodle-doo* of domestic chicken, especially at dawn and dusk, but can be shriller. Many other calls described, with a range of clucks and cackles. **SHH** Introduced (Asia) to Reunion at 200–1,200 m, especially 500–800 m; in E, especially above St-Benoît with occasional records elsewhere. Mainly in dense, humid, secondary forest (often Rose-apple or bamboo thicket and abandoned *Citrus* plantations), occasionally in open pasture, grasslands and cultivated field margins. Often in coveys of 1 ♂ and 4–5 ♀♀. Rarely observed, and usually detected by voice; when located, shy and unapproachable.

Helmeted Guineafowl *Numida meleagris*   61 cm
The only large gamebird in the region, apart from Red Junglefowl, which is confined to Reunion where guineafowl are not naturalised. Part-grown, mobile young confusable with other species if ads not seen. **Voice** Extremely loud and distinctive, and familiar from domestic birds: *harsh cackling sequence of long kaaaaaa and short kak notes repeated over long periods*; also a harmonious whistled two-note call, *buck-wheat*, and single *cheenk*. **SHH** Presumed introduced (Africa) to Madagascar (but may be native there), Comoros, Mauritius (very rare in SW alone), Juan de Nova; elsewhere, occasional free-ranging domestic birds. Mainly in lowlands, locally up to 1,600 m in Madagascar, in forest, savanna and grassland but absent from grasslands of C plateau of Madagascar. On Comoros, known mainly from Anjouan, rarely Grande Comore, possibly extinct on Mayotte, no records on Mohéli. Usually in pairs or flocks of 10–20, but up to 150. Shy, runs when alarmed.

PLATE 22: MALAGASY REGION: QUAILS AND BUTTONQUAILS

Common Quail *Coturnix coturnix* 17 cm
Most often found and identified by ♂ advertising call. When seen, several confusion species in Madagascar and especially Reunion. Madagascar Partridge *c.*50% larger. For distinctions from other 'true' quails and buttonquail, see those species. **Voice** Advertising call of F *whIT...whit-whIT* as in rest of range; also *kreee* when flushed. **SHH** Eurasia and Africa. Presumed native on Comoros and Madagascar, probably introduced to Reunion, and vagrant to granitic Seychelles. In Madagascar, mainly in C, rarely N, highlands above 750 m; on Comoros and Reunion, often associated with mountain areas but also in lowlands. Found in open grassy savanna, grassland, cultivated areas and open tree-heath. When flushed, flies low for short distance before dropping into dense cover and running away; usually impossible to see on ground and difficult to flush again.

Harlequin Quail *Coturnix delegorguei* 17 cm
Usually detected and identified by ♂ advertising call. When seen, confusable with two species: Madagascar Buttonquail (which see) and Common Quail. ♂ distinctive if seen well; even darkest Common Quail has whitish belly, pale flank-streaks and less bold face pattern. ♀ very similar to ♀ Common Quail, but has darker underparts with more prominent spotting on upper breast and belly, and undertail-coverts more tawny-buff like breast and flanks; upperparts duller, less streaked with more prominent cross-barring; in hand shows plain outer webs to primaries, barred in Common Quail. **Voice** Advertising call of F *whit-whit-whit*, rarely *whit-whit* with *notes of equal length and evenly spaced*; also a low *peet* when flushed. **SHH** Africa and Madagascar, where locally common in lowland (unlike Common Quail) open habitats such as savanna, croplands, dry paddies and waterbodies, and degraded thorn scrub. Behaviour as for Common Quail.

King Quail *Coturnix chinensis* 14 cm
Distinguished from Common Quail and Jungle Bush Quail by its smaller size; see also Madagascar Buttonquail. Flushed groups or pairs usually include at least one distinctive ad ♂. In lone ♀, *barring extends much further down underparts and onto flanks than in all other species*. Contact call, sometimes also heard when flushed, distinctive. **Voice** Advertising call an *explosive, rich, downslurred crowing whistle*, *tchiew*, repeated 2–3 times. Soft contact calls *muiii-muiii-muiii*. Also a long, rasping hiss and a series of sharp *cheep* notes. **SHH** Introduced (Asia) to Reunion, where restricted to lowlands to 600 m; mainly in W, also NE & S, best known around St-Paul and other littoral lagoons. Uses dry grassland, croplands and savanna but often in damp habitats, unlike other quails. Coveys usually contain <10 birds.

Jungle Bush Quail *Perdicula asiatica* 17 cm
Rather plain upperparts and *entirely plain cinnamon-brown underparts of* ♀, and *black-and-white-barred underparts and cinnamon supercilium and throat of* ♂, all unique. **Voice** Advertising call of ♂ a musical trill *didididididididid*, with even tempo and pitch, only first and sometimes last notes lower, starting quietly but intensifying; may recall a falcon. Harsh notes given in alarm or in coveys. **SHH** Introduced (Asia) to Reunion; mainly in W below 600 m, with scattered records elsewhere. Uses wide range of dry, open habitats, cultivated or not. Coveys consistently large, often 12–25 individuals.

Madagascar Buttonquail *Turnix nigricollis* 15 cm
Related to waders, but more quail-like; however, slimmer, with longer neck, *yellow eye and longer, thinner bill than any quail*, and smaller than all except King Quail; *pale wing-panel* shows in more erratic, less direct flight with short glides. When flushed, often lands in sight or can be relocated on ground, unlike quails. **Voice** Song *a very low-pitched humming lasting 3–8 s*, starting softly and dying away or ending abruptly. Also a quiet, short *chut...* repeated for prolonged periods. **SHH** Endemic to Malagasy region. Found throughout Madagascar and its offshore islets, and Reunion, except highest mountain-tops and dense forest. On Glorieuses and Juan de Nova, native or introduced status not known. Often abundant in open grassy, sandy or bushy habitats and forest clearings (including barren grasslands of C Madagascar), also penetrating primary open-canopy forest types including deciduous and spiny forest. Duller ♂ undertakes most breeding duties.

PLATE 23: MALAGASY REGION: MESITES

Brown Mesite *Mesitornis unicolor* — 30 cm
Distinguished from all other ground-dwelling birds of Malagasy rainforest, especially Madagascar Crested Ibis, Madagascar Wood Rail, Madagascar Turtle Dove, couas and ground-rollers, by *horizontal carriage, uniform brown coloration*, short bill and long thick tail. Pattern of head and neck somewhat variable but not shown to be linked to sex or geography. **Voice** *Low, mournful piew.. piew...piew..piew.. piewpiewpiew, or crucrucrucru cru cru criew creiw* or as a duet *crucrucrucru ..cruiew dool creiw criewdool criewdool*; contact call low toneless *ksst* or *crrt*. **SHH** Endemic to Madagascar; only in primary E rainforest to *c*.1,000 m, where rare; very locally in degraded edges of primary forest. Usually in pairs or family groups, on ridges or steep slopes, in dense vegetation; shy and hard to see. **VU**

White-breasted Mesite *Mesitornis variegatus* — 31 cm
From all other terrestrial birds of W Malagasy forest, especially Madagascar Crested Ibis, Madagascar Turtle Dove and couas by *horizontal carriage, whitish supercilium, rufous breast-band, dark-spotted breast and straight bill*. Does not normally overlap with other mesites, but very rare observations in lowland rainforest, at sites where Brown Mesite recorded, indicates need for careful identification; similar size and shape, but coloration and voice very distinct. Like Brown, pattern of head and breast, especially breast-band, variable but not shown to be linked to sex or geography. **Voice** ♂ sings solo, a rising *whoorreeeeeee*, also more often a duet *wheee-titititwheeeee-tititiiti-wheeeeeeeeee..., rising and becoming more frantic* over 10–20 s. Contact call *ksccht* or *zk*; also bubbling *bub-bub-bub-bub-bub*. **SHH** Endemic to Madagascar. Patchy distribution in W in Menabe region N of Morondava, Ankarafantsika N to Ankarana and Analamera and Daraina dry forests in far NE, and in lowland CE, at Ambatovaky and Betampona; only known sites mapped. Commonest in intact forest, shy and hard to see near habitation. Usually in pairs or family groups, moves slowly, turning leaves over with bill, making rustling noise in dry leaf litter. **VU**

Subdesert Mesite *Monias benschi* — 32 cm
From all other terrestrial birds of SW Malagasy spiny forest, especially Madagascar Turtle Dove, couas and Long-tailed Ground-roller, by *relatively long, decurved bill, pale supercilium, reddish (C) or black-spotted (F) breast*. **Voice** *Unique chorus song combination of trills and whistles*, made by several birds at once. Contact call quiet *nak*; also a bubbling purr. **SHH** Endemic to Madagascar; found only in spiny forest in SW, between Fiherenana R and Mangoky R, where patchy but sometimes common; usually in groups of 2–8, rarely more than 2♀♀ per group. Mesites are remarkable and ancient birds restricted to Madagascar, only distantly related to any other birds, and this species is perhaps the most extraordinary of them all. **VU**

PLATE 24: MALAGASY REGION: FLUFFTAILS AND RAILS I

Madagascar Flufftail *Sarothrura insularis*   14 cm

Tiny size and *strongly streaked plumage* distinctive; ♂ with *contrasting chestnut head and tail*, ♀ more strongly streaked and barred than ♀ Slender-billed Flufftail. Occurs alongside much rarer Slender-billed, which is however confined to marshes; see that species. Baillon's Crake only slightly larger but adults distinctive with grey underparts, and juv shows range of obvious plumage differences including barred flanks. Other rails are much larger, although chicks of other species could be confused if accompanying ads not seen. Voice unique among Malagasy rails. **Voice** Territorial song loud, high-pitched and ringing, carrying up to c.800 m, lasting c.20s: *1–2 trilled notes followed by several loud double notes, then a long series of single notes, accelerating and decreasing in pitch, become faint towards end*; drr—drr—KEEKEE—KEEKEE—KEEKEE—KEE—KEE—KEE—KEE—KEEK—KEEK—KICK—KICK—kick-kick-kick-kick-kikikikikikikiki.... Other vocalisations represent excerpts from this call. **SHH** Endemic to Madagascar, where common in humid areas, using wide range of degraded and intact habitats, wet and dry: secondary brush including areas dominated by heath vegetation and large ferns, grassland at forest edges, clearings and herbaceous marshes; occasionally in forest, when usually but not always close to edge. Extremely secretive and difficult to flush, but vocal.

Slender-billed Flufftail *Sarothrura watersi*   15 cm

Differs from Madagascar Flufftail by *longer, finer bill, much longer toes, and drabber, plainer plumage*: ♂ lacks streaking of Madagascar Flufftail, and ♀ plumage also quite unlike intricate brown-and-black-barred plumage of the commoner species. Requires distinction also from Baillon's Crake, although numerous plumage differences exist. Voice highly distinctive. **Voice** Territorial song composed of simple sequences of a *medium- to low-pitched knocking or croaking sound, given singly (repeated rapidly) or as doublets or triplets*: triplets *tokotok-tokotok-...* or *ug-ugug...ug-ugug...*; doublets *gu-duk gu-duk gu-duk...* or *GOO-goo* (alternate syllables accentuated), or single-note series *tok-tok-tok-tok-....*; starts slowly then speeds up and becomes more rhythmic. Rhythm may recall a galloping horse. **SHH** Endemic to Madagascar, where restricted to wetlands of E region between Tsaratanana and Andringitra massifs at 900–1,800 m. Known sites alone mapped; poorly known and may occur higher and/or lower, and in other humid areas. May belong in own genus, *Lemurolimnas*. **EN**

Sakalava Rail *Amaurornis olivieri*   19 cm

Distinctive, elegantly plumaged small rail with *brilliant yellow bill and red eyes and legs*; other plain-coloured rails in Madagascar are all much larger. Duller juv is still distinctive. **Voice** Frequent *tic–tic* or *tic-tic-tic huaww* vocalisations accompanied by upward-flicks of tail. Call *truwrurururururururururu* lasting 4–6 s heard frequently in morning and at dusk as birds prepare to roost. Contact call between adult and young a medium-high *check* or *teck*, slightly clicking, every 1–2 s. **SHH** Endemic to Madagascar, where known from a few sites in W lowlands between L Kinkony and Mangoky R, particularly L Kinkony and Mandrozo and the Besalampy and Antsalova regions; only known sites mapped. Rare and threatened: appears to be restricted to marshes with dense beds of *Phragmites* reeds and floating vegetation over which it walks to feed. Often tame. **EN**

Baillon's Crake *Porzana pusilla*   18 cm

Comparable in size only to the slightly smaller flufftails *Sarothrura*, both of which have entirely different plumage, lacking heavily marked upperparts and barred flanks of this species at all ages. Ad with *blue-grey underparts even more distinctive*. For separation from vagrant small crakes, see those species (Plate 104). **Voice** Not described in Madagascar; elsewhere advertising call of ♂ given mainly at night, a hard, dry rattle lasting 1–3 s, repeated every 1–2 s, audible >250 m away. Brief versions given in courtship flights. Also loud, energetic, rapidly repeated grating notes, often in a phrase *kraa-kraa-kraa cha cha cha cha cha*. Alarm call a sharp *tac* or *tyuik*. **SHH** Europe to Australasia to Africa, and Madagascar, where resident; no records in rest of Malagasy region. Frequents marshes, freshwater lakeshores, or brackish wetlands with islets of reeds and sedge, edges with dense aquatic vegetation; sometimes rice paddies. A typical small wetland-living crake, very difficult to observe: occasionally flies to surmount an obstacle or to escape danger; otherwise offers only fleeting views in open before disappearing.

PLATE 25: MALAGASY REGION: RAILS II

Madagascar Wood Rail *Mentocrex kioloides* — 28 cm
A large, rufous-brown forest rail with a *grey face*. Does not overlap with Tsingy Wood Rail. White-throated Rail, sometimes seen in forest, is larger, has longer, red-based bill; face lacks extensive grey, has white undertail-coverts and extensive white throat-patch; voice very different. Brown Mesite is largely plain brown, lacking striking markings on face and rear underparts shown by wood rails, and has a long tail. **Voice** *A distinctive, long sequence of medium- to high-pitched, penetrating whistles* chuchuchuchuchuchuchuchu or pyupyupyupyupyupyupyu... lasting up to 30 s, sometimes more; also low-pitched bubububububub... lasting 10–30 s, plus various shorter notes, metallic clinking, clucking and mewing, including a repeated, sneezing *tsik* or *chilk* or *chilyik* in alarm. **SHH** Endemic to Madagascar, where widespread in E forests including a few isolated patches, sea level to 1,960 m but most below 1,000 m. Paler race *berliozi* in humid NW. Restricted to floor of undisturbed rainforest and contiguous second growth, occasionally in trees when flushed.

Tsingy Wood Rail *Mentocrex beankaensis* — 30 cm
The only *Mentocrex* in dry W of Madagascar. White-throated Rail occurs in this region but is not normally forest-dwelling although may forage there. *Largely cinnamon-rufous head* of Tsingy Wood Rail recalls White-throated; latter has longer, red-based bill, white undertail-coverts and extensive white throat-patch. **Voice** Poorly known. Contact a low, continuous *dud-dud-dud-dud*..., repeated every 0.5–1.0 s; sometimes quiet, or louder when agitated; also a loud high *snick*, apparently an alarm. **SHH** Endemic to Madagascar, in dry forest of Beanka and Bemaraha limestone karst (*tsingy*) massifs, in CW. Feeds on forest floor, often tame. Population and range both small but habitat often inaccessible; threatened by habitat loss, especially due to fire, and hunting. **NT**

Madagascar Rail *Rallus madagascariensis* — 25 cm
From other rails in Madagascar by *long slim bill and all-dark plumage with white undertail-coverts*. White-throated Rail is larger with shorter, stouter bill and conspicuous white throat contrasting with dark chestnut head. Madagascar Wood Rail also larger, more uniformly rufous, with short bill, grey head and small white throat-patch. Call distinctive. **Voice** Calls frequently by day from dense cover. Calls include short *chiik* or longer *crriiiiik*, almost disyllabic, rising slightly; shorter, more clicking *crik* or *tik*; and a *low-pitched, quiet knocking gokgokgokgokgokgok*... slowing at end; In apparent chorus, a *loud, rolling, whinnying or galloping bidlibidlibidlibidlibidli*... lasting 5–10 s, dropping towards end; sounds like a duet between two birds giving the same whinnying but out of sequence; often leads to responses from neighbouring birds. **SHH** Endemic to E Madagascar, from sea level to at least 1,300 m, perhaps higher, in marshland dominated by grasses and sedges, including small marshes covering a few ha. Threatened by habitat destruction. Usually shy and difficult to observe, flushing reluctantly, with short flight. **VU**

White-throated Rail *Dryolimnas cuvieri* — 35 cm (nominate), 30 cm (*aldabranus*)
The commonest marsh-living rail in Madagascar and the only one on Aldabra. From other rails by *long thick bill, much more extensive white throat* (often invisible on wood rails), white undertail-coverts, and completely different voice. No confusion species on Aldabra, where is it the only resident rail. **Voice** Call in Madagascar *loud, metallic k-leeeeee*, clicking at start and dropping at end; also a very low *gug...gug...gig*... Song, often given by two or more birds, *a loud phweeeeeeiiii or creeeeeeeiiii*, a characteristic sound of wetlands in Madagascar, often given repeatedly in crescendo. On Aldabra, various short grunts, clicks, whistles and purrs at any time; song a series of loud whistles; pairs often duet slightly out of sequence. **SHH** Endemic to Malagasy region. Common and widespread in Madagascar from sea level to high mountains and on many small islets; mainly in wetlands including mangroves. Locally also in adjacent forest and occasionally entirely dry areas. Also Aldabra (race *aldabranus*) where limited to Polymnie, Malabar (Middle) and Picard Is and Ile aux Cèdres; uses all habitats on these islands, mainly dense, dry scrub but also mangroves. Nominate race vagrant to Comoros (Mayotte; may breed); once also Mauritius, and probably Reunion. Often fairly bold, sometimes seen walking through open vegetation or on bare ground, flicking tail. Racial differences in plumage are modest, but shorter wing renders Aldabran birds flightless; could be treated as a species (Aldabra Rail).

PLATE 26: MALAGASY REGION: GALLINULES AND JACANA

Purple Swamphen *Porphyrio porphyrio* — 46 cm
Huge size, purple plumage, massive red bill, red frontal shield and red legs distinctive; imm duller but still distinctive. Sympatric Allen's Gallinule has roughly similar pattern but is much smaller, with greenish-blue frontal shield. **Voice** Very loud calls an unforgettable feature of W Malagasy marshes; typically a loud, *low-pitched elephantine trumpeting parp... parp... parp...*; also a variety of other loud notes on a similar or lower pitch, a low *crrrek* and a range of other creaking, groaning and grunting noises. **SHH** Scattered populations across Old World. In Malagasy region, confined to Madagascar up to 800 m; locally common in CW lowlands, scarce elsewhere. Frequents freshwater or brackish marshes with rich aquatic vegetation, often with *Phragmites* reeds, large sedges and *Typha* bullrushes. Seen in pairs or loose groups, especially in morning or at dusk, usually near cover, but at times far out in open. Egyptian, sub-Saharan African and Malagasy populations sometimes separated as African Swamphen *Porphyrio madagascariensis*.

Allen's Gallinule *Porphyrio alleni* — 26 cm
Like Purple Swamphen but much smaller with *small bill, blue-green frontal shield, and no pale blue on face or throat*. Brown imm like imm Common Moorhen, but overall paler and browner with pale feather edges on upperparts, and lacks white streaking on flanks of moorhen at all ages. **Voice** A loud, hooting *aangh*, a rapid, repeated *wekwekwigawigawigawiga* and a clicking *tek tek tek tek tek tek terk..terk...terk cherk...cherk* dropping in pitch. **SHH** Sub-Saharan Africa and Madagascar, where widespread but scarce up to 750 m, commonest in CW lowlands. Undertakes long but poorly understood migrations; rare migrant on Comoros, recently breeding on Mayotte, and vagrant throughout Seychelles and Rodrigues. Found in larger freshwater marshes, lakes and river mouths, especially with water lilies and other floating vegetation interspersed with dense marsh vegetation. Walks on floating aquatic plants; generally shy.

Common Moorhen *Gallinula chloropus* — 35 cm
Bold and conspicuous. From ad Allen's Gallinule by *slaty plumage, red frontal shield, white line on flanks and pale undertail-coverts* (white on Seychelles, buff elsewhere). Imm Allen's Gallinule buffy to whitish below, without grey, has some greenish in wings and pale scaling on back. Ad Red-knobbed Coot has white bill and frontal shield, imm lacks whitish line on flanks and pale undertail-coverts of Common Moorhen. See vagrant Lesser Moorhen (Plate 104). **Voice** Race *pyrrhorhoa* gives a hoarse, *reedy trumpeting, ark...arr...kar-kekar* or *kar-kek-kakarr*, deeper than African or European birds; other loud calls include *curruc*, single or repeated *kek or kerrk, preeeuk!* and a brief, piercing *pree-ee*. **SHH** Wide global range; endemic race *pyrrhorrhoa* almost throughout region except Seychelles, where Asian *orientalis*. Common to 2,300 m in Madagascar, in nearly all wetland habitats, primarily inland fresh waters, occasionally brackish or saline, adjacent grassland, sometimes even woodland. In Seychelles often in dry wooded habitats; resident granitic Seychelles but has colonised Bird (2006) and vagrant Amirantes.

Red-knobbed Coot *Fulica cristata* — 40 cm
No closely similar species. Ad has all-dark plumage with *white bill and frontal shield*; red knob often hard to see in the field. Most similar to Common Moorhen, which has red (not white) frontal shield, white line on flanks and pale undertail-coverts, whereas coot is *all dark*. See moorhen for distinction of imm. Far more inclined to form flocks swimming in open water than moorhen, and less often walks on dry land. **Voice** A low *berp, boof* or *wheep*, and *a higher, slightly trilled kik or krrik*, repeated. **SHH** Africa and S Europe; widespread but scarce, patchy and local in Madagascar on shallow fresh or brackish water with blocks of aquatic vegetation, typically on lakes and open water of larger marshes.

Madagascar Jacana *Actophilornis albinucha* — 30 cm
Extremely distinctive at all ages. **Voice** Vocal: a medium-pitched, trilling *trrrt...trrrt...trrt*, often from several birds at once in apparent display. Also a sharp, nervous *kreeeeeee*, repeated at diminishing volume, particularly during quarrels. **SHH** *Graceful action, walking over floating or waterside vegetation*, recalls rails and certain waders. Endemic to Madagascar, where found in freshwater wetlands especially with water lilies *Nymphaea* and sometimes exotic Water Hyacinth *Eichhornia crassipes*: large and small marshes, lakes and ponds with open water and floating vegetation, fringes of slow-flowing rivers; sea level to 750 m, mainly in lowlands of N & W; locally in E.

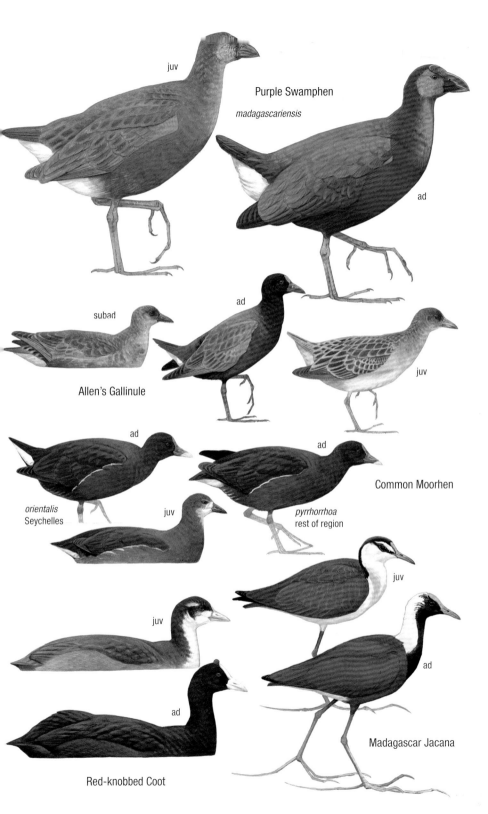

PLATE 27: MALAGASY REGION: MISCELLANEOUS SHOREBIRDS

Greater Painted-snipe *Rostratula benghalensis* 24 cm
From Madagascar Snipe by *pale stripe up side of breast* (resembling a harness) and *white or buff eye-patch*, distinctive at some distance. In flight, snipe shows pale wingbars on tips of wing-coverts, whereas painted-snipe has *plainer wings with more golden spots on wing-coverts* (especially on ♀) and barred flight feathers. Vocalisations distinctive, often the main means of detection. **Voice** Medium- to low-pitched, rhythmic *whup* or *pawhup* repeated every 1.0–1.5 s for minutes on end, at night or early in morning. **SHH** Africa, Asia and Australasia. In Malagasy region, restricted to Madagascar, where widespread in herbaceous freshwater wetlands including rice and other flooded fields, marshes and lakeshores typically with islands of dense low vegetation and open water, coastal marshes and brackish lakes, but not extensive *Phragmites* or *Typha* beds, from sea level to 1,500 m; no records in arid SW.

Crab-plover *Dromas ardeola* 41 cm
Black-and-white plumage and large black bill extremely distinctive in all plumages; only vagrant Pied Avocet (Plate 105) is remotely similar. **Voice** Vocal: harsh or mournful, shrieking or barking, but nearly always loud, *druuet*, *kerrui* or *ha-how* with chattering or trembling quality; single notes or in long series. Alarm call *witik! witik!* **SHH** Breeds on coasts of N Indian Ocean; non-breeding range to Indian Ocean coasts and islands from W India through breeding areas to Mozambique. In Malagasy region, mainly on coralline Seychelles and NW Madagascar; does not breed but dependent young often present. Found on coasts (never inland) throughout, on sandbanks, sand- and mudflats, estuaries, beaches, mangroves, lagoons and coralline islands, often in groups.

Black-winged Stilt *Himantopus himantopus* 38 cm
Uniquely distinctive, with *fantastically long pink legs and fine, black bill*. Alarm calls distinctive and are a sound familiar to observers approaching their shallow wetland haunts. **Voice** Simple *kek* or *kraak*, repeated in long sequences, often becoming higher or lower; frequently the first bird to alarm-call, as well as the most persistent. **SHH** Widespread in Old World tropical and temperate regions. In Malagasy region, restricted to Madagascar, mainly on coasts and coastal plains of N, W & S, and scattered sites elsewhere although generally not on C plateau. Vagrant Seychelles.

Madagascar Pratincole *Glareola ocularis* 25 cm
The only common pratincole in the region; a distinctive tern-like wader. From three vagrant pratincoles (Plate 105) by much *darker plumage with white streak below eye, chestnut lower breast*, short and shallowly forked tail, and white belly; plainer juv has similar underparts pattern. **Voice** Vocal: calls *usually variations on vik vik, vikavik, vikvikvik*, tern-like but sometimes rising or falling and with whinnying quality, often while feeding on the wing. **SHH** Breeds only in Madagascar, possibly only E & NW; recorded from sea level to 1,500 m throughout apart from far SW. Vagrant elsewhere in region. Present late Sep–Mar, migrating to E Africa (Somalia to Mozambique) for austral winter. Usually on short grass (often saline or sandy coastal grassland and airfields), and a range of open habitats, often near water, including rivers and coasts with emergent rocks and dead trees, on which often rests and may nest.

Madagascar Snipe *Gallinago macrodactyla* 31 cm
The only true snipe resident in the region or recorded in Madagascar: heavy-bodied wader, usually seen when *flushed from densely vegetated freshwater swamps*, appearing all dark brown but heavily streaked, with *very long straight bill*; see Greater Painted-snipe. Similar to numerous other snipe species, of which four Palearctic species (Jack, Great, Pintail and Common Snipes) recorded as vagrants, so far only on Seychelles; see Plate 107. **Voice** When flushed and occasionally on ground, *a sharp, reedy or hoarse checka-kep... kepkep... kep...*. Aerial display sounds consist of rapid sequence of notes, *wikwikwikwikwikwik...*, lasting 2–4 s in swooping display-flight which may crescendo and then descrescendo; also emits a deep *whor-whor-whor*, often at night and in level, as well as swooping, flight. **SHH** Endemic to Madagascar, in wetlands of C plateau and E at all altitudes: prefers dense vegetation in large or small marshes and swamps dominated by grasses and sedges, other flooded fields, edges of watercourses and occasionally weedy rice paddies. **VU**

PLATE 28: MALAGASY REGION: PLOVERS I

Pacific Golden Plover *Pluvialis fulva*   26 cm
Most similar to Grey Plover, which is larger and bulkier, has black axillaries and whitish (not brown) rump. Grey Plover is typically spangled whitish on upperparts, whereas Pacific Golden has *yellowish-brown spangles,* but both are very subject to wear and fading (especially imm in Apr–Sep) and may appear confusingly plain and brown. Call quite unlike Grey Plover's. **Voice** Loud 1–2 syllable calls, usually in flight: *plaintive, melodious klee-ye, or tlu-ee,* with emphasis on second syllable; also a sharp *chuwit.* **SHH** Breeds in Arctic; non-breeding range S Asia to Pacific; in Malagasy region, annual migrant in very small numbers to granitic Seychelles; vagrant elsewhere.

Grey Plover *Pluvialis squatarola*   31 cm
The only large plover common in the region. *Large size, grey upperparts and short bill* unique in region, although when worn may appear brown; in flight, *contrasting black axillaries* diagnostic. See much rarer Pacific Golden Plover. **Voice** Very distinctive, *mournful whistle plee-uu-wee*: first syllable highest-pitched, second lowest. **SHH** Breeds in Arctic; vast non-breeding range on N & S Hemisphere temperate and tropical coasts. In Malagasy region, on nearly all coasts, including remotest islands; mostly Oct–Mar, but not rare at any time; usually in small groups. Most common on intertidal mudflats in estuaries and river deltas; also other coastal wetlands and rarely fresh water, especially near sea.

Common Ringed Plover *Charadrius hiaticula*   19 cm
Small, robust plover with *single complete breast-band and white underparts,* similar to Madagascar and Kittlitz's Plovers but latter lacks breast-band, and both have underparts washed cinnamon-buff, and black eyestripe joining breast-band on side of neck. Three-banded Plover has two breast-bands. Breeding plumage of Common Ringed has brighter orange bill-base and orange legs. See vagrant Little Ringed Plover (Plate 106). **Voice** *Liquid tooip rising in pitch.* **SHH** Breeds across temperate to Arctic zones; non-breeding range mostly sub-Saharan Africa. Race *tundrae* widespread in small numbers (rarely >50) on coasts, especially refers mudflats, beaches, estuaries and mangroves, throughout Malagasy region; most in Oct–Mar, very few all year.

Lesser Sand Plover *Charadrius mongolus*   19 cm
Both sand plover species are distinguished from other *Charadrius* as follows. Common Ringed Plover has complete black or dark brown breast-band. White-fronted and Kittlitz's Plovers have white or pale nuchal collar in all plumages. Greater Sand Plover usually markedly larger than all, but Lesser often not so. Separation of Greater and Lesser in non-breeding plumage very difficult. *Greater is heavier and proportionately bulkier, with a longer, heavier, more bulbous bill with conical tip,* longer legs and less rounded head, with overall impression recalling Grey Plover, whereas daintier *Lesser Sand Plover is a more typical small Charadrius, with rounded bill tip.* Greater has legs greenish or yellowish-grey rather than dark grey as in Lesser, but variation and staining makes this of limited value. In breeding plumage, Greater has narrower rufous breast-band. In flight, wingbar on Greater widens more on primaries more than does Lesser's, and tail shows more contrasting dark subterminal bar. Foraging Greater moves 9–10 paces between pauses to search for prey, compared to 2–3 paces in Lesser, and pauses for longer to search for prey (5–10 s vs 2.5–3.0 s in Lesser). **Voice** *Short, hard, chitik, chiktik, trik or drrit, usually in flight.* **SHH** Breeds in mountains of C & far NE Asia. Non-breeders on coasts of Africa, Asia and Australasia. Appears widespread on Malagasy coasts (intertidal mudflats, sandflats and lagoons), most Oct–Mar, a few all year, but details uncertain because of identification problems; only small numbers reported, usually fewer than Greater Sand Plover. Malagasy region birds are race *pamirensis,* a member of C Asian-breeding *atrifrons* group of races, sometimes considered a separate species from NE Siberian-breeding *mongolus* group ('Mongolian Plover').

Greater Sand Plover *Charadrius leschenaultii*   23 cm
For separation of both sand plovers from other *Charadrius* species, and for distinctions between this species and Lesser Sand Plover, see the latter. **Voice** *Short, soft, trilled trrri, often repeated 3–4 times; also chirrirrip, and a clear one- or two-note whistle*; distinctness from Lesser Sand Plover uncertain. **SHH** Breeds from Turkey through C Asia E to Mongolia. Non-breeders from Africa to Australasia. In Malagasy region, race *scythicus* or *leschenaultii* occurs in small numbers on all coasts including remotest islands, most Oct–Mar, a few all year.

PLATE 29: MALAGASY REGION: PLOVERS II

Madagascar Plover *Charadrius thoracicus*   14 cm
A very elegantly marked, small plover; similar to Kittlitz's but with strong and **broad black breast-band**, also present albeit duller greyish in juvs. Both species can have bright orange tint to lower breast and belly. Differs from Common Ringed Plover, which also has breast-band, by **head pattern similar to Kittlitz's**: black eyestripe and white supercilium both extend backwards to form hind-collars, the former also joining black breast-band. Slightly but significantly heavier and longer in wing and tarsus than Kittlitz's Plover. **Voice** Short, medium-high *pip* or *prip*, repeated every 2–3 s; also *pipipipreeeet*, or *pipipipipriieeeooit*; single note repeated rapidly 3–6 times, then short trill on same note or slightly modulated, lasting 2–5 s. **SHH** Endemic to Madagascar, where almost always outnumbered by Kittlitz's Plover; rare (population *c*.3,100 birds) but readily found in correct habitat and range. Mainly occurs, and only breeds, on sparsely vegetated habitat, especially grassland, around open lakes and mudflats; also uses coastal mudflats (wet or dry), saltmarshes, edges of temporary or permanent pools (typically brackish), lakes and mangroves, and shrimp farms. **VU**

Kittlitz's Plover *Charadrius pecuarius*   13 cm
Small plover, similar in appearance to Madagascar Plover especially when viewed from behind and breast not visible; Kittlitz's is *smaller and lacks black breast-band*. Head patterns similar, but distinct from other sympatric plovers, with dark brown crown surrounded by white stripe and black frontal bands. Juv has whitish forehead, supercilium and hind-collar; may resemble White-fronted Plover and Lesser Sand Plover but has longer legs, darker lesser coverts and pale hind-collar, while brown patches at sides of breast variable but never approach full breast-band of Madagascar Plover. **Voice** Quiet, short *trit* or *tick* and plaintive *tu-lit* or piercing, trilled *tritritritrit...* in flight. When breeding, *krewwwwwwww* call (as for White-fronted Plover) used in courtship or chasing away other plovers from nesting or brooding area. **SHH** Africa and Madagascar, where breeding resident and intra-island migrant; most below 950 m, occasional up to 1,400 m including around Antananarivo but commonest along W & S coasts and far NE. Uses coastal and inland saltmarshes, dry mudflats, sandy or muddy riverbanks, alkaline grasslands with short vegetation, and temporary pools and mangroves; sometimes rice paddies and inland wet grassland.

Three-banded Plover *Charadrius tricollaris*   18 cm
Distinctive among Malagasy plovers, resident or migrant: *small and dark, with dark grey face, chin and throat, two dark bands across breast separated by white, and red eye-ring*. Juv fairly similar to ad, and thus easily distinguished from congeners of all ages. In flight, upperparts very dark, with much narrower wingbar than any other *Charadrius* plover in Madagascar. **Voice** At rest, on taking flight or in alarm, gives a fluty, shrill, plaintive *teewit, teewit, teewit* or *tuuieet*; also *eek-eek–eek*. **SHH** Africa (mainly E & S) and Madagascar, where most widespread in E, W & N, typically inland but not above 1,400 m and rare on C plateau including Antananarivo. Vagrant to Comoros. Mainly found at freshwater wetlands, particularly margins of streams and rivers, muddy ponds in open country, wet rice fields and artificial lakes; not usually on coast. Rarely gregarious. Malagasy race *bifrontatus* differs strongly from African form and may be a species (Madagascar Three-banded Plover).

White-fronted Plover *Charadrius marginatus*   17 cm
The most strikingly *pale, small plover* in the region. Ad usually distinctive, and often strikingly marked; *extent and brightness of rufous or cinnamon on upperparts variable*. The other three breeding species all have prominent dark markings, apart from juv Kittlitz's Plover; latter is darker overall, with darker legs and hint of pale supercilium reaching hindneck. Among migrant visitors, most similar to Lesser Sand Plover, which is considerably larger, darker and has more extensive dark markings on breast-sides; Greater Sand Plover usually larger still. **Voice** Soft *pwit*, *woo-et* or *whiiitt* in flight or on ground, often when joining a group. Ad courtship call a sharp *krewwwwwwww*, also used to expel intruders from nesting or feeding area. **SHH** Africa and Madagascar, where a breeding resident recorded on all coasts and inland up to 1,500 m; common on open coasts where few other shorebirds occur. Rare visitor to Comoros; no breeding confirmed. Found in open, sandy habitats; also visits saltmarshes, mudflats, pools and mangroves. Inland birds particularly associated with rivers. May be the only wader species in its habitat, especially on sandy beaches, lakeshores and riverbanks. Madagascar population usually treated as endemic race *tenellus*, but differences from E African population not clear.

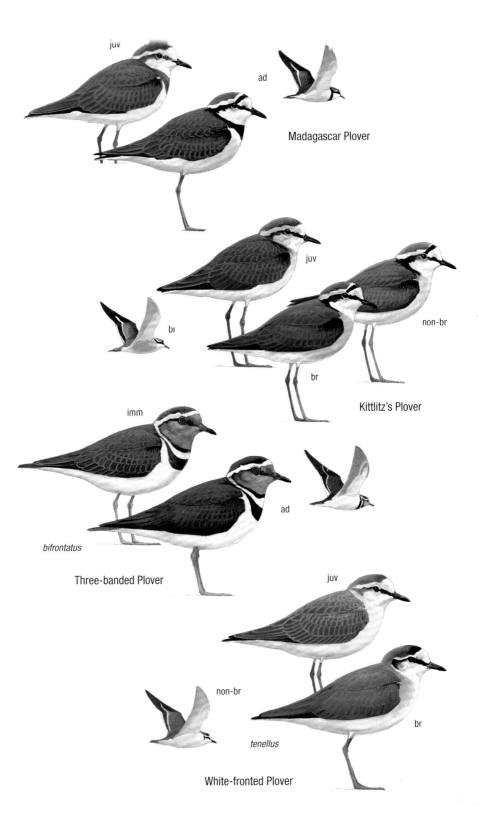

PLATE 30: MALAGASY REGION: LARGE WADERS

Whimbrel *Numenius phaeopus* — 41 cm
Bar-tailed Godwit has straight bill and generally paler plumage. For distinctions from Eurasian Curlew, see that species. Whimbrel is much commoner than either. **Voice** *Twittering or clear, rapid keekeekeekeekee*, or *queequeequeequee*, 3–15 notes, often heard in flight. Also, a single or slowly repeated medium-high *queeee* or *kreeeee*, slightly rising, similar to Eurasian Curlew. **SHH** Breeds across Arctic. Non-breeders almost worldwide S of 35°N, concentrated in tropics including Malagasy region, where race *phaeopus* common (other races possibly vagrant) and almost ubiquitous on coasts, including remote islands, nearby freshwater areas, inland along river systems to 750 m. On open coast and coralline islands, rocky shores, short coastal grassland and coastal lagoons, and, particularly, intertidal mudflats and mangroves. Most occur Oct–Mar but not uncommon year-round.

Eurasian Curlew *Numenius arquata* — 52 cm
From Whimbrel by *larger size, very long, more evenly downcurved bill, absence of dark lateral crown-stripes and pale supercilium*, and overall paler plumage. **Voice** Loud *quoi-quoi* or drawn out *coor-wi* or *curlEE*, *kyoi-yoi-yoi-yoik* and hoarse *kahiyah-kahiyah* when flushed. **SHH** Breeds in temperate and subarctic regions of Old World; non-breeders in Europe, Africa and Asia. In Malagasy region, widespread but rare visitor to coastal mudflats, sandy edges, estuaries and mangroves, Oct–Mar. Most appear to be E race *orientalis*.

Bar-tailed Godwit *Limosa lapponica* — 38 cm
From Whimbrel and Eurasian Curlew by *slightly upcurved bill*; Common Greenshank smaller, shorter-billed and greyer. See vagrant Black-tailed Godwit (Plate 108). **Voice** Harsh *kirruk, kvittett* or *ki-ki-ki* in flight; not very vocal. **SHH** Breeds in N Palearctic and W Alaska. Non-breeders from Britain S to South Africa and E to Australasia. In Malagasy region, birds assumed to be race *taymyrensis* (formerly included in W race *lapponica*) annual visitors to all coasts, especially W Madagascar and Aldabra, in small numbers. Birds resembling E populations, probably race *menzbieri* (formerly in E race *baueri*), recorded on Seychelles and Reunion. Coastal mudflats, estuaries and mangroves, often preferring sheltered bays over more exposed coasts.

Common Greenshank *Tringa nebularia* — 32 cm
The only large *Tringa* common in region; see smaller, much rarer Marsh Sandpiper. Bar-tailed Godwit shares *long straight bill and white rump extending up back*, but is larger, with longer bill, browner, less grey-and-white, plumage and very different call. **Voice** Loud, mellow *tyu-tyu-tyu* or *chew-chew-chew*, up to ten notes, very often three. **SHH** Breeds across Eurasia in taiga and forest zones. Non-breeders in S Europe, Africa and Asia to Australasia. In Malagasy region, found on coasts throughout, more locally inland (in Madagascar, to 1,400 m). Uses a wide range of shallow-water habitats, especially intertidal mudflats, estuaries and mangroves, but also saltpans, floodwaters, lakes, ponds, rivers and rice paddies. Most occur Oct–Mar but not uncommon year-round.

Marsh Sandpiper *Tringa stagnatilis* — 25 cm
Like small Common Greenshank, but much rarer. Marsh Sandpiper more elegant: smaller (30% shorter-winged and about half as heavy), with *thinner, straight bill* (slightly upcurved in Common Greenshank) *with less extensive pale base, and proportionately longer legs*. See Terek and Wood Sandpipers. **Voice** A sharp *yup* or *tchip*, much deeper than Common Greenshank; also a single or rapidly repeated *tew* or *keeuw*, higher-pitched, thinner and less ringing than Common Greenshank. **SHH** Breeds at mid-latitudes of Palearctic. Non-breeders in Africa, Asia and Australasia, but in Malagasy region a rare visitor mainly recorded in W Madagascar freshwater wetlands, Nov–Mar; vagrant Seychelles, Comoros and Mauritius.

Wood Sandpiper *Tringa glareola* — 20 cm
Unique combination of *square white rump, no wingbar in flight, greenish-yellow legs, spotted back and fairly conspicuous supercilium*; lacks white trailing edge and upturned bill of Terek Sandpiper, or flickering flight of Common Sandpiper. See vagrant Green Sandpiper (Plate 108). **Voice** Loud, single note repeated rapidly, three or more times, *jif-jif-jif*; as greeting or alarm call. **SHH** Breeds in boreal and subarctic zones of Palearctic. Non-breeders in Africa, Asia and Australasia, mainly in tropics. In Malagasy region, rare migrant mainly to Madagascar (most on or near W coast) and Seychelles (granitic islands and Amirantes). Vagrant Mauritius and Reunion. Most on freshwater or brackish lakes near coast but may use other wetlands including rice paddies; vagrants sometimes forced on to mudflats.

PLATE 31: MALAGASY REGION: SMALLER WADERS

Terek Sandpiper *Xenus cinereus*   25 cm
Distinctive, with *short, usually orange legs, gently but noticeably upturned bill*, and usually frenetic feeding activity, running at high speed with head lowered. In flight, diffuse *white trailing edge to wings* unique in region, and call distinctive. Common Sandpiper has shorter, straight bill, greyish- or yellowish-green legs; and stiff-winged flight showing obvious pale wingbar. **Voice** A whistled, rippling *2–5 syllabled hühühühühü*, or more melodious, whistled *teeuu-duey, duey, wi-wi-wi-yu*. **SHH** Breeds in boreal to subarctic Old World; non-breeders on tropical and subtropical coasts. In Malagasy region, mainly Oct–Mar, most in W Madagascar, also regular in E Madagascar, granitic Seychelles, Comoros and Mauritius; elsewhere rare visitor or vagrant. Strictly coastal, on mud- and sandflats, in estuaries, sheltered bays and rivers, sometimes nearby freshwater; often in mangroves.

Common Sandpiper *Actitis hypoleucos*   20 cm
Confusable with Wood, Terek and Curlew Sandpipers. Differs in flight by f*lickering action on bowed wings, conspicuous white wingbars* and olive-brown rump, and call; on ground by white showing between brownish breast-sides and bend of wing. *Persistent bobbing* on land also distinctive. Wood has longer legs, plain upperwing and white rump; Curlew Sandpiper has downcurved bill and white rump. **Voice** Loud, *high-pitched tsee-wee-wee*, usually in a descending series, often repeated, usually in flight. May sing in non-breeding areas: rapidly repeated, rising *kittie-needie*. **SHH** Breeds across temperate to subarctic Old World; non-breeders to Africa, Asia and Australasia. Most widespread wader in Malagasy region, found Oct–Mar in almost all wetland habitats from remotest oceanic islands to high mountains of Madagascar, although numbers often small; wanders into nearby dry habitats.

Ruddy Turnstone *Arenaria interpres*   23 cm
Very distinctive, dumpy, *short-legged and short-billed wader, with unique contrasting plumage*. Chuckling call distinctive. **Voice** Low, rolling *trrrrpup, trrrrrk* or *trrrrp*; also hoarse, medium-pitched, staccato *tewk... tewk...*, extending into excited prolonged trill *tirrrrrrrrrrrrrrk*. **SHH** Breeds in Arctic lowlands of Old and New Worlds; non-breeders on almost all tropical and temperate coasts; Malagasy range includes remotest islands, but rare inland or even on fresh water near coast. Uses many habitats: rocky shores, sandy beaches, mudflats, mangroves, saltpans, sand and gravel beaches, estuaries, coralline islands, short coastal grassland and exposed reefs.

Sanderling *Calidris alba*   21 cm
From all other *Calidris* species by *striking wingbars*, call and often behaviour. Most in Malagasy region are in juv or ad non-breeding plumage, when *white appearance of face and underparts* is distinctive; darker lesser coverts may show as blackish patches at carpal joint. Very pale waders chasing waves on sandy beaches are likely to be Sanderlings. See smaller Little Stint. Curlew Sandpiper has much longer bill and white rump. **Voice** *Call twick twick given in flight*; foraging flocks may give a steady twittering, or quiet trill in chases. **SHH** Breeds in high Arctic; non-breeders on coasts almost worldwide. In Malagasy region Sep–Mar, some all year, on mudflats and lagoons but not inland; also sandy beaches shunned by most other waders except White-fronted Plover.

Little Stint *Calidris minuta*   15 cm
From Sanderling by smaller size; in non-breeding plumage, Little Stint *has darker grey upperparts and thinner wingbar in flight*; in breeding plumage, *whitish chin and throat*, and yellowish or cream mantle lines (Sanderling is plainer). Juv shows much more *conspicuous warm rufous fringes to mantle and scapulars*. See also vagrant *Calidris* species (Plate 110). **Voice** A rapid *stit-it*, and occasionally purring calls *trurr* or similar. **SHH** Breeds in Arctic tundra of Old World; non-breeders mainly Africa and SW Asia. In Malagasy region, scarce on most coasts, particularly mudflats and brackish or freshwater wetlands of W Madagascar, Oct–Mar; on islands, very local.

Curlew Sandpiper *Calidris ferruginea*   22 cm
Long, downcurved bill and prominent white rump in flight unique among region's small waders. Juvs in Sep–Oct distinguished by peachy wash on breast and delicately scaled upperparts. Non-breeding plumage darker above than Sanderling. Begins to acquire distinctive breeding plumage before departure, Mar–Apr. **Voice** Distinctive trilling *chrrut* often given in flight. **SHH** Breeds in high-Arctic tundra; non-breeders widespread in Old World. Commonest migrant wader in Malagasy region, on most coasts, especially open coastal mudflats of estuaries and deltas; also mangroves, lagoons, saltpans and muddy freshwater lakeshores, occasionally inland. Gregarious. Oct–Mar, some remain (in non-breeding plumage) all year.

PLATE 32: MALAGASY REGION: SKUAS

South Polar Skua *Stercorarius maccormicki*   53 cm
From very similar Subantarctic Skua by *smaller head, narrower wings and finer bill* and tarsi. Light-morph South Polar has *paler brown head, neck and underparts, and prominent pale nuchal collar*, all contrasting with much darker rest of plumage. Ad and imm dark-morph South Polar have more *uniform blackish upperparts, faint nuchal collar, colder dark greyish-brown head and neck*, contrasting with blackish underparts and underwing-coverts; often has paler and more *prominent patch at base of upper mandible*. Juv South Polar distinguished from all ages of Subantarctic by pale to medium-grey head, neck and underparts contrasting with blackish rest of upperparts and underwings. **Voice** Silent in region. **SHH** Breeds Oct–Mar in Antarctica; non-breeders disperse to equator and beyond. In Malagasy region, confirmed only on Seychelles but probably overlooked.

Subantarctic Skua *Stercorarius antarcticus*   62 cm
For separation from South Polar Skua, see that species. Both distinguished at all ages from dark-morph ad or dark imm Pomarine and Arctic Skuas by *larger size, heavier build, large head and neck, broad wings, all-dark bill and much larger white patches on upper primaries*. Juv and imm Kelp Gull less dark and lack white primary patches. **Voice** Silent in region. **SHH** Race *lonnbergi*, breeds mainly subantarctic islands, a fairly common visitor, singly or in groups of up to six, especially in areas rich in seabirds. An opportunistic predator, scavenger and kleptoparasite, targeting species up to size of Red-footed Booby. Flight powerful, steady and laboured, but capable of great speed and agility when chasing other birds. [Alt: Brown Skua]

Pomarine Skua *Stercorarius pomarinus*   46 cm + 9 cm breeding tail-streamers
Imm smaller skuas are very difficult to separate. Pomarine is *thickset, thick-necked and more barrel-chested* than Arctic, with *broader-based wings* (inner wing appears wider than distance from rear edge of wing to tip of tail) and heavier bill. *Central tail-feathers blunt-ended and broad*, barely projecting beyond rest of tail. *Double pale patch on underwing*, created by pale bases to primaries and primary coverts, typical of imm Pomarine, rare in Arctic. Juv usually darker overall than Arctic, which sometimes is paler on head and belly. Ads in breeding plumage have *unique long, blunt, twisted central tail-feathers*, unlike straight, pointed feathers of Arctic; younger birds and non-breeding ads may show hint of this. Juv and imm Kelp Gull less dark and lack white primary patches. **Voice** Silent in region. **SHH** Breeds in Arctic; non-breeders widespread to Southern Ocean but few records in Malagasy region; vagrant N to S Seychelles. Often kleptoparasitic, attacking victims up to size of boobies but especially terns and petrels. [Alt: Pomarine Jaeger]

Arctic Skua *Stercorarius parasiticus*   41 cm + 8 cm breeding tail-streamers
Intermediate in size between Pomarine and Long-tailed Skuas, flight often rather fast and falcon-like; see those species for distinctions. Imm Arctic has *central tail-feathers pointed* and *projecting (slightly: 1–3 cm) more* than in Pomarine; longer and more pointed in adults. **Voice** Silent in region. **SHH** Breeds in Arctic and subarctic; non-breeders widespread to Southern Ocean. Rare in Malagasy region; vagrant Seychelles. Status elsewhere uncertain with few records (widely scattered). Strongly parasitic, often chasing terns and petrels. [Alt: Parasitic Jaeger]

Long-tailed Skua *Stercorarius longicaudus*   38 cm + 18 cm breeding tail-streamers
Similar to Arctic Skua, but *smaller and more graceful with small head, slender bill, narrow wings and tail base*, and *light, buoyant (even tern-like) flight*, sometimes *dipping to sea surface*; impression is less menacing than Arctic Skua. Ad distinctive: in breeding plumage, *pointed central tail-feathers project far beyond tail tip*, and in all plumages has ashy grey-brown wing-coverts unlike darker brown of other species, underwings *uniformly dark without pale primary patch*, and shafts of only two outermost primaries whitish (3–8 in Pomarine and Arctic). Juv typically greyer, less brown than Arctic, although darkest birds similar; protruding central tail feathers usually as in Arctic (1–3 cm, sometimes longer) but more round-tipped, and bill has paler base. Older imm has creamy feather tips and bars, and darker areas, increasing confusion with Arctic Skua; *structure, flight, pale-barred rump and grey coloration* are best features of subad Long-tailed. **Voice** Silent in region. **SHH** Breeds in Arctic and subarctic; non-breeders to Southern Ocean. In Malagasy region, probably a rare migrant or vagrant, except in S Mozambique Channel where apparently common in southern summer. The most oceanic and least parasitic skua, usually feeding by tern-like surface-seizing and aerial-dipping, but parasitises smaller seabirds. [Alt: Long-tailed Jaeger]

PLATE 33: MALAGASY REGION: GULLS AND TERNS I

Kelp Gull *Larus dominicanus* 60 cm
The *only large gull* resident in region or recorded in Madagascar, *much larger than Grey-headed Gull* and unlikely to be mistaken for any other resident species. Brown imm confusable with skuas, but lacks white primary flashes of larger skuas. Madagascar endemic race *melisandae* differs from S African and S Indian Ocean island populations by size, bill shape and reduced white in primary tips. See vagrant Lesser Black-backed Gull (Plate 111). **Voice** Loud *kiyock-kiyock* or *kok-kok-kok....* **SHH** S Hemisphere coasts. Apparently resident but uncommon in SW Madagascar; rarely >10 together. Rare vagrant on Reunion possibly from other populations. A typical large gull, feeding opportunistically on sandy beaches, rocky shores, marine islets, lakes and lakeshores; seen on town seafronts (Tolagnaro) but not as strongly commensal in Madagascar as elsewhere in range.

Grey-headed Gull *Chroicocephalus cirrocephalus* 41 cm
The *only small gull* resident in the region. *Grey hood* of ad in breeding plumage diagnostic. In size and overall colour, more similar to large terns, but these have narrower and more pointed wings, longer and finer bills, forked rather than rounded tails and usually more buoyant flight. For distinctions from vagrant small gulls, see Common Black-headed Gull (Plate 111). **Voice** Quiet but may utter a harsh *kyaah*. **SHH** South America, sub-Saharan Africa and Madagascar, where widely scattered. Commonest at L Alaotra and in W lowlands; elsewhere erratic. Inhabits freshwater lakes and marshes with open water, brackish areas and estuarine mudflats.

Caspian Tern *Hydroprogne caspia* 54 cm
The largest and most heavily built tern, with *dagger-like bill, red tipped black*. Never shows extensive white on crown or forehead. In flight *has extensive black on outer primaries*, particularly striking on underwing, and shallow fork to short tail. Juv has dull orange-red bill. Greater and Lesser Crested Terns are much smaller, at all ages with yellowish and orange bills respectively, duller in imm but never red or orange-red. **Voice** A loud harsh *krowk* or *kaar-arr*. **SHH** Scattered populations worldwide. Breeds Madagascar, Aldabra and Europa; regular Cosmoledo and Astove; vagrant Glorieuses and granitic Seychelles. Coastal and not very gregarious. Flight powerful and direct with slow, shallow wingbeats, wings held stiffly. Hovers and plunge-dives in shallow water.

Lesser Crested Tern *Thalasseus bengalensis* 39 cm
A large, elegant tern with rounded head, neat crest and *straight orange bill*. Non-breeder has black band behind eye to nape and pale grey upperparts, rump and tail centre. Juv has pale yellow bill, faint carpal bar, brown outer primaries and indistinct brown tips to wing-coverts forming dark bars. Greater Crested Tern is heavier and longer-winged, with more angular head and heavier, drooping, yellow bill; juv has rather plain forewing and coverts. See also vagrant Sandwich Tern (Plate 112). **Voice** A high-pitched, rasping *kik-kirek*. Also a *kee-kee-kee*. **SHH** Common non-breeding visitor from N Indian Ocean (also Mediterranean and Australasia) to Madagascar, Seychelles, Comoros and Mozambique Channel islands, mainly Dec–Jun but some all year; vagrant elsewhere. Coastal and gregarious, sometimes in flocks of thousands. Flight fast and graceful, bill held horizontal, plunge-diving vertically.

Greater Crested Tern *Thalasseus bergii* 49 cm
Bigger and less elegant than Lesser Crested Tern, with *larger, more angular head*, shaggier crest and *larger, more drooping yellow bill*. Ad breeding has *narrow white band between black cap and bill*. Ad non-breeding has forehead and forecrown more extensively white, bill paler yellow. Upperparts and tail pale grey in usual race *thalassina*, slate-grey in vagrant *velox*. Imm has *dark outer primaries and strong contrast on coverts*. See also vagrant Sandwich Tern (Plate 112). **Voice** In flight a soft *whep-whep*. Feeding birds give a continuous *keek-keek* and a harsh *krow*. At colonies, gives cawing *korr* on ground. Also a rasping *kerrack* and high-pitched screaming *kree-kree*. **SHH** Pacific and Indian Oceans and SW Africa; breeds (race *thalassina*) Madagascar, coralline Seychelles, Juan de Nova and possibly Glorieuses. Non-breeders present all year and disperse widely in region, although rare on Mascarenes. Race *velox* vagrant from Red Sea and N Indian Ocean to Seychelles. Coastal, often in small parties. Elegant, bounding flight, plunge-diving in shallow water or dips to surface.
[Alt: Swift Tern]

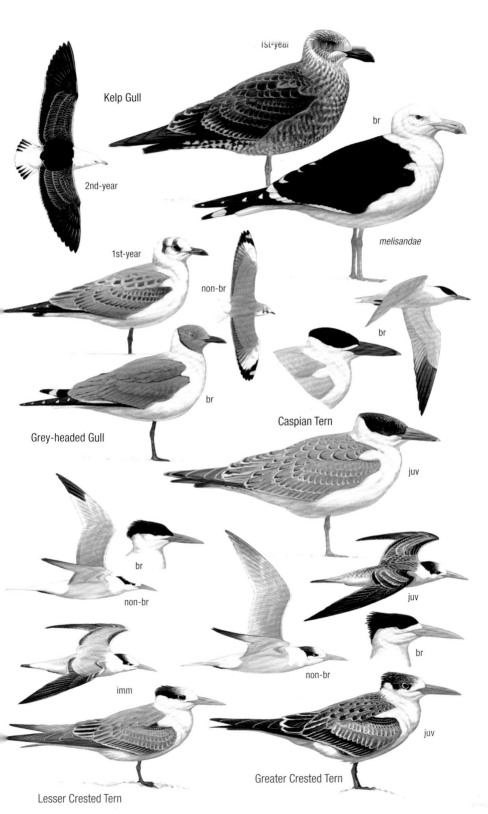

PLATE 34: MALAGASY REGION: TERNS II

Roseate Tern *Sterna dougallii*   38 cm

Ad *much whiter than Common Tern*, especially on underwing, lacking broad dark grey trailing edge to outer primaries of Common; Roseate has *outer 3–5 primaries all dark greyish*; greyer upperparts on Common contrast with white rump and tail. Bill longer than Common, and legs brighter orange. Proportionately short-winged, flying fast with *shallower and stiffer wingbeats* than all other terns and has characteristic angled plunge-dive. Ad breeding has *rosy wash on breast, long tail-streamers*, and long, slender bill; nominate race usually has largely black, red-based bill; race *arideensis* a black-tipped red bill. Imm and non-breeding ad have dark carpal bar as in Common Tern, but this and other grey areas somewhat paler. Juv has more extensive blackish on head and prominent black-scaled pattern on back. See also vagrant White-cheeked and Antarctic Terns (Plate 112). **Voice** *A distinctive chiw-ik*, alarm a loud rasping *krak*; when defending territory a rapid repetitive *kek-kek-kek*. **SHH** Breeding populations scattered worldwide; *arideensis* (validity uncertain) breeds Seychelles and St Brandon, and W Madagascar; races elsewhere in region uncertain. Non-breeding movements unknown but probably not far from colonies. Local in the region; Common Tern much more widespread. Pelagic and gregarious, but rarely in large flocks.

Black-naped Tern *Sterna sumatrana*   30 cm

A distinctive, medium-sized, very pale tern, from Roseate and Common by *narrow, clear-cut band running from eye to eye*, broadest on nape, contrasting with *very pale grey upperparts* (appearing almost pure white in bright sunlight), lacking contrast with white underparts. Juv also has whitish head with poorly defined, greyish-brown eyestripe, sepia-brown mantle mottled grey-brown; bill dull yellow at first, darkening to black with age. Fairy Tern is pure white but has no black on head other than immediately around eye. **Voice** A high-pitched repeated *kee-ik*. Alarm call *cheet-cheet-cheeter*. **SHH** Tropical Indo-Pacific; W Indian Ocean race *mathewsi* (possibly invalid) breeds Seychelles (Aldabra, Cosmoledo, African Banks, St Joseph's Atoll, Farquhar, St François and possibly elsewhere), and vagrant to granitic Seychelles, Madagascar, Comoros and Glorieuses. Frequents coral atolls, feeding inside the lagoon or close to shoreline, dipping to surface.

Common Tern *Sterna hirundo*   35 cm

Often confused with the more localised Roseate Tern (which see). Usually seen in non-breeding and imm plumages, with *grey back and wings, white forehead, dark carpal bar, black bill* and no tail-streamers; imm may show pale orange base to bill and dark secondary bar. Ads moult into summer plumage before leaving in Apr–May and have greyish underparts and long tail-streamers. Whiskered Tern, resident in Madagascar and vagrant elsewhere has very shallow tail fork and uniform grey rump/tail; see that species. See also vagrant White-cheeked and Antarctic Terns (Plate 112). **Voice** A drawn-out *keee-yaah* or *keee-rrr* emphasising first syllable. **SHH** Breeds N latitudes, spending boreal winter mainly S of Tropic of Cancer. In N of region, mainly present on passage Sep–Oct and Mar–Apr; elsewhere most Oct–Mar but some year-round. Coastal and gregarious, often abundant in W Madagascar and frequent elsewhere. Rather languid flight; plunge-dives, sometimes hovering first.

Fairy Tern *Gygis alba*   28 cm

Ad is *pure white, graceful and delicate* with translucent wings and large *slightly uptilted black bill* with blue base; Roseate and Black-naped Terns are pale grey above with black head markings, and lack Fairy's distinctive large-headed appearance and big-eyed look. Juv has sparse brown or grey mottling on grey-washed upperparts, black spot behind eye and blackish shafts to outermost primaries. **Voice** At nest gives low, slightly quacking *kwe kwe kwekrekrekre kwe kwe kwe… kwe*. A low purring chuckle when perched. Also a mechanical clicking and a distinctive, buzzing *byowp* when excited. **SHH** Pantropical, breeding throughout Seychelles, Rodrigues (Ile Cocos and Ile aux Sables), Agalega and St Brandon; vagrant to Madagascar, Reunion and probably Comoros and Mauritius. Buoyant, graceful flight, often in pairs low above sea, twisting and turning, sometimes hovering. [Alt: White Tern]

PLATE 35: MALAGASY REGION: TERNS III

Little Tern *Sternula albifrons*   23 cm

Little and Saunders's are the smallest terns in the region, also characterised by *yellow bill with black tip* in breeding plumage; very similar to each other but obviously smaller than all but the slightly larger and usually greyer 'marsh terns' (*Chlidonias* species), which have very different flight and obvious plumage differences at all times. For distinctions from Saunders's Tern, see that species. **Voice** A harsh *ket-ket*. **SHH** Breeds W Africa, Eurasia to Australasia, spending boreal winter S to South Africa. Vagrant to Seychelles, Apr; probably under-recorded throughout region but status blurred by presence of often indistinguishable Saunders's Tern; see map for that species. Both species fly with rapid deep wingbeats and much hovering.

Saunders's Tern *Sternula saundersi*   23 cm

Only safely separable from Little Tern in breeding plumage (in region around Apr or Sep) when *smaller white square forehead-patch extends only to eye; on Little, white patch extends behind eye as supercilium*. Pale grey upperparts lack contrast with rump and tail; on Little, paler rump contrasts with white tail. Legs yellow-olive; in Little, orange-grey or brown, sometimes tinged yellow. Distinctions disappear in non-breeding plumage, although black mask averages slightly broader than in Little. In all plumages, outermost primaries more extensively black (3–4 primaries black with black shafts) than Little (2–3 black primaries with pale shafts) but this is very difficult to see in the field. **Voice** A hoarse, grating *kip*, *wip*, or *tchijjick*; similar to Little Tern but perhaps less 'urgent'. **SHH** Breeds Red Sea to Sri Lanka, spending boreal winter to south. Little/Saunders's Terns occur throughout Malagasy region coasts S to Toliara (SW Madagascar), but identification rarely certain; ? on map shows occurrence of unidentified birds. Saunders's positively identified in NW Madagascar and Seychelles (mainly Sep–Apr) and probably occurs at least across N of Malagasy region, but southern limit uncertain. Favours sheltered coasts, where it is fairly gregarious. Energetic, swift and purposeful flight, plunge-diving to surface.

Whiskered Tern *Chlidonias hybrida*   26 cm

Ad breeding differs from all except vagrant White-cheeked Tern (Plate 112) *by dark grey underparts and black cap contrasting with white cheeks and undertail-coverts*. Ad non-breeding and imm recall Common and White-winged; see also vagrant Black Tern (Plate 112). From Common by shorter wings, *thick short bill*, grey rump and tail with shallow fork, and long dull red or black legs. Flight more hesitant, dipping over water. Juv similar to juv White-winged, with blackish ear-coverts and rear crown reaching almost to mantle, dark trailing edge to upperwing and distinctive grey-and-brown, variegated 'saddle' contrasting with greyer wings, but bill thicker, and rump grey not white, not contrasting with back. **Voice** Alarm a loud, rasping *kerch*. When breeding calls include *kiriririck*, *kek*, *kee-kee*; and a soft, low *kura-kura-kiu*. **SHH** Breeds across Palearctic, spending boreal winter to south; also local resident S Africa and Madagascar, mainly in W. Vagrant to Seychelles Oct–Nov and Mar–Apr, probably also Reunion. Favours inland sites and occasionally coastal mudflats. Steady flight, wingbeats more regular and deeper than congeners, dipping to surface or making shallow plunge-dives.

White-winged Tern *Chlidonias leucopterus*   23 cm

Smaller and shorter-winged than Common and Roseate Terns. Ad breeding very distinctive; moulting birds show diagnostic remnants of contrasting black underwing-coverts. Non-breeding ad from Whiskered by *slimmer shorter bill, contrast between grey upperparts and white rump*, smaller black patch behind eye, grey-flecked crown usually separated from eye-patch by white supercilium, dark central nape stripe, and white hind-collar. Juv has dark saddle like Whiskered but plainer brown, contrasting more with paler mid-wing panel, white (not grey) rump, narrow blackish carpal bar and faint dark trailing edge; see also vagrant Black Tern (Plate 112). **Voice** Alarm or contact call a loud churring *kerr* or *kreek-kreek*. **SHH** Breeds E Europe to China, spending boreal winter mainly in sub-Saharan Africa. Annual on Seychelles mainly Oct–Apr, vagrant to rest of region but possibly overlooked. Prefers freshwater habitats but also estuaries, sheltered coasts, lagoons and rubbish dumps. Flight swift, agile and bouncy with slightly stiff wingbeats. Dips to surface, but does not plunge-dive. [Alt: White-winged Black Tern]

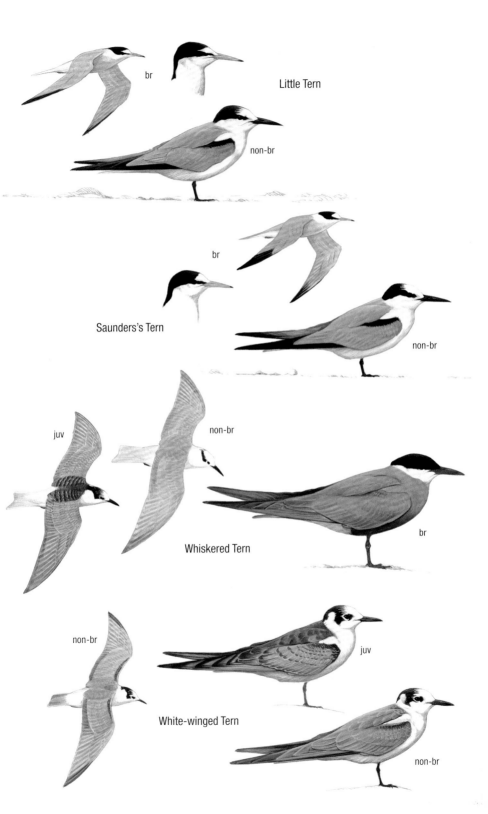

PLATE 36: MALAGASY REGION: PELAGIC TERNS AND NODDIES

Bridled Tern *Onychoprion anaethetus* 35 cm

A medium-sized, slim-winged tern. Very similar to Sooty Tern, but *smaller, paler, browner, less uniform above and shorter-winged*; beware browner appearance of Sooty in strong sunlight. Usually shows narrow white collar. *White forehead extends behind eye* as a narrow white supercilium accentuating black line through eye and *black loral stripe continues to gape with even width*; on Sooty, rounded white forehead-patch is restricted to front of eye with black loral stripe narrowing towards bill. Flight-feathers above slightly darker than upperparts; on Sooty Tern, same blackish colour. Contrast between white coverts and dark flight-feathers on underwing less distinct than in Sooty Tern. Non-breeder somewhat paler with pale streaks and fringes on crown and back. Juv has shorter, entirely grey-brown tail, brown upperparts broadly tipped white or cream and mainly whitish underparts, greyer on flanks and sides of breast, unlike almost all-dark juv Sooty, which has white tips only on belly and upperparts. **Voice** A yapping *whup-whup* at colonies. At evening communal roosts a trilling *dirrr-ip*. **SHH** Pantropical, but scarce and local in Malagasy region: breeds locally off W coast of Madagascar (possibly now only Barren Is) and Seychelles (Récif, Aride, Cousin, Cousine and some smaller islands). Pelagic outside breeding season, occurring throughout Seychelles to Comoros; probably only rare vagrant to Mascarenes (does not breed). Flight buoyant and graceful low over sea. Hovers and dips to surface, occasionally plunge-dives without submerging.

Sooty Tern *Onychoprion fuscatus* 43 cm

A large, long-winged tern, *black above, white below with long tail-streamers*. Juv lacks tail-streamers, upperparts and scapulars mainly blackish-brown tipped cream, underparts blackish-brown tipped whitish on lower belly. Sometimes retains smoky-grey underparts until time of first breeding. Similar to Bridled Tern but overall more black and white, larger and longer-winged; see that species. **Voice** A nasal *er-wakey-wake*. Colonies extremely noisy. **SHH** Pantropical, breeding Madagascar (several small colonies in NW, N & E), Seychelles (major colonies at Bird, Aride, Récif, Desnoeufs, Farquhar and Cosmoledo, and several smaller colonies), Mauritius (large colony Serpent I), Rodrigues, Outer Islands (major colonies Glorieuses, Juan de Nova and Europa; also St Brandon and Agalega). Highly pelagic outside breeding season and gregarious at all times. Strong bounding flight.

Lesser Noddy *Anous tenuirostris* 33 cm

A medium-sized all-dark tern, smaller and slenderer than Brown Noddy in direct comparison, with slimmer bill. Ad has *grey crown and nape, and whitish-grey forehead, extending around front of eye and becoming dark on lores*, unlike Brown's more restricted grey crown sharply demarcated at the lores. Dark plumage areas more blackish or greyish, less brown, than Brown Noddy: very dark brown upperparts have slight greyish cast, tail blackish-brown and underparts charcoal-grey. In flight shows evenly dark underwing (pale-centred in Brown Noddy). Juv has browner plumage than ad, with pale cap more contrasting (more like Brown Noddy). **Voice** At nest medium-pitched *arrrk-arrk-* or *eeek-eeek-* or lower-pitched *ugugug*. A mechanical buzzing *byowp* when excited. **SHH** Tropical Indian Ocean. Nominate race breeds W Indian Ocean, including Seychelles (world's largest colony at Aride; also other granitic islands and Amirantes), Mauritius (Serpent I), Rodrigues, Outer Islands (St Brandon, Agalega). Regular on Madagascar off NW & NE coasts, and Comoros, but few data here and in rest of region. Mainly pelagic outside breeding season and gregarious at all times. Flight more buoyant than Brown Noddy with more rapid wingbeats, and less likely to settle on sea. Hovers and dips, occasionally splashes, but does not plunge-dive.

Brown Noddy *Anous stolidus* 41 cm

A medium-sized, all-dark tern with a *whitish cap*, larger and stockier than Lesser Noddy, with heavier bill; see that species. In flight has *shorter broader wings than most terns*, the upperwing blackish-brown with browner wing-coverts and dark brown underwing with greyish centre. Broad, blackish-brown, *wedge-shaped tail* appears pointed when closed and during moult can even appear forked for a short time. Juv darker, pale crown less distinct and more restricted, with indistinct pale fringes to feathers of upperparts and wing-coverts. **Voice** At nest low mechanical rattling *groooo* or *arrrhh* or *groo-groo-groo-groo*, repeated 3–4 times. **SHH** Pantropical, breeding on islets off Madagascar mostly in E (several colonies), throughout Seychelles, Mauritius (Serpent I), Reunion, Rodrigues, Outer Islands (Glorieuses, Agalega and St Brandon); regular visitor to Comoros. Pelagic, but some present at breeding colonies year-round. Gregarious, feeding in flocks, often with other terns. Flight purposeful and direct, appearing falcon- or skua-like. Hovers, then snatches prey from surface.

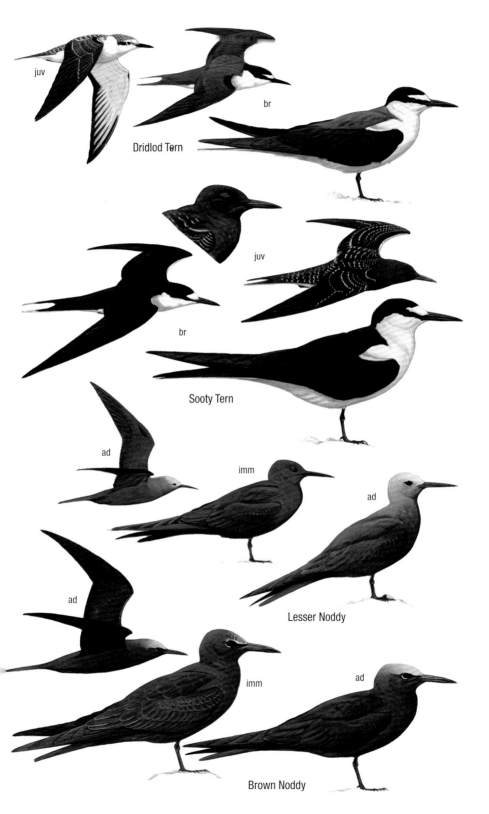

PLATE 37: MADAGASCAR: SANDGROUSE AND PIGEONS

Madagascar Sandgrouse *Pterocles personatus* — 35 cm
Large sandgrouse; no other species known or likely in the region. From all other Malagasy birds by stocky body, with thickset profile and **broad pointed wings, pale panels in upperwing** and **rapid, powerful flight**. On ground, black mask of ♂ contrasts with mainly rufous-tan body; ♀ is striped brown on upperparts. Juv similar to ♀ but often smaller and duller with very pale buff wings, barred black. **Voice** *A clattering, rapid ag-ag-ag-ag-ag-ag-ag* or *ak-ak-ak-ak-ak-ak-ak*, often in flight. Usually many birds calling continuously. **SHH** Endemic to Madagascar; in W & S from N of Mahajanga (Maromandia) to Tolagnaro; may occur rarely in far NW but absent from C plateau. In open dry areas below *c.*1,000 m, in morning and evening often flying in groups of up to 40 to water, where hundreds may gather.

Feral Pigeon *Columba livia* — 33 cm
The familiar urban pigeon. Very variable; rufous form from Madagascar Turtle Dove by lack of black spots on neck and *lack of contrast between grey head and rufous back*. **Voice** A long, moaning *o-o-orr*, repeated monotonously, or a hurried, rambling *coo-roo-ooo-ooo*. **SHH** Commensal worldwide including most of region; in Madagascar rare far from towns. Often in flocks, when variation between individuals is marked.

Madagascar Turtle Dove *Nesoenas picturata* — 28 cm
Ad has unique combination of **grey head and rufous body**. Juv lacks rufous, with darker brown upperparts, dull brown-grey breast and buff wash on lower breast. **Voice** A low, cooing *cooo-coo-huh* or *coooo-coowuh*, each note higher than the previous. **SHH** Endemic to region (although possibly also native to Diego Garcia) where fairly widespread. In Madagascar (nominate) in most wooded habitats including primary forest, degraded areas, plantations and gardens; scarce on C plateau. Feeds mainly on ground, rising with noisy wingbeats, flight swift, powerful and usually low.

Namaqua Dove *Oena capensis* — 28 cm
Only *small dove* in Madagascar. **Long graduated tail** and reddish underwings highly characteristic in flight. Juv browner with greyer crown, faint dusky bars on mantle and primaries have chestnut tips; juv ♂ can show blackish breast feathers with fawn tips. Madagascar endemic race *aliena* ♂ darker on upperparts than African birds; less black on face and pale grey (not black) undertail-coverts. **Voice** A medium-high *woou-huh*, slow and rather mournful, second note higher, louder and shorter. **SHH** Sub-Saharan Africa and Arabian Peninsula; race *aliena* resident and common in N, W & S Madagascar, below *c.*1,500 m in dry areas; scarce on C plateau and rare in E. Vagrant to Seychelles, race not confirmed but one possibly *aliena*.

Madagascar Green Pigeon *Treron australis* — 32 cm
From other pigeons by **grey-green upperparts, pale wing markings** and yellow-orange legs. In poor light or in flight, green not always obvious but pale yellow wingbars striking and tail longer than Madagascar Blue Pigeon. Imm duller with grey bill. Race *xenius* in W & S is paler and more yellowish than nominate, upperwing-coverts tinged bluish-grey, with mauve patch at bend of wing more extensive. **Voice** Remarkable song like other *Treron* pigeons, a *fast medley of whoops, whistles, trills and yaps*, *tiwiorrr... whip... towiorrr... whip-por-whip koorr*, and variants. **SHH** Endemic to Madagascar and Comoros. Common in Madagascar in lowlands below *c.*1,000 m. In primary forest, more common in W & S than in E. Rare or absent from most of C plateau. Both races common in degraded forest, other wooded areas, cultivation, plantations and gardens, usually near fruiting trees. Highly mobile; usually in small flocks. Comoros race *griveaudi* (Plate 70) may be a separate species.

Madagascar Blue Pigeon *Alectroenas madagascariensis* — 26 cm
From other pigeons by **dark blue coloration**, in flight by **short red tail** and broad dark unbarred wings. Neck and throat silvery blue-grey. Juv browner than ad, glossed greenish-blue, underparts sooty-brown, less red on tail, bare area around eye smaller. **Voice** Rarely heard. Low muted cooing only call known. **SHH** Endemic to Madagascar; common in rainforest, scarce in degraded forest and plantations in E & humid NW and forests on C plateau; rare in W (Ampijoroa, Bemaraha, Toliara area). Often perches in top dead branches of canopy tree.

PLATE 38: MADAGASCAR: PARROTS

Grey-headed Lovebird *Agapornis canus*   14 cm
Tiny parrot; from all other species by **grey head** to upper breast in ♂ and **green body**. ♀ more uniform green, with green underwing-coverts (black in ♂), upperparts browner. Juv similar to ♀ but has darker bill, grey hood of ♂ suffused green. Race *ablectaneus* (W & S) darker green above than nominate, less yellow below, ♂ with grey area darker and bluer. **Voice** High-pitched, single *chik orchilp*, and more complex, slightly lower *tswik* or *tsiwik*, singly or many calling together. Song a pleasant twitter of similar notes, *chickitwick-psick-tsiwick-tsikichick*; occasionally also squawks or whistles. **SHH** Endemic to Madagascar; introduced to other islands, now only on Comoros where possibly native. Mostly in lowlands, especially W & S, patchily in E & N; occasional on C plateau. In flocks in wooded areas.

Greater Vasa Parrot *Coracopsis vasa*   50 cm
From Lesser with difficulty by larger size, *deeper (usually pale) bill with straight edge to (darker) bare skin at base*; forehead profile in straight line with bill – see figure; large area of bare skin around eye more conspicuous than Lesser. ♀ *in breeding season may have bare orange head*. Bill usually pale in ad, dark in imm. Race *drouhardi* (N, W & S) paler than nominate. **Voice** A varied loud squawking and whistling, *mostly lower than Lesser*; *kwa... kwa... kwa...*, single *kark* or *kia-kark*, also *kwieirk*; also in flight a high *weet*; ♀ song consists of lower rasping notes: *kowerr-kweeaaw-wheeaw-wheei-kaarrk-whiioo-whioo-wark-wieer-kweeooer*. **SHH** Endemic to Madagascar and Comoros; common in Madagascar lowlands to *c.*1,000 m, in forest, degraded areas; also plantations and cultivation. Sometimes in large flocks; compared to Lesser, scarcer in rainforest and commoner in degraded areas and S; not known to mix with Lesser. Usually in groups of 4–10.

Lesser Vasa Parrot *Coracopsis nigra*   40 cm
From Greater with difficulty by smaller size, *smaller bill (usually dark) with rounded edge to bare skin at base*; small area of bare skin around eye. Juv has duller brown plumage and darker bill than ad. Race *libs* (S, SW & CW) paler than nominate (N & E). **Voice** *Clear koo... keeh... kiweek* and variants, higher than Greater; also loud, low, harsh dropping *kakakakerkerker* given possibly as alarm on take-off; song, loud rhythmic *koo-ker-kee koo-ker-kee-koo-ker-kee ker-kuh* (often at night), more repetitive and higher than Greater. **SHH** Endemic to Madagascar and Comoros; common in Madagascar to *c.*2,000 m, in forest, degraded areas, in nearby plantations and cultivation; compared to Greater, rarer far from forest, commoner in rainforest and scarcer in S; largely absent from C plateau. Graceful flight with more rapid wingbeats than Greater.

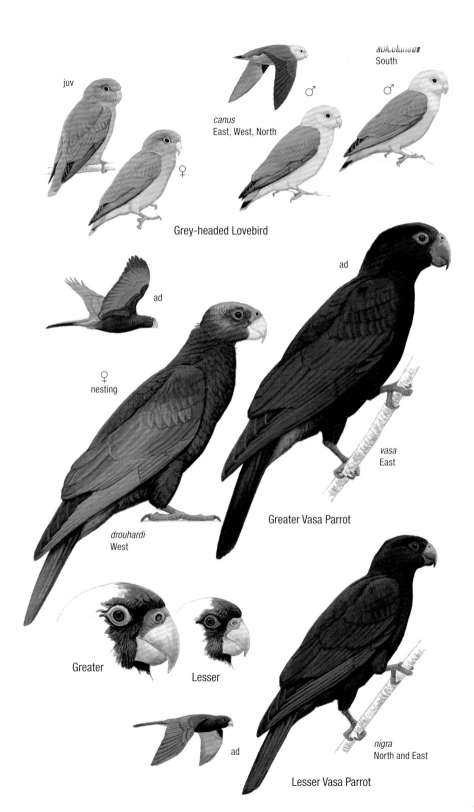

PLATE 39: MADAGASCAR: COUAS I

Crested Coua *Coua cristata* — 42 cm
Very different from other couas apart from Verreaux's, from which distinguished by larger size, *shorter, pale-tipped crest*, larger blue patch around eye and *orange tinge to breast*. Juv duller, with eye-patch absent or reduced, but always some pale orange on breast. **Voice** Low double alarm *kokok* or *pupup*, or *kok-rrr* or *kokokok-rrr*. Also loud screeching alarm *kreek kreek kreek*, preceded by a low *bubububububub*. In evening, *'check-in' call* (each bird calling single sequence of 3–7 notes) a loud descending *chooik... chooik... chooik... chooik* or *chaik... chaik... chaik*. **SHH** Endemic to Madagascar; in primary and adjacent degraded forests (to c.1,000 m) in drier parts of N, and patchily in E, S limit around Farafangana (nominate race), common all through W (race *dumonti*, paler especially on underparts) and S, especially inland (race *pyropyga* with chestnut undertail-coverts; recently recognised as a species by BirdLife International). Sometimes in cultivated areas if little hunted; scarce or absent from coastal spiny forest. Distinctive, large, dark race *maxima* known from one specimen taken 1948 near Tolagnaro (SE); possibly a species, but validity needs confirmation, and if valid may be extinct.

Verreaux's Coua *Coua verreauxi* — 36 cm
Thinner and smaller than Crested, *lacking any pink below*; ad has *longer, blackish-tipped crest often held pointed forward* over forehead. Juv duller, eye-patch reduced. **Voice** Similar but higher-pitched and more rasping than Crested; quiet high cooing *koor... koor... koor*, also a single high-pitched *kriik*. *'Check-in' call higher, quicker and harsher*: descending *krooooik... krooik... krooik...* or *kriik... kriik... kriik*. **SHH** Endemic to S Madagascar, in primary spiny forest along coast between Fiherenana R and Mandrare R; rare more than 20 km inland and less common in E of range. **NT**

Blue Coua *Coua caerulea* — 49 cm
From all other couas by *entirely blue plumage*; bright, bare eye-patch edged black. Juv has some brown feathers, lacks gloss, eye-patch reduced. **Voice** Contact call spluttering, reedy *prrrt... prrrt... prrrt*, often heard in rainforest. Also a low grunting *mrrh... mrrh*, a quiet *bok... bok... bok...*, and cooing, purring *ptoooooooor*. 'Check-in' call descending like Crested but simpler and longer, 10–15 notes, *whik... whik... whik.... whik... whik... whik...* or *cuak... cuak... cuak... cuak*. **SHH** Endemic to Madagascar; common in rainforest and locally in degraded forest, in E & NW (to c.2,000 m) and in wetter parts of forest in far N & NE except Amber Mountain. Absent from C plateau.

Red-fronted Coua *Coua reynaudii* — 39 cm
Smaller with thinner black bill than Red-breasted Coua, *ad with greenish on breast and reddish cap*. Juv has yellow bill, dull brown cap, feathers of upperparts tipped rufous, bare area around eye duller and smaller. **Voice** *Long, irritated rattle krrrrrrrrrrrrrrrrrrrr*, or harsher *kkkkkkkkkkkkkkkk*, getting stronger; *loud, short low kwaa*, clicking at start; at distance, only *aaaah* audible. Also scolding *tswick*, or *whick*, quiet, low popping *bopbopbopbopbop* and a quiet *guurrr*. **SHH** Endemic to Madagascar; in rainforest from NW to far SE, usually commoner at mid-altitudes; also in isolated rainforest fragments in NE, on C plateau and along coast. Often in dense understorey, on ground or in trees.

Red-breasted Coua *Coua serriana* — 42 cm
Darker, larger, longer-legged and thicker-billed than Red-fronted; *breast red and cap olive*. Juv duller, bill initially pale, skin around eye feathered; top of head olive-brown more uniform with mantle. **Voice** Clear, abrupt *chio-quip*, clicking at start of second syllable; quiet *crrooo*, and *thkkkrummm*, a quiet rattle followed by humming. *Song loud, mournful single keyou or kio* lasting c.1 s, dropping sharply and repeated once every few seconds; sometimes an extended *kiiioooo*. Similar Red-fronted Coua *kwaa* call is gruff and much less clear. **SHH** Endemic to Madagascar; fairly common in N part of rainforest S to about Manakara, below c.1,200 m; also in forest fragments around Daraina, but not on C plateau. Usually terrestrial.

PLATE 40: MADAGASCAR: COUAS II

Red-capped Coua *Coua ruficeps*   42 cm

From Coquerel's and Running Couas by *longer legs and bill*, more black on face, *upper breast and sides of breast pinkish-lilac*, merging with cinnamon lower breast, flanks and belly; thighs, vent and undertail-coverts deeper cinnamon-brown. Coquerel's and Giant are darker below. Nominate has rufous cap. Race *olivaceiceps* duller, cap brown, underparts paler, upper breast lilac-grey, lower breast, flanks to undertail-coverts cinnamon-buff. Juv duller, with pale tips to upperwing-coverts. **Voice** Loud rattling *dzer dzer dzer dzer dzer* or *creee creee creee creee*; deep hum, rattling bill while calling *brrrrrrr*; quiet *grr grr grr*; low dropping *dudududududu* and short, descending trill *brrrip* recalling Blue Coua. Song varied; in S and as far N as Bemaraha, a *hysterical rising, then dropping quee quee quee quee quii quii kou kou kou*, 5–20 notes; also similar but less hysterical *who who who who wher wher coo coo coo*; in NW (Ampijoroa) a similar but shorter *whrioouu cou cou*, first note whirring and dropping, last notes much lower. **SHH** Endemic to Madagascar. Race *olivaceiceps* in S from Mangoky R (SW) to Mandrare R (SE), nominate in NW (Analalava to Betsiboka R), intermediates or nominate in CW (Betsiboka R to Mangoky R). Mostly terrestrial in spiny or deciduous forest and adjacent scrub; often in more degraded areas than Running or Coquerel's. Sometimes treated as two species (*olivaceiceps* 'Brown-capped Coua') but broad area of intermediates in CW.

Coquerel's Coua *Coua coquereli*   42 cm

Top of head, side of neck and upperparts brownish olive-green; upper breast creamy with deep chestnut lower breast and belly. From Running and Red-capped Couas by dark belly and undertail, *from Giant Coua by smaller size and bill*, less black on face (especially below eye) and small or no pink spot on ear-coverts. Juv duller and browner, top and side of head dark olive-brown, bare face-patch without black borders. **Voice** Calls *quia-cooo*, or *eya-coo*, first note rising, second much lower, similar to Giant Coua but higher-pitched and less rasping; also quiet *cooo* or *ooo* and harsh *kakakakaka* in alarm. *Song cor-quee quee quee* or *cou qua qua qua qua*, often with a lower, slightly reverberating *coo* at end. **SHH** Endemic to Madagascar; common in primary W forest from Zombitse-Vohibasia (SW) to far NW and also dry forests of NE (Daraina, Analamera). Reported further S on Mahafaly plateau. Mostly terrestrial.

Running Coua *Coua cursor*   37 cm

From Red-capped by shorter legs and bill, *large pink area at rear of facial skin*, and narrow black surround to facial skin; from Coquerel's and Giant Couas by *extensively whitish underparts*. Duller pale grey above than other species. Juv duller, facial skin dull black lacking border, tips of upperwing-coverts, remiges and mantle cinnamon-buff with narrow blackish subterminal lines. **Voice** Call similar to Coquerel's; short, medium-high *cri-ooo*, more croaking than Coquerel's. Song a loud, high-medium 3–6-syllable *kwio kwio kwio kwio croo croo croo*, or *gweer gweer gweer gweer gulk-gulk-gulk* or *cri cri cri cri croo croo croo*, the first few notes dropping slightly and the last few notes much lower. Lacks initial rising notes of Red-capped and slower than Coquerel's, with more low notes at end. Sometimes a simpler *weer gulk-gulk-gul-gu*. **SHH** Endemic to S Madagascar, common but patchy in spiny forest and adjacent slightly degraded areas from Mangoky (SW) to just W of Tolagnaro. Mostly terrestrial.

Giant Coua *Coua gigas*   60 cm

Larger and *stronger-billed* than other couas; from Coquerel's by thicker bill tip and larger pink spot at rear of facial skin. Darker below than Running and Red-capped. Juv duller, facial skin dull blue. **Voice** A *loud rasping ey-yoooo*, highly distinctive; rising then drops very rapidly. Also low *oooo* or *gooo*. *Song very loud, kiugiugugugu*, the first note high and shrill, last few very low and dropping. **SHH** Endemic to Madagascar. Scarce and patchy in W & S, from Betsiboka R to just N of Tolagnaro, in primary deciduous, spiny and riverine forests; also locally in humid littoral forest (NE of Tolagnaro), and various forest types of western C plateau (Isalo, Zombitse-Vohibasia and Ambohijanahary). Scarce or absent in coastal spiny forest. Mostly terrestrial.

PLATE 41: MADAGASCAR: CUCKOOS AND NIGHTJARS

Madagascar Coucal *Centropus toulou*   40 cm (♂), 43 cm (♀)
Completely distinctive. Ad breeding *mainly black except for chestnut wings*. Non-breeder duller, dark greyish-brown, streaked and spotted whitish on head, mantle and breast. ♀ significantly larger than ♂. Juv has blackish upperparts, speckled cream on mantle and crown, rufous wing barred dusky-brown, dusky-brown underparts, streaked on upper breast. **Voice** A repeated hollow *pop... pop... pop...* also loud, toneless rattle, *kakakakakakekekek*, and single, toneless *kisschheerrrrk* or *tish* dropping at end. *Song, distinctive far-carrying rapid low popping, popopopopopopopopopo popopopopopewpew*, falling and slowing at end. Usually a duet in overlapping sequence at lower or higher registers. **SHH** Endemic to Madagascar and Aldabra. Nominate race common throughout Madagascar to *c*.2,000 m in forest, dense secondary scrub, plantations, cultivation and wooded areas, not in open grassland; common in reedbeds. Often mobbed by smaller birds.

Thick-billed Cuckoo *Pachycoccyx audeberti*   36 cm
From Madagascar Cuckoo by *lack of barring on underparts*. Juv has top of head white with dark brown blotches, side of head white, browner on mantle and scapulars with white fringes, remiges and wing-coverts broadly tipped white. Nominate race endemic to Madagascar; differs from African races by less extensively pale bill and more white fringing on wing-feathers. **Voice** In Africa (also heard once in Madagascar) includes *kloo kloo kla kla kla klo kloo kloo...* or rolling *krukrukrukrukru*, possibly ♀ call, and *repeated, loud, whistled oui... yes-yes*, or *were-wick*. **SHH** Sub-Saharan Africa and Madagascar. Only seven documented records in Madagascar, most in CE rainforest below 1,000 m; also Isalo (SW) in riverine forest; presumably overlooked but hard to explain rarity of records.

Madagascar Cuckoo *Cuculus rochii*   27 cm
The only 'typical' (genus *Cuculus*) cuckoo common in Madagascar. Rather hawk-like, especially in flight, but wings relatively short and pointed, tail long and rounded, and lacks hooked 'raptor' bill. Juv browner than adult with whitish mottling, yellow bare parts duller. See also vagrant Lesser and Common Cuckoos (Plate 113), not yet confirmed in Madagascar. **Voice** ♂ song, a very common sound in forests in austral summer, 3–4 repeated, mellow notes followed by one lower *pyu-pyu-pyu-too*, *pow-pow-pow too* or *chow-chow-chow-tu*; sometimes last note omitted or only slightly lower than preceding. Presumed ♀ call high, slow trill *pepepipipipipepepe*. **SHH** Breeds only in Madagascar, most migrating to E & C Africa in Apr–Sep; a few present all year. In forest and wooded habitats; also plantations, cultivation; scarce above *c*.1,500 m, rare on C plateau and in far S. Hard to see; sometimes joins mixed-species flocks.

Collared Nightjar *Gactornis enarratus*   24 cm
A highly distinctive and beautifully marked nightjar, with owl-like facial discs. From Madagascar Nightjar by *lack of white on wings* and rufous collar; also *large dark spots on crown*. **Voice** Song unknown. Alarm call a repeated soft, liquid *kow, keeow* or loud *tsrrr*. **SHH** Endemic to Madagascar; in rainforest, mostly at low and mid-altitudes, and in far N in dry forest (Ankarana, Daraina); unconfirmed reports in W. Sleeps during day on ground, often huddled in pairs. Unusually for a nightjar, nests on low shrub or tree-fern, and forages by sallying from low perch near clearing, not in prolonged flight.

Madagascar Nightjar *Caprimulgus madagascariensis*   22 cm
A typical small nightjar; from Collared by lack of both dark spots on crown and rufous collar, and by *conspicuous white on wings*; from potential vagrant European Nightjar (Plate 114) by much smaller size. **Voice** Song, heard very frequently, *loud single note followed by somewhat lower, short trill tip-tttttrrrr*. Call *pow-pit, wow-up, wow-wow, tpit-woau, wau, wa-pit*, or *ku-wuh*, often several birds calling simultaneously, in flight or perched. Also, possibly this species, a long, low, glugging or clucking trill lasting 10–30 s, moving up and down scale and becoming faster and slower. **SHH** Endemic to Madagascar and Aldabra; widespread in Madagascar in towns, villages, wooded areas, plantations and cultivation but patchy above *c*.1,500 m and absent from interior of rainforest and open savanna. Feeds in flight, often over water.

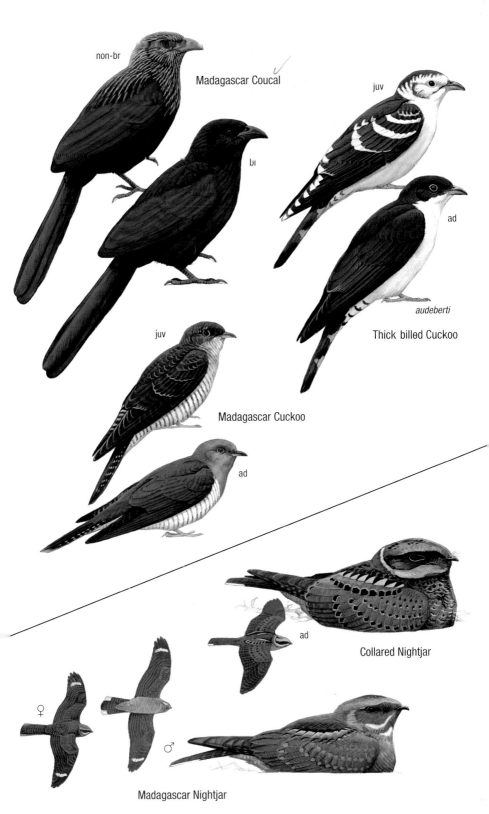

PLATE 42: MADAGASCAR: OWLS

Madagascar Red Owl *Tyto soumagnei*   29 cm
Much more *orange-rufous overall* than Barn Owl, particularly below, but care required as Barn Owls vary and can be pale orange below; also *smaller, lacking grey fringes above, call distinctive*. From other owls by lack of strong streaking or barring below. **Voice** Loud, toneless, medium-/high-pitched, hair-raising *kcchhhhrrrrrrrr*, like Barn Owl but *dropping markedly at end*. **SHH** Endemic to Madagascar, elusive and rarely recorded; in rainforest and locally in dry forest in far N; feeds in adjacent degraded areas and cultivation. Roosts in dense vegetation, often in banana or Traveller's Tree. **VU**

Barn Owl *Tyto alba*   35 cm
Larger than Madagascar Red Owl, with noticeable *grey fringes to many wing and back feathers*; usually paler below, less orange; care required due to orangey underparts of some Malagasy birds. **Voice** Loud, toneless hissing *ksscchh*, *on one note*. Hisses and snores when at rest and makes loud bill-snaps when disturbed at or near nest. **SHH** Practically worldwide; in region, absent from Mascarenes and coralline Seychelles. Widespread in Madagascar (race *poensis* as in Africa) including towns, but commonest in cultivation and wooded areas; usually absent in dense rainforest interior, but present in fragmented or open forest and edge; less common above *c.*1,800 m.

Madagascar Scops Owl *Otus rutilus*   23 cm
Occurs in brown and rufous morphs (latter most common). From other Malagasy owls by *small size*, *yellow eyes*, short ear-tufts. W race *madagascariensis* (sometimes treated as a separate species, Torotoroka Scops Owl) from nominate with great difficulty by narrower barring on flanks, greyer crown and more conspicuous central streak on wing-coverts; tail slightly longer; more easily separated by voice, although some songs intermediate. **Voice** Nominate song *tu..tu..tu..tu..tu..tu..tu*, up to 30 quite high, simple, short whistles; *madagascariensis* shorter sequences (5–10) of lower short trills *prrp..prrp..prrp..prrp..prrp*, but much variation between the two, especially in N, SE & C plateau. **SHH** Endemic to Madagascar. Nominate race common in E rainforest and adjacent areas. Race *madagascariensis* in S, W & C plateau, including Antananarivo; common in forest, towns, plantations and cultivation; absent from savanna.

White-browed Owl *Ninox superciliaris*   26 cm
From all other owls by *white eyebrows and barred not streaked breast*, eyes dark. **Voice** A low *disyllabic hoot pooorow*, rising at end, often repeated, and usually followed by loud medium-low barking *warkwakwakwakwakwakwak...*, first note usually lower; similar to Madagascar Long-eared Owl but lower and lacks crescendo. **SHH** Endemic to Madagascar; common below *c.*800 m in W, N & S, also patchily in lowland E, in primary forest and adjacent degraded areas, scrub and less commonly in cultivation. Genetic evidence shows closer relationship to genus *Athene* than *Ninox*, but may belong in its own genus.

Madagascar Long-eared Owl *Asio madagascariensis*   34 cm
From Marsh Owl by *long ear-tufts, streaked breast* and barred back and tertials. Juv streaked paler buff above than ad, underparts paler; when first fledged, face black and body covered with creamy-white down. **Voice** A *loud wugh... wugh... wugh...* or *harkh... harkh... harkh*, 10–50 notes, speeding up and becoming higher, often slowing at end; also single notes repeated at 2–5 s intervals; lacks creaky tone of Madagascar Crested Ibis, higher-pitched and more varied than White-browed Owl. Juv begging call toneless screech *krrrrrch-krrrrch*, usually two notes, unlike Barn Owl. **SHH** Endemic to Madagascar, fairly common in primary forest in NW, N & E, usually in riverine forest in W & S; also in degraded forest and plantations on plateau (including Antananarivo) and E. Not in savanna or on high mountains.

Marsh Owl *Asio capensis*   34 cm
From Madagascar Long-eared Owl by *uniform back and breast*; much larger, longer-winged and darker than Barn and Madagascar Red Owls. Madagascar endemic race *hova* larger and plainer on wings and tail than African birds. **Voice** Not vocal; croaking *creeouw* when flushed; also faster, repeated croaks *quick quark quark*. **SHH** Sub-Saharan Africa to Morocco; in region only in Madagascar, where patchy in wetlands, rice fields and towns, most common on C plateau including C Antananarivo; absent from high mountains and far S.

PLATE 43: MADAGASCAR: SWIFTS

Madagascar Spinetail *Zoonavena grandidieri* 12 cm
From Little Swift by slimmer build, *less contrasting and smaller pale rump*, less dashing flight; from possible vagrant Mascarene Swiftlet (Plates 82/86) with great difficulty by diagnostic tail-spines; Spinetail shows more prominent black shaft-streaks on rump and underparts, shorter tail and slightly larger size but these characters are hard to judge in the field; either species may show a shallow, or no, tail-fork. Palm Swift has long forked tail and narrow backswept wings. **Voice** Short *chip... chip... chip...* or *zip... zip... zip...*, often in chorus or as short trill *zrizrizri*. **SHH** Endemic to Madagascar (nominate race) and Comoros. Commonest in E Madagascar, usually over forest or adjacent degraded areas or cultivation, scarce in W & S, absent from C plateau. Flight rather slow and erratic; in small groups, sometimes with Palm Swift.

African Palm Swift *Cypsiurus parvus* 16 cm
From Madagascar Spinetail by paler coloration, lack of pale rump, longer, swept-back wings and, especially, *long, pointed, deeply forked tail*, usually held closed. Madagascar endemic race *gracilis* darker above and paler below than E African race. **Voice** Short, vowel-less, scratchy notes *tchh tchh tchh* sometimes in long sequence. **SHH** Sub-Saharan Africa, Madagascar, Comoros; vagrant Aldabra group. In Madagascar, common in warm lowlands, scarce in coastal areas of S; usually not far from palms, around cultivation and villages; sometimes in large numbers over wetlands or salt flats; scarcer over primary forest. Usually flies around treetop height, sometimes with other swifts.

Alpine Swift *Tachymarptis melba* 20 cm
Very large brown swift; from all others by *white belly* and chin (latter hard to see). Madagascar endemic race *willsi* smaller than E & S African birds; brown areas much darker. **Voice** Rolling trill *kirritirrirr*. **SHH** Breeds S Eurasia, mainly spending boreal winter in Africa, where also resident populations, as in Madagascar, where sparse and erratic, more regular on C plateau and mountains. Vagrant to Comoros (race unknown) and Seychelles (probably migratory nominate race). Occurs over forest, degraded areas and cultivation, often flying high with African Black Swift. Often near cliffs where it may breed. Very strong, powerful flight, sometimes slow.

African Black Swift *Apus barbatus* 16 cm
Dark swift with paler secondaries and inner primaries; *lacks white belly of Alpine Swift* and white rump of Little; tail more forked than latter. Palm Swift is smaller and slimmer, with much longer deeply forked tail and proportionately longer, more swept-back wings. Madagascar endemic race *balstoni* smaller and darker than E & S African races. See vagrant Common Swift (Plate 114). **Voice** High screaming trill, *zzzzziiieeewwww*, dropping at end, often in chorus. **SHH** Sub-Saharan Africa, Madagascar and Comoros. Fairly common but erratic throughout Madagascar; breeds around inland and sea-cliffs, and rocky offshore islands. Malagasy region birds may constitute a separate species, Madagascar Black Swift *Apus balstoni*.

Little Swift *Apus affinis* 12 cm
Small stocky swift with *bright white rump* and chin; from other swifts by *short notched or square tail*; from Madagascar Spinetail by thicker body and blacker plumage. **Voice** A shrill, rippling trill, tailing off at end. **SHH** NW Africa to India and sub-Saharan Africa; in Madagascar very local, breeding in small numbers in Antananarivo, rare in N & SE; vagrant to Seychelles. Usually in groups with other swifts or Mascarene Martin. Race or races in region unknown.

144

PLATE 44: MADAGASCAR: KINGFISHERS AND ASITIES

Madagascar Malachite Kingfisher *Corythornis vintsioides* — 13 cm
The only regularly occurring **blue-backed** kingfisher in the region. Pale blue rump and back contrast with darker wings and tail. Bill black at all ages. Potential vagrant Malachite Kingfisher *C. cristata* from sub-Saharan Africa has solid blue back concolorous with wings and tail, and red bill (black in imm). **Voice** High *tsip* or *tsrip*, often repeated, *tsip-ip* or *tsrip-ip*; also flatter *chipchipchip..chipchip*, sometimes given by 2+ birds together. **SHH** Endemic to Madagascar (race *vintsioides*) and Comoros; vagrant Aldabra. Common near water below *c.*1,600 m, on coasts, along rivers, around lakes and even in towns; scarce in dense forest and absent from driest parts of S.

Madagascar Pygmy Kingfisher *Corythornis madagascariensis* — 13 cm
The only **red-and-white** kingfisher in the region. **Voice** Single, penetrating *chick, chiuk, tsiewk* or *tsiewk-ik*, repeated irregularly; song similar, given from top of rainforest tree. **SHH** Endemic to Madagascar; fairly common in rainforest and degraded edges below *c.*1,200 m, scarcer in W forest, usually in more humid areas, as far S as Isalo and Zombitse-Vohibasia, from where race *dilutus* described but not reported recently and validity uncertain. Rare far from forest and in S. Not strictly tied to water; often hunts on forest floor.

Velvet Asity *Philepitta castanea* — 15 cm
Breeding ♂ distinctive: entirely **velvety black** apart from small golden-yellow spot at bend of wing and large bright lime-green caruncle above eye from rear of ear-coverts to forehead; non-breeding ♂ has caruncle much reduced and black plumage marked with yellow scales. ♀ from Schlegel's Asity by **pale streak across face, lack of yellow on undertail** and lack of pale eye-ring, plus larger size. **Voice** Quiet, *squeaky, repeated whee-doo*, with emphasis on first syllable; also long series of simple *weet* notes, often produced by lekking ♂♂; shorter and monosyllabic *wit* or *wheet* calls while foraging, also thin, high *seeeeee*. **SHH** Endemic to Madagascar; fairly common in rainforest below *c.*1,500 m; sometimes higher but absent from montane forest and C plateau. Often alone or in family groups, near fruiting understorey shrubs; sometimes with mixed-species flocks.

Schlegel's Asity *Philepitta schlegeli* — 14 cm
♂ distinctive: **black head, yellow underparts and large green-and-blue caruncle**; non-breeding ♂ olive-green above, wings and tail brownish, mottled darker on breast and upper belly, brighter **clear yellow on undertail-coverts**. ♀ from Velvet Asity by yellowish undertail, pale eye-ring, smaller size, shorter and thicker bill. **Voice** Calls include single squeaky notes. Song very distinctive; quiet but penetrating, rapid, brief (*c.*1.5 s) *hu-hu-hu-hi-hi-hu*, first 4–5 notes rising and last 2–3 dropping, from hidden perch or high in canopy. **SHH** Endemic to Madagascar; patchy and scarce in W forest, from far N (Andavakoera) to just S of Tsiribihina; commonest in pinnacle karst massifs (Bemaraha, Namoroka) to SW of Betsiboka, and humid parts of NW forests including Ampijoroa/Ankarafantsika N to humid NW (especially Manongarivo), where overlaps with Velvet Asity. Usually in canopy or subcanopy, often near flowering trees, singly or in small family groups. **NT**

PLATE 45: MADAGASCAR: BEE-EATER, ROLLER, CUCKOO-ROLLER AND HOOPOE

Olive Bee-eater *Merops superciliosus*   30 cm
Only common bee-eater in region; from vagrant Blue-cheeked (Plate 115) by overall duller colours, *lack of blue around supercilium, and chestnut cap*. Juv duller green and lacks tail-streamers. From vagrant European (Plate 115) by lack of yellow on throat and blue below. **Voice** *Sequence of rapid, hollow and rolling medium-pitched notes*, *krioor krioor krioor* or *kriew kreiw kriew*, often repeated and in chorus from flocks; also single *prioop*. **SHH** Sub-Saharan Africa, Madagascar and Comoros; vagrant to Aldabra. Common mostly below *c*.1,300 m, in open wooded habitats, often near water, on coasts, in cultivated areas, plantations and gardens; scarce in rainforest, migrant over plateau in Apr–May and Sep–Oct. Usually in flocks, sometimes large. [Alt: Madagascar Bee-eater]

Broad-billed Roller *Eurystomus glaucurus*   29 cm
A stocky, large-headed roller, mainly **cinnamon above** and deep lilac below. From Madagascar Kestrel and potential vagrant European Roller (Plate 115) by turquoise tail, broad yellow bill. Ground-rollers live in forest interior; most are terrestrial, with weak flight and highly distinctive plumage. Nominate race breeds only in Madagascar, and is larger with darker greyer underparts than African races. **Voice** A loud, croaking or rattling *gagagagaggrikgrikagrika... grik... grik* or *karr kriik kriik karr krikrikrikrikrikrikukrikukrikukukuk... uk... uk... uk*. Many variants, longer and shorter, with single notes given when mobbing intruders. **SHH** Sub-Saharan Africa; breeding visitor to Madagascar, passage migrant on Aldabra group, Comoros and Mozambique Channel islands; vagrant to Mauritius, Reunion and Seychelles W of Aldabra group; in non-breeding season in C & E Africa. Common and conspicuous in wooded areas, plantations, cultivation and towns of N, W & S to *c*.1,600 m, patchy on C plateau; fairly common in E but absent from rainforest interior. Arboreal and aerial feeder, with dashing falcon-like flight.

Cuckoo-roller *Leptosomus discolor*   50 cm
Very distinctive. In flight, like floppy raptor but with *deep, sweeping wingbeats*. When perched, from all other large birds by *spotted breast* (♀) or *dark iridescent upperparts with black bands on head* (♂). **Voice** Song high-pitched, fluid, rolling *riieeew reew rew rew*, or *crrrieeewww riiieew rew rew*: 3–4 (sometimes five) notes, the last 2–3 descending; sometimes second note higher-pitched than first, and more trilled when more excited; often given by 2+ birds together. Call a sequence of 2–10 *pupupupupup* notes; also a short *pup*. **SHH** Endemic to Madagascar (nominate race) and Comoros. Fairly common in E & W forests below *c*.2,000 m, scarcer in adjoining degraded areas and wooded savanna, and in cultivated land and plantations; in S only regular in gallery forests, and scarce on C plateau, apparently a migrant. Noisy and conspicuous when displaying, discreet and hard to see when foraging. Several birds may follow each other during display, flying over forest; otherwise usually in pairs.

Madagascar Hoopoe *Upupa marginata*   32 cm
Very distinctive even in briefest views, pattern and structure unique among resident Malagasy birds. In flight, shows **broad rounded black-and-white wings**. ♀ slightly smaller than ♂, with white areas tinged beige. Juv similar to ♀ but crest and bill shorter, and duller. See vagrant Common Hoopoe *Upupa epops* (Plate 115). **Voice** Completely unlike other hoopoes. Call a low harsh growling *raghr*, given infrequently, and a low-pitched, raucous hissing *chrrshhh*. Song a *low cooing trill rrrrrrooow*, dropping slightly, lasting 1.5–2.5 s, repeated every 3–10 s. **SHH** Endemic to Madagascar; common in N, W & S, in forest, wooded areas, degraded areas, cultivation and plantations; scarce on C plateau (including Antananarivo) and in open areas of lowland E below 1,600 m. Usually in pairs, may follow couas and mesites in forest. Often treated as conspecific with Common Hoopoe but completely different voice, and plumage and size differences from nearest other hoopoes (in Africa), support separation.

PLATE 46: MADAGASCAR: GROUND-ROLLERS

Short-legged Ground-roller *Brachypteracias leptosomus* — 34 cm
From other ground-rollers by **whitish supercilium and breast-band**; and usually in trees. Juv duller, browner above than ad, with fainter buffish breast-band, flanks and lower breast mottled (not barred) brown. From White-browed Owl by breast-band and longer bill. **Voice** Call soft dove-like *coor* or *kroo-kroo* or single *poop*. Song short low booming *boop* or slightly more modulated *mwop* lasting *c*.0.5 s or less, repeated 5–30 times every 1–5 s (usually *c*.1 s), most frequently around dawn, *from perch 5–30 m up*; lower and usually more rapidly repeated than Scaly or Pitta-like Ground-rollers. Also rarely a single *boop* followed by *doodoodoodoodoodoodoodoodoo*, a rapid descending series of short hoots. **SHH** Endemic to Madagascar. Very patchy in rainforest below *c*.1,500 m; often hard to detect as spends long periods immobile in subcanopy. Makes long flights to catch food, sometimes to ground. **VU**

Scaly Ground-roller *Geobiastes squamiger* — 29 cm
Very distinctive; from all other terrestrial birds by **scaled plumage, rufous collar and long pink legs**. When flushed, shows pale bluish tips to outer tail; lacks pale wing-flashes of Pitta-like and Rufous-headed Ground-rollers. Juv extensively rufous on head and upperparts, with ill-defined black moustache. **Voice** Call a brief muffled *ko-uh*. Alarm a hissing *kwish-sh* decreasing in volume. Song loud, mellow, *mooop* or *booop* or more modulated *mowooop*, lasting *c*.1 s, repeated every 5–30 s; higher-pitched and more widely spaced than Short-legged, lower and longer than Pitta-like and Rufous-headed; usually at dawn, from near ground. **SHH** Endemic to Madagascar, patchy in rainforest below *c*.1,000 m; commoner at low altitudes and rare in adjacent degraded forest. Shy and discreet; hard to find, usually solitary. Runs along ground, stopping at regular intervals, sometimes flies short distances. **VU**

Pitta-like Ground-roller *Atelornis pittoides* — 27 cm
White throat permits identification even at distance. When flushed, shows whitish wing-patch like Rufous-headed Ground-roller, but **blue on head** usually evident. Juv throat pale, speckled brown. **Voice** Brief, explosive *trrrt* and hissing *kashhhhh*; also low *kowk* when flushed. Song *clear, single, short, whup* or *mwup*, slightly barking, lasting *c*.0.5 s, repeated at 3–10 s intervals, from around 2–3 m up, mostly around dawn; lower and less disyllabic than Rufous-headed and higher than Scaly or Short-legged. **SHH** Endemic to Madagascar; fairly common in rainforest including Amber Mountain (N) and Ambohitantely and Ankaratra (C plateau) to *c*.1,500 m, locally in montane rainforest, especially drier (western) edges where Rufous-headed absent; also in degraded fringes of rainforest and rarely in nearby plantations; secretive, runs along ground, pausing frequently.

Rufous-headed Ground-roller *Atelornis crossleyi* — 26 cm
Smaller than Pitta-like and Scaly Ground-rollers, with **reddish head and pale spot in wing** obvious when flushed. Juv has duller head, lacks black triangle on throat and has purple tinge to nape. **Voice** Call soft clucking *bok*. Also a sneezing *ketch*. Song *clear, short pop* or *pwop* or *paup* or *glop*, slightly disyllabic when close, lasting *c*.0.5 s, repeated every 2–10 s from a perch 2–5 m off ground mostly around dawn; higher and more disyllabic than Pitta-like. Occasionally extended to sequence of similar calls, *whop-whop-whop-whot-whot-whot*, slightly descending. **SHH** Endemic to Madagascar; in wetter, east-facing parts of montane rainforest above *c*.1,000 m, commonest at 1,200–1,750 m; rare in degraded areas or drier fringes; absent from Amber Mountain and C plateau forest fragments. Secretive; runs along ground, stopping regularly. **NT**

Long-tailed Ground-roller *Uratelornis chimaera* — 47 cm (♂), 39 cm (♀)
Wonderfully distinctive. From terrestrial couas by smaller size, **grey upperparts and black breast-band**. ♀ shorter-tailed. Juv has narrower gorget, especially in ♀. **Voice** Calls low *boo-boo-boo-boo*; also scratchy bisyllabic *too-tuc too-tuc...*, possibly an alarm. Song low popping *too-tuc too-tuc too-tuc* followed by short wing-snapping (while flying up to perch) and then low, loud, descending hooting, *boo-boo-boo-boo-boo*, from perch 1–2 m up, at dawn or dusk. **SHH** Endemic to SW Madagascar, in a narrow band of spiny forest between Mangoky R and Fiherenana R. Within this area, in low coastal scrub, degraded and primary dense spiny forest and closed-canopy deciduous forest; absent from cultivation or open areas. Walks or runs on ground, singly or in family groups; secretive and graceful. **VU**

PLATE 47: MADAGASCAR: SUNBIRD ASITIES AND SUNBIRDS

Common Sunbird Asity *Neodrepanis coruscans*   10 cm
From Yellow-bellied Sunbird Asity with difficulty by *duller yellow flanks brighter than mottled olive belly* and breast, *longer, more decurved bill*, ♂ has yellow fringes on wing feathers and more wedge-shaped caruncle, *thickest portion well behind and above eye*. From ♀ sunbirds by much more decurved bill, yellow flanks, and hyperactive behaviour. **Voice** *Call rapid series of 10–15 high-pitched squeaky hisses*, very characteristic and penetrating, starting closely spaced, becoming less so; often followed by widely spaced series of similar notes; more hissing and forceful than Yellow-bellied. **SHH** Endemic to Madagascar, in mid-altitude rainforest at *c*.500–1,500 m, commonest around 1,000 m; visits flowering trees at forest edge, scarce in degraded areas adjacent to primary forest; recorded but rare in C plateau forest fragments (Ambohitantely); absent from littoral forest and drier parts of rainforest. Very hyperactive, rarely stopping for more than a few seconds; visits flowering trees in forest, sometimes in groups; occasionally mobs intruders and (especially in cool season) joins mixed-species flocks.

Yellow-bellied Sunbird Asity *Neodrepanis hypoxantha*   10 cm
From Common Sunbird Asity with difficulty by *brighter yellow underparts, brightest on throat*, and *shorter less decurved bill* (hard to judge); ♂ has *uniform blue upperparts, lacking yellow fringes* on wing-feathers, and *large blue caruncle* with central green panel *extending much further in front of eye* than in Common. Smaller than sunbirds with much brighter underparts. **Voice** *Calls weaker and higher than Common*, lacking the hissing, squeaky quality; could be mistaken for tree-frog; single, very high-pitched forceful *tip* repeated at *c*.1 s or longer intervals, sometimes repeated more rapidly, in similar rhythm to Common. Wings make whirring noise in flight, especially ♂. **SHH** Endemic to Madagascar, in montane rainforest usually above *c*.1,200 m; commonest in dense mossy forest around 1,600 m, absent from drier areas and C plateau. Incredibly nervous and hyperactive; visits flowers in forest especially parasitic mistletoes *Bakerella*; mobs intruders including people. **VU**

Madagascar Green Sunbird *Nectarinia notata*   15 cm
From Souimanga Sunbird and sunbird asities by *long, strong bill*, larger size and generally darker plumage; ♀ *heavily streaked darker below* and has *pale supercilium*, ♂ *lacks pale on belly*, although moulting birds can be paler on belly and resemble ♂ Souimanga more. **Voice** Usual call distinctive, high-pitched insistent *chert chert chert chert*, or *tjup tjup tjup tjup*, slightly hoarse, repeated every 2–10 s. Song (rarely heard) complex warbling phrases. **SHH** Endemic to Madagascar and Comoros; in Madagascar throughout, to *c*.2,050 m, but more common in lowlands; in forest, degraded areas, plantations, cultivation and gardens; scarce in S.

Souimanga Sunbird *Nectarinia souimanga*   11 cm
From Madagascar Green Sunbird by smaller size, shorter slimmer bill, and in ♂ *pale belly* (dull yellow in nominate, whitish in race *apolis* in dry S & SW) and reddish breast band; ♀ *lacks heavy streaking and pale supercilium* of Madagascar Green. From ♀ sunbird asities by larger size, thicker bill (especially tip), lack of clear yellow on flanks (Common) or breast (Yellow-bellied). ♂ *apolis* differs from nominate in having little yellow on belly, underparts below sooty band creamy-white; metallic upperparts brighter green. ♀ paler, less yellow below and greyer above. **Voice** Call a *single plit* or *chit* or *chip* in flight; also high-pitched, drawn-out, peevish (or mewing) *chweee*. Song high-pitched twittering *plit... chit... chirrit ... chirrititiwitittirrit*; last phrase very complex and sometimes extended to 1.0–1.5 s. **SHH** Endemic to Madagascar, Aldabra group and Glorieuses. Very common in all wooded habitats; possibly seasonal in parts of S (visiting flowering trees). Often in mixed-species flocks.

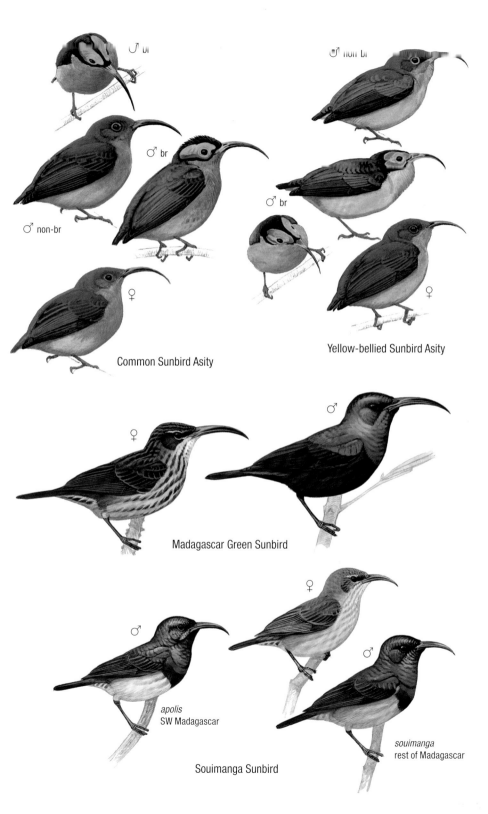

PLATE 48: MADAGASCAR: LARK, MARTINS, SWALLOW AND WAGTAIL

Madagascar Lark *Mirafra hova*   13 cm
The only lark on Madagascar; from potential vagrant larks (Plate 117) by *streaked breast and small size*, from vagrant pipits (Plate 117) by stouter bill and broad pale supercilium. **Voice** Calls *chirrup* and short *chit*. Song very variable, given in flight; high-pitched, slightly squeaky, rhythmical *chiriip-chit-it-per-chrrip* or *chriiit... pitit... chirreep* or simpler *chirrii-ee*, slightly higher at end, repeated every few seconds. **SHH** Endemic to Madagascar, very common in open habitats everywhere especially on C plateau; absent from densely wooded areas and forest. Usually singly or in pairs. Genetic evidence shows closer relationship to sparrow larks (genus *Eremopterix*) than to *Mirafra*, but further study needed.

Mascarene Martin *Phedina borbonica*   15 cm
From Brown-throated Martin by *heavier build and clear streaks on throat, underwing-coverts darker than flight-feathers*, and head and neck stronger; see vagrant Sand Martin (Plate 116). Madagascar Spinetail and African Palm Swift (Plate 43) have very different flight, closing wings much more rarely and wingbeats much stiffer. **Voice** Very characteristic; medium- to high-pitched *bubbling or buzzing crreeeww*, also similar rising *crisssww* or *criissspp*, or buzzing *gzeee-gzeee* or *phree-zz* call; also a shorter, medium *chep* or *cherp*, both often given in chorus. Song, in flight or perched, a complex warble. **SHH** Endemic breeder in Malagasy region (Madagascar, Mauritius, Reunion) with some movement in austral winter to SE Africa; vagrant to Seychelles. Race *madagascariensis* is common in Madagascar, with wide movements but some present all year. Found up to *c*.2,000 m, often around wetlands including rice fields, rivers, lakes and on coast; also in cities and towns. Away from SW, where Barn Swallow occurs, the only hirundine seen regularly in lowlands.

Brown-throated Martin *Riparia paludicola*   12 cm
Smaller and slimmer than Mascarene Martin and *lacks streaks on throat, with weak hesitant flight*, lower to the ground. Underwing more uniform, and head smaller and more rounded. See also vagrant Sand Martin (Plate 116). **Voice** Very different from that in Africa; medium-pitched, *mechanical, bubbling trill titititetetetet*, dropping at end, lasting 1.5–2.0 s. also short medium-low *pit-a click*, last note very mechanical, and a quiet *tew*. **SHH** Sub-Saharan Africa, SE Asia to Philippines; in region resident only in Madagascar (race *cowani*) where only in E highlands above *c*.500 m, commoner above 1,000 m, in open grassland and around wetlands, not usually in towns; vagrant to Seychelles (Asian race *chinensis*; see Plate 116). Taxonomy of Asian, African and Malagasy populations deserves further research; could be >1 species.

Barn Swallow *Hirundo rustica*   18 cm
Slimmer and longer-winged than Mascarene Martin, ad has *blue upperparts, long tail-streamers and white windows in tail*, and red throat with dark breast-band. Imm duller, pale buff on chin and throat, and shorter tail-streamers; tail still much more deeply forked than resident martins. **Voice** Contact call *witt-witt*, sometimes run together as series of *tswitt* notes. Song, a more prolonged bubbling twitter, not recorded in Malagasy region. **SHH** Breeds Eurasia and North America, spending boreal winter in sub-Saharan Africa and South America. Migrant to Madagascar and Seychelles, in Madagascar commonest in SW, N of Toliara, where up to several hundred roost in reedbeds in austral summer; very scarce elsewhere.

Madagascar Wagtail *Motacilla flaviventris*   19 cm
The only wagtail known from Madagascar; from potential vagrant wagtails (Plate 118) by combination of *black breast-band and yellow belly*. Sexes similar, ♀ slightly paler, juv with less conspicuous breast-band and supercilium. **Voice** Calls loud mellow *tsreee-teeeew* or *teee-tseewew*, or *tsree-teweew*. Song, often in duet *tsee-tsweeuuw, treeeew, tsweeweeuw* and similar, in irregular sequence, some notes trilled. **SHH** Endemic to Madagascar, common throughout except in S and possibly in NW around Nosy Be, where scarcer. Often around water, along coasts, in rice fields, along streams in forest, also common in built-up areas. In S restricted to rivers and nearby cultivation.

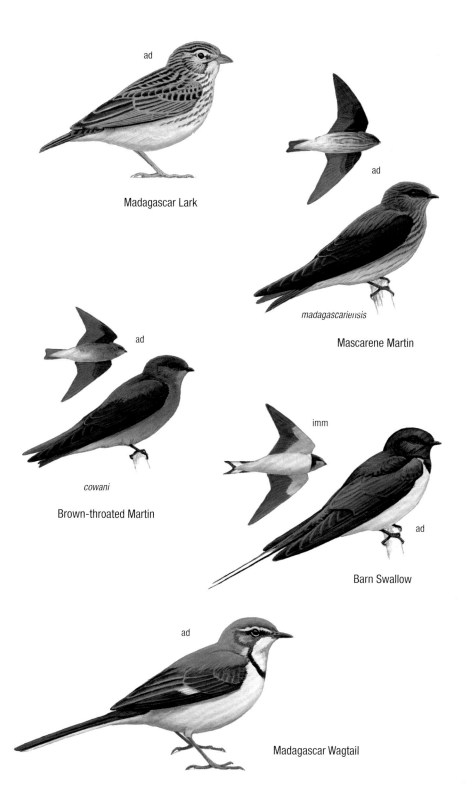

PLATE 49. MADAGASCAR: ROCK THRUSHES

Forest Rock Thrush *Monticola sharpei*   16 cm

♂ from Littoral Rock Thrush by weaker bill, *richer blue on head and extensive red on tail*; ♀ from Madagascar Magpie-robin and Littoral Rock Thrush by reddish tail and buffy-orange breast streaked whiter. ♂ of Amber Mountain race (*erythronotus*, formerly considered a separate species) has *reddish back* and wholly red breast, ♀ *lacks whitish streaks on breast*; ♂ from Bemaraha (CW) may appear intermediate with more red on breast than nominate, and those in SW (formerly, and sometimes still, treated as a species, 'Benson's Rock Thrush *M. bensoni*', best known from Isalo) like nominate but *paler and greyer*, ♀ *with darker streaking on breast*. Imm similar to respective ad, with pale fringes to feathers of back, head and breast, wider on breast. **Voice** Song variable across range, unclear if differences consistent, but typical versions follow: a *simple, quiet tew-de or tew tew tew teer teer*, a more complex *tew-deer ter di-eer* or *tew teweier teweier teweier* (Eastern *M. s. sharpei*), or *whetew whetew whetew whetew... poreee poreee whetew whew whew whew* (Isalo; scratchier and more prolonged than forest birds), *tew tereee weeweeuh* (Bemaraha), *whit whit whit whiti whitiooweooweoo* (Amber Mountain). Contact call repeated quiet *weed*. *Alarm low tactactactac* or *wheet-tock-tock*. **SHH** Endemic to Madagascar; in rainforest, commonest above *c.*1,000 m, also at Amber Mountain (N), Bemaraha (W) and Isalo (SW) and adjacent rocky massifs in W & SW, even reaching far SW near Toliara. Inconspicuous and rather inactive, not shy, in forest interior, edges of rocky or shrubby areas especially at high altitude or in karst or sandstone massifs. Genetic analysis does not support recognition of more than one species, but plumage variation considerable and Amber Mountain birds particularly distinctive. Observations in W & S still valuable, including Isalo, Zombitse, Toliara (Sept Lacs region), Bemaraha (especially differences N & S of Manambolo R); note tone of blue on head, extent of red on breast and any rufous on mantle of ♂, and overall tone (buffy/greyish) and extent and colour of streaking on breast of ♀.

Littoral Rock Thrush *Monticola imerinus*   16 cm

Both sexes from Forest Rock Thrush by *strong bill, dark tail with no rufous*, and overall greyer plumage; ♀ has whitish breast lightly streaked greyer, lacking rufous or buff. Juv has narrow pale fringes to feathers of upperparts, breast feathers widely edged pale brown. **Voice** Song a *high-pitched, hurried mixture* of falling whistles, trills and chacking notes, *crrieer crier whit-sic treeeir whit-sic crieerr treeoeer ttttttr* or *treeerrr tirrri tirri tirri... treeer siirsiirsiir...*, also *tac*, often repeated as alarm, and low *tschrrr*. **SHH** Endemic to coastal SW Madagascar, from about Mangoky R to near Fort-Dauphin, but rare N of Onilahy R and E of Cap Ste Marie. Found in *Euphorbia* scrub and edges of spiny forest; usually perched on shrubs or mounds, singly or in family groups.

Forest Rock Thrush

Littoral Rock Thrush

PLATE 50: MADAGASCAR: MAGPIE-ROBIN AND STONECHAT

Madagascar Magpie-robin *Copsychus albospecularis* — 18 cm

♂ has at least three plumage forms; all have head, breast and wings black with a violet-blue gloss and white lesser-coverts. Nominate (NE) has a *black belly and short black tail*; *inexspectatus* (SE) similar with whitish belly and undertail-coverts; *pica* (W & C plateau) is *larger and longer-tailed, with white belly and lower breast*, white greater coverts, *outer fringes to tertials and two white outer tail-feathers*. ♀ has similar pattern but greyish-brown not black; *pica* and *inexspectatus* more rufous-brown on upperparts and flanks. Juv has upperparts dark brown, mottled pale cinnamon-brown, flanks cinnamon-buff, belly and undertail-coverts buffy-white, and dark-mottled white wing-patch. ♀ from rock thrushes by white in wing, longer tail, usually held cocked, and from African Stonechat by larger size and longer tail. **Voice** Calls apparently similar in E & W. *Song disjointed fluty notes* interspersed by higher whistles *tsiiir... tseeetseirtseetseetsew... tseetsiritichiweee... teeesiritchiweetseesewee...* or *seechirrew... tseeteechirrew tsiiiiiii chirri* or *whit-eeeer whit-eeeer...*, being very variable. Alarm a low *tchee* or *chizz*, or very high *tsiiiiiiiii*, difficult to locate. **SHH** Endemic to Madagascar; E forms in rainforest or margins, from sea level to *c.*1,500 m; nominate mostly in N half of E, intermediates in centre and *albospecularis* in S. Race *pica* common in and outside forest in W & S and rarely on C plateau in forest relicts and plantations to *c.*1,500 m. Terrestrial or in understorey, often in pairs or family groups, not shy and sometimes tame. Differences between race *pica* and races *albospecularis/inexpectatus* appear sufficient to separate the former at species level.

African Stonechat *Saxicola torquatus* — 13 cm

A small upright chat, ♂ with *black hood, wings, tail and mantle* (edged brown), *conspicuous white wing-patches* and bright dark chestnut upper breast. Three races endemic to Madagascar: *ankaratrae* (C Madagascar to W coast) similar to widespread *sibilla* but significantly larger; *tsaratananae* (Tsaratanana Massif) highly distinctive, ♂ with variable black streaking on breast, obscuring rufous patch; lacks pale rump. ♀ similar to ♂ but black replaced by mid-brown, streaked dark brown and underparts buff-brown. In flight shows white wingbar and white uppertail-coverts. Juv very scaly, gradually resembling ♀ with age but paler and more streaked on head, mantle and breast. From other resident species by small size, short tail, habit of perching atop bushes in open. See vagrant Whinchat (Plate 119). **Voice** Song complex, rapid mixture of high warbles and grating notes *trkwhirtrikwhirpirchk*. Call a characteristic *whiiit-tck-tck* or *wheet-chak-chak*. **SHH** Sub-Saharan Africa, parts of W Africa, SW Arabia and Malagasy region, where resident on Madagascar and Comoros. Common in open shrubby habitats, especially at higher altitudes (*c.*1,000 m and above), but also to sea level particularly in E Madagascar. Usually alone or in family groups, perching on tall shrubs and feeding on ground. Research needed to assess taxonomic status of Malagasy stonechats, which appear genetically distinct from, although look similar to, some African races.

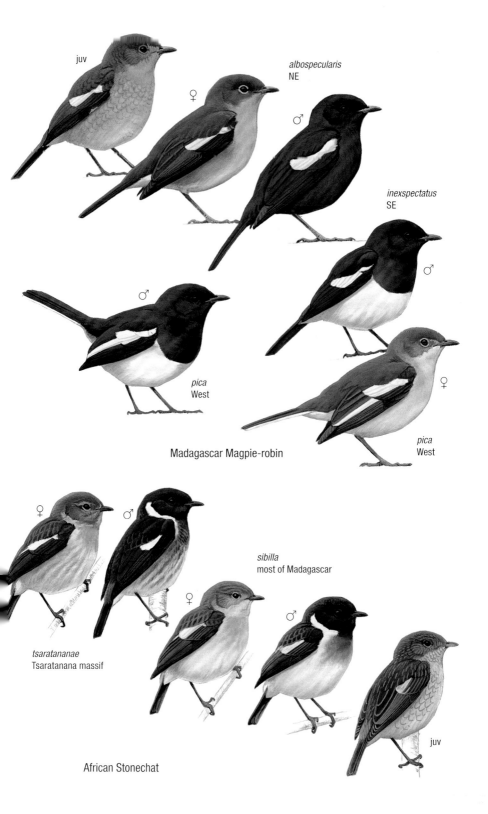

Madagascar Magpie-robin

African Stonechat

PLATE 51: MADAGASCAR: JERIES AND TETRAKAS I

Common Jery *Neomixis tenella*   10 cm
From other jeries and smaller tetrakas by *clear grey nape, pale legs and lower mandible, yellow throat*, weak supercilium and eyestripe, and (in ad) pale eye. Nominate (N, NE & NW) bright green above with grey nape and clear yellow throat, lacking conspicuous streaking. CW birds (*decaryi*) similar but slightly duller, birds in S half of rainforest (*orientalis*) and SW/S (*debilis*) duller olive above with *slightly stronger streaking on throat and upper breast*; *debilis* smaller and shorter-billed. Juv has dark eye and unstreaked or finely streaked breast. **Voice** High-pitched descending *seep-seep-seep*, sometimes with lower spluttering calls interspersed; song in SE, S & SW starts with this series then ends with lower, spluttering notes *pitrrr*, or even lower *stitit*. In CW, NW, N & NE, single rising *seeeip* followed by descending *seep-seep-seep* notes. **SHH** Endemic to Madagascar. Common everywhere except completely open areas and above *c.*1,600 m. In small groups with white-eyes, newtonias or other jeries, in canopy or in low scrub. Further research needed on taxonomy; perhaps >1 species.

Green Jery *Neomixis viridis*   10 cm
Short-tailed jery. From other jeries and smaller tetrakas by small size, *lack of yellow below* (sometimes a tinge on side of neck), bright green upperparts and *orange legs*. Lacks bright white eye-ring and clear yellow throat of Madagascar White-eye. Race *delacouri* (NE) slightly darker and brighter green above. **Voice** A harsh, sharp, flat *tick* or *tickititit*. Song similar to Common Jery but pitch rises: *se-see-see-si-stikit*, final phrase very high, slightly rasping; some notes, including last, may be omitted. **SHH** Endemic to Madagascar; only in rainforest, where common mainly below *c.*1,600 m, usually in canopy; sometimes on forest edge, in groups with newtonias, other jeries or white-eyes.

Stripe-throated Jery *Neomixis striatigula*   11 cm
Largest, stocky jery; from other jeries by *long, dark bill and legs, clear streaking on throat and upper breast*, and clear dark eyestripe; see Rand's Warbler. NE race *sclateri* very similar to nominate, but greener above and on ear-coverts. W/S race *pallidior* paler, with shorter bill and longer legs as well as distinct voice. **Voice** In E, a quiet *tsiep*; in W/S, *tsick*, *click* or *tsilik*. Song in E alternates *two sweet phrases*, ascending then descending; *too-too twee-twee twee dree dree see see...see-see dree-dree-dree-dree-twee twee-tcheechee*; sometimes truncated and then very similar to Rand's Warbler, but usually ascends slightly. In W, song higher, faster, more rattling and flatter, a mix of high notes and rattles: *see-see-see see-see-trchchch-sisisisisisi-tchchchch*. **SHH** Endemic to Madagascar. In E (nominate and *sclateri*), in or near primary rainforest below *c.*1,600 m, in canopy, often in mixed-species flocks; in S & SW, not in primary forest, but common in adjacent scrub and in primary spiny forest. Sings for long periods from bare treetops, or, in W/S, shrubs. E and W/S populations may be separate species.

Rand's Warbler *Randia pseudozosterops*   12 cm
Small tetraka, *grey above and unmarked pale grey below*, unlike all jeries. From Common Newtonia by *striking pale supercilium* and whitish underparts, strong, slightly decurved bill, and behaviour. Some have bright pale orange lower mandible. **Voice** Short, high-pitched *tick* or *schip*, repeated, or high chittering made by several individuals. *Song rapidly repeated tew notes, strengthening and slowing slightly*, sometimes with final, lower *tu* note; recalls Stripe-throated Jery, but smaller range and variation: *titititew-tew-tew-tew-tew-tew tew tew*. **SHH** Endemic to Madagascar; fairly common in rainforest up to *c.*1,200 m. Usually in groups in mixed-species flocks; often feeds along horizontal or angled branch looking underneath alternate sides. Sings for long periods from canopy trees, often near Stripe-throated Jery.

Cryptic Warbler *Cryptosylvicola randrianasoloi*   11 cm
Small, long-tailed tetraka, usually at high altitudes, unlike similar species. From jeries by *unstreaked throat*, slightly larger size, *longer tail, dark legs*, and pale pink lower mandible; from Wedge-tailed Tetraka by whitish belly, dark eyestripe and lack of grey ear-coverts. **Voice** Scolding *tsick* or *chick-ess*, sometimes in short series. Song loud, *a single harsh note followed by 5–12 rasping notes on same pitch*, chick *tss-tss-tss-tss-tss-tss*, fading away. **SHH** Endemic to Madagascar; rather scarce in E rainforest, mainly above 1,000 m, commonest above 1,600 m; in mixed-species flocks and singing from canopy.

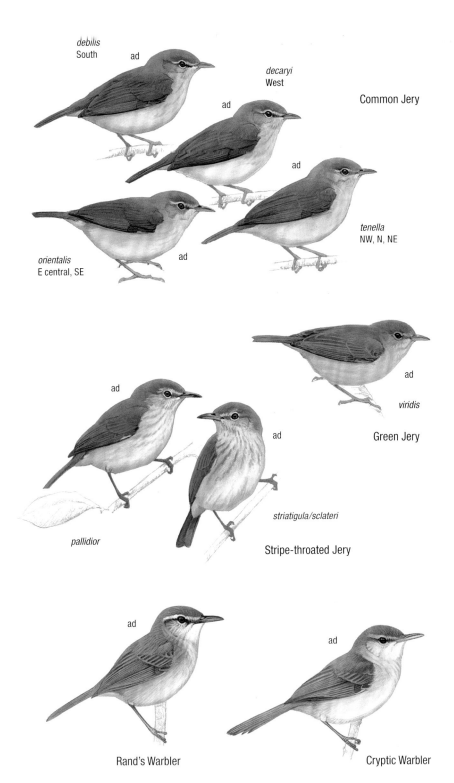

PLATE 52: MADAGASCAR: WARBLERS AND TETRAKAS II

Madagascar Cisticola *Cisticola cherina* 11 cm
Tiny open-country warbler, from all others by *small size, streaked upperparts and white tips to short tail*. **Voice** Song a very high-pitched, metallic *tint... tint... tint* or *pik... pik... pik... pik...*, repeated at 0.5–1.0 s intervals; also a high-pitched, metallic, repeated *triit* or *chiit* or *chreep*, slightly trilled and modulated. Both given in song-flight, and in duet when may be delivered more rapidly. Alarm note *tic*. **SHH** Endemic to Madagascar, Cosmoledo, Astove and Glorieuses. Common throughout Madagascar, except in forest; especially but not only near water.

Grey Emu-tail *Amphilais seebohmi* 17 cm
Small marshland warbler. From Madagascar Brush Warbler and Madagascar Swamp Warbler by *small size, streaking on back and 'decomposed' tail* as in Brown Emu-tail (Plate 54). **Voice** Call a low grating *chip... chip*, rather quiet and unobtrusive. Song a complex duet; low *chip chip chip chipgrr*, then high, tuneful warbling trill *pipipipipi* or *titititititi*, followed by a slightly lower, slower trill *twewewewe*; duet finishes with low chipping noises. Frequently only first and second elements given. **SHH** Endemic to Madagascar; in sedge swamps in or adjacent to rainforest, not in reedbeds, above 700 m; mostly above 1,000 m. Very secretive except when singing, and only sings briefly.

Thamnornis *Thamnornis chloropetoides* 15 cm
Pale understorey tetraka. Greyish above, whiter below, with *narrow green fringes to wing-feathers* and more marked supercilium than Subdesert Brush Warbler; also *pale tips to outer tail-feathers*. Juv undescribed. **Voice** Call an *excited, rubbery trill*, speeding up then slowing down, *trrrrreeeeeeetrrrrrrrrrreeeeee*, lasting up to 1 min; song often follows, a long series moving slightly up and down scale *teeteeteetewtewtewteeteetewtewtewtewteetee*. **SHH** Endemic to spiny forest in SW Madagascar. Usually near ground, but sings from tall *Didierea* or other tree; often in family groups or with newtonias.

Madagascar Brush Warbler *Nesillas typica* 18 cm
Solid ground-layer warbler. E birds *dull mid-brown*, barely differing in measurements and subtle colour tones; *wispy tail long*, unlike Dark Newtonia; from Madagascar Swamp Warbler by weaker bill, less obvious streaking. W/N race *obscura* darker with variable yellow spot at front of supercilium (often absent). Juv similar but even duller. See also Subdesert Brush Warbler. **Voice** Noisy; calls a *coarse, medium-low trkkk... trrrkk... trrrk...* or *checkacheck... checkacheck...*, also a long, chacking trill *trrrrrrrrrrrrrrrrrrrk*. Calls of race *obscura* less forceful and quieter than nominate, but voice not fully described. **SHH** Endemic to Madagascar and Comoros. In rainforest to 2,750 m, in understorey and shrub layer, plantations and garden in E, C plateau, S to Isalo and Analavelona; also rare in W around Ankarafantsika/Ampijoroa. Race *obscura* only in widely spaced pinnacle karst massifs in W & N (especially Ankarana, Namoroka and Bemaraha); with distinctive appearance and habitat, it may be a separate species.

Subdesert Brush Warbler *Nesillas lantzii* 17 cm
Pale ground-layer warbler; very like Thamnornis but lacks greenish edges to wing-feathers and pale tips to outer tail-feathers; from Madagascar Brush Warbler in narrow zone of overlap by *overall much paler coloration and call*. **Voice** Harsh, high ticking, similar in rhythm to Madagascar Brush Warbler but much higher-pitched, *tik... triiik... tikikikik... tkkkkkkkkkkkkk*. **SHH** Endemic to Madagascar, very common in S coastal scrub from Mangoky R to just E of Tolagnaro; less common inland and further N along W coast. Singly or in family groups.

Madagascar Swamp Warbler *Acrocephalus newtoni* 18 cm
Solid 'reed warbler'; from Madagascar Brush Warbler, Subdesert Brush Warbler and Thamnornis by *stronger bill, finer more discrete breast streaking and stronger supercilium*; lacks green edges to wing feathers. **Voice** Call deep, fluid *chep* or *cherp*, sometimes *chrep* or *chrepachep*, *chepchepchep*. Song long, slow, varied, tuneful, and deep, e.g. *chup-e weer weer, tuweep, tweep tweep weepweepweep, tuwherero, tuwherero, churrep, churrwerreo, chu-weer, lululululu...*, and opening phrases often repeated with low trill in place of full song. **SHH** Endemic to Madagascar; widespread. Usually singly or in family groups in reedbeds, montane vegetation and dry scrub usually near water, to around 2,500 m.

PLATE 53: MADAGASCAR: TETRAKAS III

Dusky Tetraka *Xanthomixis tenebrosa*   15 cm
Dumpy short-tailed tetraka. Ad from Spectacled (with some difficulty) by *earth-brown upperparts tinged greenish*, narrower yellow throat and eye-ring, darker breast-band, *shorter tail and relatively longer and stronger legs; posture upright*. From juv White-throated Oxylabes (with yellowish throat) by eye-ring and shorter tail, although juv oxylabes has part-grown tail so care needed. Juv even darker and duller, lacks eye-ring and yellow throat, but normally seen with ad at this age. **Voice** Call forceful, descending *seet... seet... seetseetseetseet*; metallic, piercing, unlike trilling, spluttering calls of Spectacled Tetraka; like Common Sunbird Asity but lower. Also a quieter *tsit*. Song unknown. **SHH** Endemic to Madagascar, very rare in NE rainforest, below 1,000 m; no regular sites known. Usually on ground, sometimes in low shrubs, in family groups. **VU**

Spectacled Tetraka *Xanthomixis zosterops*   16 cm
Medium-sized understorey tetraka. From others by *bright yellow throat, upper breast and spectacle*, yellow-greenish breast and belly, and *mostly orange bill*; from Dusky by *longer tail*, and from juv White-throated Oxylabes by greenish upperparts and yellow eye-ring. Juv lacks strong head pattern of ad and yellow is less bright. Amber Mountain race *fulvescens* lacks yellow below, upperparts greyish-green. **Voice** Calls squeaking *chich... chich... chich, or tsit... tsit... tsit...*, when moving, also spluttering *ptrrrrr*. Song *ptrrrr... chich chich chich chich*, running slightly up and down scale. **SHH** Endemic to Madagascar; commonest tetraka, but only in rainforest; mostly below 1,200 m, in family parties and mixed-species understorey flocks. On saplings in understorey, sometimes on ground. Several races described in addition to *fulvescens* but all perhaps best treated within nominate, birds from most humid areas being darker and brigher yellow/green.

Appert's Tetraka *Xanthomixis apperti*   15 cm
Slim, elegant tetraka. From Grey-crowned (no known overlap) by *pale supercilium and pale orange not yellow underparts*. Smaller and shorter-billed than Long-billed. Juv lacks bright colours and is more uniform. **Voice** Call, *sharp trill prrp... prrp*, also *sit... sit... sit* or *tsit... tsit... tsit*. Song *sit...sit... sit... sisisiseeseeseesipsip*, middle notes harsh and final notes harder. **SHH** Endemic to Madagascar, only in SW, in deciduous forest at Zombitse, Vohibasia and in submontane forests at Analavelona, 500–1,300 m; rare in adjacent spiny forest at sea level. In low understorey and on ground, in family parties and mixed-species flocks. **VU**

Grey-crowned Tetraka *Xanthomixis cinereiceps*   14 cm
A slim, colourful tetraka. From other rainforest tetrakas by *blue-grey cap and cheeks, white throat, green upperparts and pale yellow underparts*; bill relatively short and weak. Juv lacks grey cap and head is uniform greenish, throat yellowish. **Voice** Call a high, *spluttering trill ptrrrr*, also weak, very high *seeseesee* or *sitsitsit*, sometimes *tsit-tsit*. Song very high *sisisiseeseeseesi*, rising and becoming stronger, then slowing and dropping slightly; very similar to Appert's but less penetrating and slightly shorter. **SHH** Endemic to Madagascar; fairly common above c.1,000 m in rainforest, most frequent above 1,500 m. Often climbs mossy trunks; on understorey saplings or in low montane forest canopy, in mixed flocks. **NT**

Long-billed Tetraka *Bernieria madagascariensis*   19 cm
A large, lanky tetraka. From others and Madagascar Brush Warbler by *large size, dull olive upperparts, dark line to eye* and weak creamy supercilium; *long bill* (longer in ♂). W race *inceleber* is paler and greyer than nominate. **Voice** Calls low and harsh, different from *Xanthomixis* tetrakas: alarm a low *treeck treeck treeck* or *chick chick*, contact call rattling *chickikik* or *chickik*; also quiet *chep...chep*, the latter often while foraging; song in W low, fluting *pew-tu peo-pew tchew-irit*; in E simpler, *whi-wiew-tu tu tu whew*; often introduces *chack* or *chep* note or series before or between phrases. **SHH** Endemic to Madagascar, in all native forest except S; fairly common in rainforest to c.1,200 m (nominate), commoner in W deciduous forest and scarce in N part of S spiny forest (*inceleber*). In understorey and lower canopy, in family groups and mixed-species flocks with other tetrakas.

PLATE 54: MADAGASCAR: TETRAKAS IV

Wedge-tailed Tetraka *Hartertula flavoviridis*   13 cm
Small, long-tailed understorey tetraka, larger than jeries, differing also by *all-yellow underparts, grey ear-coverts, long wedge-shaped tail* (often splayed) and sharply pointed, grey bill. **Voice** Series of very high-pitched *see* notes, or descending, *spluttering trill ptsrrrt*. Song a sequence of similar *see* notes, trills and whistles, in rapid succession with short gaps over several minutes. **SHH** Endemic to Madagascar; scarce and patchy in E rainforest at 100–1,600 m, commonest at *c*.1,000 m. Usually in understorey flocks with other tetrakas; hangs upside-down from dead leaf clumps. **NT**

White-throated Oxylabes *Oxylabes madagascariensis*   18 cm
Medium-sized terrestrial and understorey tetraka; distinctive *rufous cap and white throat*, longish tail. Juv confusing, youngest ones olive-brown all over; after post-juvenile moult has pale yellow throat; see *Xanthomixis* tetrakas. **Voice** Call high, downslurred trill, *ssshhrewwww*; also a quiet *tsit*, only audible when close. Song a duet, a *short loud whistle, up and down scale, whit-treet tirooet teeoo whireeet, and a loud, wooden, rattling trill* coming in halfway. Sometimes given independently, not as duet. **SHH** Endemic to Madagascar. In dense understorey of rainforest to *c*.1,600 m; commoner at mid-altitudes. In mixed-species flocks and family groups, with Dark Newtonia and other tetrakas.

Madagascar Yellowbrow *Crossleyia xanthophrys*   15 cm
Medium-sized terrestrial tetraka. *Striking yellow supercilium and throat.* Juv has short buffish supercilium and diffuse buffish band across lower cheeks. **Voice** Call *penetrating tsip or tsiop*; also *tser-sit*, spluttering *tsee-tsee-tsee* and quiet *shrrrr*. Song high, weak, descending *tsit-tsit tseeer tseetsee tsit tsit*. **SHH** Endemic to Madagascar, only in montane rainforest above *c*.1,000 m, commonest above 1,500 m. In family groups within mixed-species flocks; hops on ground and in low shrubs. **NT**

Brown Emu-tail *Bradypterus brunneus*   15 cm
Small, forest understorey warbler. From other warblers *by warm brown, streaked upperparts, grey ear-coverts* and 'decomposed' tail almost lacking barbules (only shaft and barbs); juv entirely dark tawny-brown with fully formed tail. **Voice** Call spluttering *ptrrrrrr*. Song duet; in N & C, initial *high-pitched see-see-see followed by twiwiwi*, or faster *twirra*, slightly higher. In SE, harsher *tch-tch-tch* followed by high-pitched *chirrrrr*, then slow, low *wi-wi-wi*. **SHH** Endemic to Madagascar; usually found singly in dense undergrowth within montane rainforest from *c*.1,000 m to 2,500 m; very secretive.

166

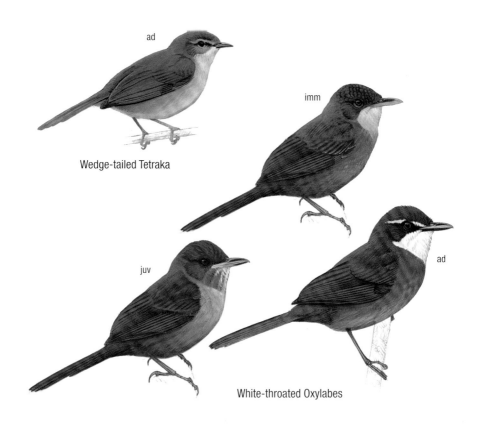

Wedge-tailed Tetraka

White-throated Oxylabes

Madagascar Yellowbrow

Brown Emu-tail

PLATE 55: MADAGASCAR: WHITE-EYE AND VANGAS I (NEWTONIAS)

Madagascar White-eye *Zosterops maderaspatanus*   11 cm
From jeries and tetrakas by *bright green upperparts, conspicuous white eye-ring* and clearly demarcated yellow throat. Juv eye-ring less well marked or absent (only in fledgling), brighter green above. **Voice** Call *nasal tew, ter* or *ter-tew* notes, and peevish rising *wrrrri* or *trrrrri*. Song a medium-pitched rapid warble, *tewteerweerteerteerpiwitewteerteertewtewpiwi*, often given around dawn. **SHH** Endemic to Madagascar (nominate), Comoros, Aldabra Group, Glorieuses and Europa. In Madagascar, very common and gregarious in native forest, scarcer in S, also in plantations, gardens; in large monospecific groups (up to 50) or mixed-species flocks, to *c*.2,300 m. Madagascar birds treated here as monotypic, but tend to be paler and larger in drier environments.

Archbold's Newtonia *Newtonia archboldi*   12 cm
Small vanga, from very similar Common Newtonia by *rufous patch on forehead* and around eye, difficult to see; *legs and bill dark*. Juv brighter orange on breast, with strong rufous wingbar and secondary edges. **Voice** Call scolding *tchiewtchiewtchiew*. Song rapid *low-pitched melodious whistle*: *pitit-chew whetit*; *chitipew-chewit*; *chitipew-chewhew*. **SHH** Endemic to Madagascar; fairly common in spiny forest, also W deciduous forest S of Morondava. With Common Newtonia and jeries in mixed-species flocks within low canopy of spiny forest.

Common Newtonia *Newtonia brunneicauda*   12 cm
Small vanga, from other small warbler-like birds *by grey upperparts*, pale orange on breast, *white iris, pale legs* and lower mandible. Race *monticola* (Ankaratra Massif) slightly darker greyish olive-brown above than nominate (elsewhere), richer brownish-buff below; larger. Juv with faint chestnut wingbar and edges to tertials. **Voice** Call harsh *kchh kchh kchh...* or *bzz bzz bzz....* Song a rapid rattling *chikuchickuchikuchicku* or *kitrikitrikitrikitrik*, often in chorus. **SHH** Endemic to Madagascar; common in all native forest (in rainforest to *c*.1,600 m); absent from relict patches on plateau. Common in scrubby spiny forest but otherwise absent from degraded or secondary forest/plantations. Forages like warbler in canopy, sometimes lower, in family groups and flocks with jeries.

Dark Newtonia *Newtonia amphichroa*   12 cm
Dull understorey vanga; from Common Newtonia by *olive-brown upperparts and dull grey-brown underparts; iris pale brown, but looks darker*, from Dusky Tetraka by lack of yellow throat and eye-ring, and smaller size. Juv rufous-brown all over, with darker iris. **Voice** Call penetrating *shreep*, alternating with quiet *pit..pit..pit*. Song a *loud rolling, rather low and irregular rapid warble*, *chiraweetaraweetarachiraweetaraweetarawicharatowicharawitara*. **SHH** Endemic to Madagascar in rainforest, commonest between 800 and 1,200 m in areas of dense understorey. In mixed-species flocks or in family parties, usually within 1 m of ground.

Red-tailed Newtonia *Newtonia fanovanae*   12 cm
Slim canopy vanga, more colourful than other newtonias; from Common Newtonia and other canopy species by pale orange patches on breast-sides, *extensively reddish tail*. From very similar female Red-tailed Vanga *by lack of pale eye-ring, slimmer bill, pale wing-panel*. Juv unknown. **Voice** Calls unrecorded; song distinctive, confident descending whistles, *whit-whit-whit-whit-whit-whit-whit-whit* (or *fui-fui-fui*), often alternated with a similar series of doubled notes *whit'se-whit'se-whit'se-whit'se-whit'se-whit'se*. **SHH** Endemic to Madagascar, where rare (or only locally common) in canopy of lowland rainforest below 1,000 m, in mixed-species flocks with jeries and Common Newtonia; not yet recorded from Masoala. **VU**

PLATE 56: MADAGASCAR: VANGAS II

Red-tailed Vanga *Calicalicus madagascariensis*   14 cm
Small stocky vanga, ♂ from Red-shouldered Vanga by *grey greater coverts, dark iris*, longer wings and shorter, grey legs, ♀ by dark iris and brown wing-coverts; from Red-tailed Newtonia by thicker bill and pale eye-ring, slightly larger size. Juv similar to ♀, but upperparts with a few buff feather tips and shaft-streaks; underparts more strongly washed warm brown. **Voice** Complex; calls include wooden *tk-tk trrt* or *trrk tikatik*; *chup-chup peewer*, also hissing *kschrrr* when mobbing or in display. In most of E & N, *song per-whew or whe-weew, last syllable dropping*. In W, *pew-poo-whee* or *oo-oo-whi*, last syllable higher; also *tee-teo*, or *pit-we-eer* or *plee-plee tich-che-weh*. **SHH** Endemic to Madagascar, common in canopy mixed-species flocks in rainforest (to *c*.1,200 m) and W deciduous forest (except around Mahajanga), S into N part of spiny forest to N of Toliara; absent from degraded forest and plantations.

Red-shouldered Vanga *Calicalicus rufocarpalis*   15 cm
Small stocky vanga, from Red-tailed Vanga by *pale iris*, shorter wings and longer, pinkish legs; ♂ has more rufous on wing-coverts than Red-tailed ♂; ♀ has *red on lesser coverts*. **Voice** Call a low, peevish *karr-trkkk*, ending in a rattle; also rolling *kwoiroikk*, and chanting *ksisisisisususu* generally harsher than Red-tailed. *Song tyuh-tee or pu-teer or puh-tet*, second note louder and slightly higher, less melodic than Red-tailed. Also a quieter, weaker version of song *wer-wi....* **SHH** Endemic to Madagascar, where rare in family groups or sometimes with Common and Archbold's Newtonias, in dense *Euphorbia* scrub on limestone between Toliara and Cap Ste Marie in SW. Could occur N of Toliara, where Red-tailed found nearby in spiny forest on sand. **VU**

Blue Vanga *Cyanolanius madagascarinus*   18 cm
Medium-sized, acrobatic, spectacularly plumaged vanga. ♂ distinctive; ♀ from Chabert Vanga *by bluish upperparts, lack of eye-ring* and pale buffy underparts. Juv has (short-lived) pale fringes to head, back and wing-feathers. **Voice** Remarkably limited; *harsh grating chhhk crrkcrrk* or *chh-chh-chh-ch-ch-ch*, or *chrr-crrk-crrk-crrrrk-crk-crk*; also short *scheet*. **SHH** Endemic to Madagascar (nominate race) and Comoros; in canopy of rainforest (to *c*.1,200 m), W deciduous forest; not in spiny forest or outside primary forest. Usually in mixed-species flocks or family groups; foraging behaviour often distinctive, hanging upside-down from ends of branches.

Chabert Vanga *Leptopterus chabert*   14 cm
Small lively vanga; from all others by *bluish-black upperparts, bright blue eye-ring* and bluish-white bill. Juv lacks eye-ring and has broad white fringes to upperparts and wing-feathers; often only wing fringes remain. SW form *schistocercus* has white bases to outer tail-feathers. **Voice** Distinctive call a medium-pitched, *rhythmic, harsh chechechecheche... che... che* or *tsetsetse...* often in chorus from flocks. **SHH** Endemic to Madagascar, in all native forest types (scarce in rainforest) and, more rarely, plantations and remote forest relicts to *c*.1,600 m; forages in canopy and also catches flying insects like swallow; often in groups (up to 30 or more) flying high.

White-headed Vanga *Artamella viridis*   20 cm
Large black-and-white vanga, from all others by combination of *pale head (grey wash in ♀ and juv, white in ♂) and short pale bill*. W & S race *annae* slightly longer-billed than nominate. **Voice** Calls varied and raucous; a harsh medium-pitched *graach-graaach... graach... graach*, repeated in chorus. Also loud, medium-pitched, hoarse *chahchahchahchah*. Song a melodious but slightly off-key *kiwah* or *piwerh*, rising and nasal at end; also *whii... wheaw... wheawaw...whii....* **SHH** Endemic to Madagascar; common in all native forest types, below about 1,600 m, in degraded forest, plantations, regrowth and at edges of towns and villages, especially in W & S; not on C plateau. Usually in mixed-species flocks of large vangas, especially Sickle-billed and Chabert Vangas; forages in canopy, pulling epiphytes and bark off trees.

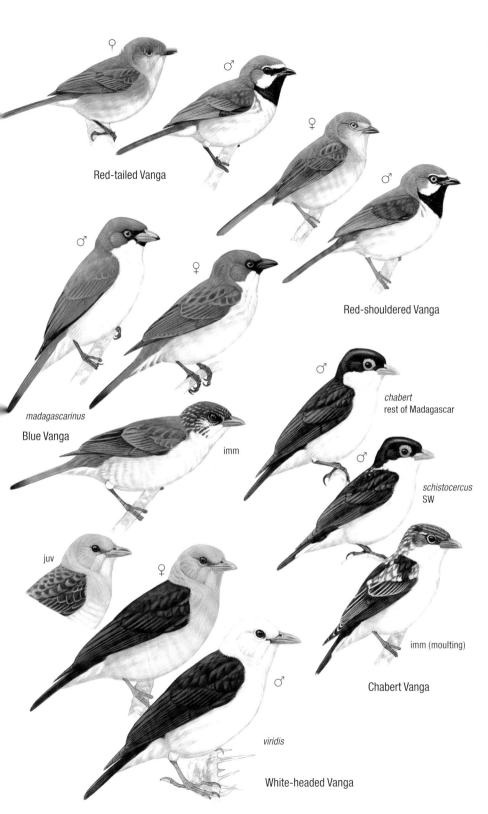

PLATE 57: MADAGASCAR: VANGAS III

Hook-billed Vanga *Vanga curvirostris*   27 cm
Strong, predatory vanga; from other large black-and-white vangas by **black wedge on rear crown, white fringes to black wing-feathers**, and solid black bill. S race *cetera* has longer and narrower bill. Extent of black on crown highly variable. Some, perhaps imm, have grey underparts and hind-collar. **Voice** Contact call a quiet *tew*, also loud, angry *karrkarrkarr...* *tewtewtewtew*; opening notes often with bill-claps. Song a **high, even whistle** *peeeeeeeeer* stronger initially, dropping slightly; difficult to place and infrequently repeated. **SHH** Endemic to Madagascar, in all native forest types, mostly below 1,200 m. Common in W deciduous forest and spiny forest, but scarce in primary rainforest; often also common in degraded edges, regrowth and plantations in E & W lowlands; scarcer in mangroves and littoral forest. Solitary and discreet, moves through dense vegetation searching for lizards and large insects.

Helmet Vanga *Euryceros prevostii*   30 cm
Spectacular large vanga; ad from all other birds by **outrageous blue bill**; bill of juv similar shape but dark with pale tip. **Voice** Vocal; contact call a **short, whistled** *phu*, higher *tseeah*, and *eesh* note. Alarm call a harsh, *treh treh treh...* or *hink hink*. Song high trill, slightly descending, *pip pi pi pi pipipipiperperpeperperper...*; notes less distinct and higher-pitched than Rufous Vanga. Also short, medium-pitched, warbled sequence lasting *c*.2 s followed by another, slightly higher sequence. **SHH** Endemic to Madagascar; only in N half of rainforest, mostly below 1,200 m, scarce but commonest at lower altitudes. Patchily distributed, and absent from many areas in this range. Often in mixed-species flocks with large vangas; perches for long periods in midstorey and catches large invertebrates and small vertebrates on leaves and ground. **VU**

Rufous Vanga *Schetba rufa*   20 cm
Distinctive large vanga; from all other vangas **by red back and whitish underparts**; W race *occidentalis* has slightly longer bill. **Voice** Varied and beautiful repertoire; descending, rippling trill, *pipipiepepepepewpewpew pew pew*; *pititwhipwhipwhipwhip*; *whit-whitoo... whit-whitoo...*; *tuituituituituituit...*, often in duet with previous phrase; frog-like *wok-wok... whow... wok-wok whow... wok-wok...*, the *whow* given in duet with bill-clapping. Also quiet *whit* or *tew* in contact. **SHH** Endemic to Madagascar; in rainforest mostly below *c*.800 m, and rather scarce; commoner in W deciduous forest but absent from degraded areas and plantations. Rarely in mixed-species flocks with other vangas and usually solitary or in family groups; perches for long periods in mid- or understorey and takes large insects or small vertebrates from ground or leaves; sometimes follows large terrestrial birds such as couas.

Bernier's Vanga *Oriolia bernieri*   23 cm
Distinctive large vanga, ♂ from other dark forest birds **by white eye and pale bill**, the ♀ is only *rufous, barred blackish* bird in the region. **Voice** Quiet. *A distinctive harsh schick,* often repeated. Also a loud *chew*. **SHH** Endemic to Madagascar; scarce below about 1,000 m in NE rainforest; only one (old) record in SE. Usually in mixed-species flocks with other large vangas, but very patchy; commoner in wetter areas in rainforest with more *Pandanus*. Excavates dead leaf clumps in branch forks, strips bark and moss from branches, often noisily. **VU**

Sickle-billed Vanga *Falculea palliata*   32 cm
Large vanga and the **longest-billed**, instantly recognisable. Juv has shorter bill than ad, but still much thinner and longer than White-headed Vanga. **Voice** Vocal. *Loud, hoarse zhe-zhe... zhe-zhe*, loud, chanting *chak-chak-chak-chak...*, *chork-chork-chork-chork...* or *kyak-kyak-kyak-kyak...*; loud *kwaaaah... kwaaah*, dropping at end, very like a baby; also low wheezing *gweeeeech* or *gweeeaaach*; often entire flock calls in unison. **SHH** Endemic to Madagascar; common in dry forest and plantations, regrowth and even the edges of towns in N, W & S. Sociable and noisy, usually in single-species or mixed flocks of vangas especially White-headed. Pulls epiphytes and bark off trees, and investigates tree holes.

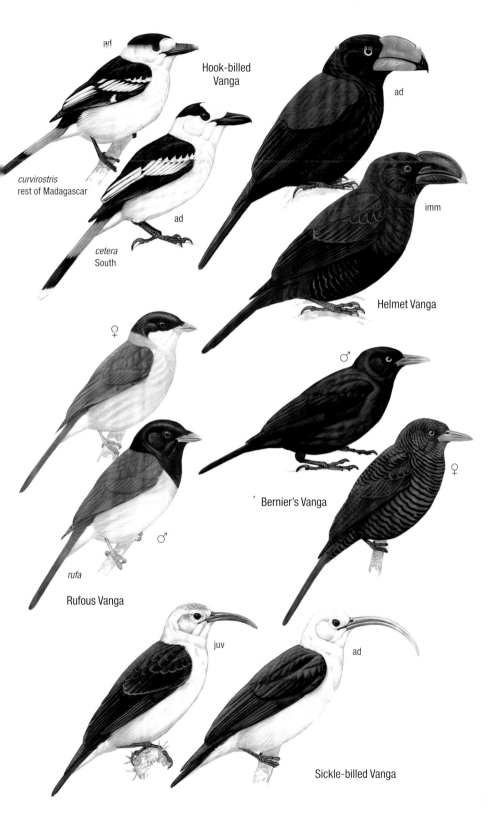

PLATE 58: MADAGASCAR: CUCKOO-SHRIKE, BULBUL AND VANGAS IV

Madagascar Cuckoo-shrike *Coracina cinerea*   23 cm
From *Xenopirostris* vangas by *slim dark bill*, less bulky body; from most Tylas Vangas by *greyish* (not olive) *back*, and lack of orange below. Tylas may have white breast, but never grey as in cuckoo-shrike. S & W race *pallida* paler, especially on breast and forehead. **Voice** Repeated *kip-kewkew... kip-kewkew* or *kipakewakewkew*, last two notes much higher; also *rapid, high-pitched piitikakewkewkew-kew-kew*, opening note high and explosive. **SHH** Endemic to Madagascar and Comoros; in Madagascar, common in native forests including mangroves, up to *c.*2,000 m, less common above *c.*1,200 m; also in adjacent degraded habitats and plantations. In mixed-species flocks with vangas, tetrakas and smaller insectivores.

Madagascar Bulbul *Hypsipetes madagascariensis*   24 cm
Distinctive *orange bill and dark cap*; darker below than Madagascar Cuckoo-shrike or any similar vanga. Juv similar albeit duller. **Voice** Very vocal; a squeaky *kee kee kee* or *whining keeeeer*, also a clucking *tut tut tut tut tut* interspersed by short falling *tchiliyu*, or given together *chup-tchilyu*. In song, extended as *chip-er-chilyu-chip-chip-chilyu-chup-eee-chup-ee-tsiliyu*. **SHH** Endemic to Madagascar, Glorieuses, Comoros and Aldabra; abundant over most of Madagascar, especially in degraded forest, plantations and gardens; less common in primary rainforest and S; often in groups of 5–15, not usually in mixed-species flocks.

Tylas Vanga *Tylas eduardi*   21 cm
Medium-sized vanga. From Madagascar Cuckoo-shrike by buffy or (rarely) whitish underparts, never grey, and greenish or brownish back, never ashy-grey like cuckoo-shrike; from Pollen's Vanga in all plumages by *narrow black bill*. W form *albigularis* has white cheeks, chin and throat, and usually paler orange on breast. **Voice** Medium- to *high-pitched, fluty pe-pi perwhit*, or *pee per-perwit* or *pe perwhit*, or *perper..whit*; the last note usually 'whiplashed'. Also a medium-pitched *tseu-chu... tseu-chu*, similar to Pollen's Vanga but less harsh. Contact call a quiet *whit..whit..whit*. **SHH** Endemic to Madagascar, in rainforest, to *c.*1,600 m, rarely higher; sometimes in adjacent plantations or degraded forest. Also rare in W deciduous forest and mangrove, most frequently around Morondava and Soalala; and C plateau forests (Ambohitantely and possibly Ambohijanahary). Usually in mixed-species flocks.

Pollen's Vanga *Xenopirostris polleni*   24 cm
Bulky vanga, from Tylas Vanga and Madagascar Cuckoo-shrike by combination of *thick, pale bill, complete dark hood* and pinkish breast (♀) or black extension from hood to mid-breast (♂). **Voice** Varied; a *loud, medium- to high-pitched tsiiiiiop or tsiiiiiep*, starting high or very high and dropping sharply; also a loud, hoarse *whorp* or *wheek*, also *tsitsichew... tsitsichew* and low warbling *wiritiew*; quiet *whitwhitwhit*, very like Tylas Vanga, and a quiet *chuck-chuck-chuck*. **SHH** Scarce in S part of rainforest, and rare or absent from most of NE; from sea level to *c.*2,000 m, occasionally in adjoining degraded forest. In mixed-species flocks with other vangas, tearing bark or moss and excavating leaf clumps. **NT**

Van Dam's Vanga *Xenopirostris damii*   23 cm
Bulky vanga, from Hook-billed, (western) Tylas, Lafresnaye's Vanga and Madagascar Cuckoo-shrike by thick *blackish bill* (darker than in Lafresnaye's), white chin (♂; difficult to see), white ear-coverts and forehead (♀), and dark slate-grey back. ♀ and juv can be buffy below, especially in N. **Voice** Loud *tsiiianh or tsiiiaaa*, starting very high and dropping sharply; also *tsiiiiep* much like Lafresnaye's Vanga, and low *peew... peew*; also *whorp*, and low slow *chirrachirrachirra*. A quiet *whit ... whit* in contact. **SHH** Endemic to Madagascar. Rare in W deciduous forest N of Betsiboka R, sometimes in mangroves in NW, and Analamera forest (far NE); only known sites mapped. Usually in mixed-species flocks with other vangas, also alone or in family groups; excavates holes in trees, breaks rotten twigs and branches. **EN**

Lafresnaye's Vanga *Xenopirostris xenopirostris*   24 cm
Large stocky vanga, from Van Dam's, Madagascar Cuckoo-shrike and Tylas Vanga by short, thick, *pale bill, grey back* and black half-hood (♀) or *hood extending below chin* (♂). Juv can be buffy on breast. **Voice** Loud *tsiiiiiop* or *tsiiiiiep*, starting very high and dropping sharply; also low, forceful repeated *chutchutchut*, or *schipschipschip*, also *chiow* or *whiep* or *chiew*, in long sequences. **SHH** Endemic to Madagascar; fairly common in spiny forests of S or locally in degraded fringes, below *c.*300 m. Usually in mixed-species flocks with other vangas, also alone or in family groups; breaks rotten twigs, removes bark and epiphytes.

174

PLATE 59: MADAGASCAR: PARADISE FLYCATCHER AND VANGAS V

Madagascar Paradise Flycatcher *Terpsiphone mutata*   18 cm + 15 cm tail (♂)
Very distinctive; both sexes from other small birds by *black hood glossed bluish*. Rufous birds are unique, as no other small, rufous passerine occurs in Madagascar's forest; Rufous Vanga is much larger. White ♂ from Ward's Vanga by very long tail and lack of black breast-band. All except white ♂ have *reddish back and underparts*. ♂ occurs in two morphs, white and rufous; in white morph, back feathers black or black tipped with white ('black-backed' birds), or white with narrow black central streaks and concealed black bases ('white-backed' birds); intermediates appear mottled grey. All ♂♂ instantly recognisable with striking elongated rectrices, and eye-catching aerobatics; ♀ always rufous and black, duller than ♂, with shorter tail and no wingbar. Juv similar to ad ♀, but paler and duller. **Voice** Scolding call a medium- to high-pitched harsh *ker* or *chrr*, usually extended *to kerkercherrrit, retret retret retretret* or other variant. Song rapid, rolling warble *kitiowitiowitiowiti*, often preceded by scolding call. **SHH** Endemic to Madagascar (nominate race) and Comoros; common in all native forest to *c.*2,000 m, adjacent plantations and degraded areas, rare in gardens.

Nuthatch Vanga *Hypositta corallirostris*   14 cm
Distinctive, small, tree-climbing vanga; from all other species by *orange bill and bluish upperparts*, ♀ duller than ♂, face washed brown. Imm ♂ resembles ♀, but brown underparts tinged blue-grey. **Voice** Calls, single *tsip* or *tsirp*; song very high-pitched, rapid trills and twitters, *tsip... tsip... tsip... tssssssiirrp... tsir-tsir... tsssirrrsiirp* or *trrrrrrrrrrrrrrrir*, slightly descending. **SHH** Endemic to Madagascar; in rainforest below about 1,200 m, sometimes in adjacent degraded forest or plantations; almost always in mixed-species flocks, where shuffles up tree-trunk then flies to base of adjacent tree.

Crossley's Vanga *Mystacornis crossleyi*   16 cm
Distinctive small terrestrial vanga; from other terrestrial species by *long bill, short tail and complex pattern of grey, white and black on face*; all ages except very young ♀ have less contrasting plumage and brown cap. Juv (accompanied by ad) share white stripe across ear-coverts. **Voice** Song given by apparent ♀ or imm ♂ a powerful, clear, high-pitched whistle *teooeer-teeee teeee*, last note slightly higher; ♂ song a *loud, penetrating, high-pitched whistle peeeeeeeeer* or *peeeeooeeer*, stronger and falling at end. **SHH** In rainforest below *c.*1,300 m, commoner at lower elevations and scarce in adjacent degraded forest; not usually in mixed-species flocks, often in family groups or alone, walking on ground.

Ward's Vanga *Pseudobias wardi*   15 cm
Flycatcher-like vanga, from other vangas and Madagascar Paradise Flycatcher by *black breast-band on white belly, and blue eye-ring*, although latter dull in juv. **Voice** Rapid flat, dry, high-pitched, *metallic trill pttttttttt*, slightly descending, sometimes followed by more urgent *pt-chichichichi*; often several birds calling together. **SHH** Endemic to Madagascar; in rainforest below *c.*1,600 m, rarely in degraded fringes or adjacent plantations; scarce or absent at low elevations (<500 m) and in NW rainforests. Occasional dispersal far from forest. Conspicuous in mixed-species flocks, flycatching from subcanopy; often in family parties.

PLATE 60: MADAGASCAR: DRONGO, CROW AND STARLINGS

Crested Drongo *Dicrurus forficatus* — 28 cm
Only all-black bird in Madagascar with *forked tail and tuft over bill*; juv has whitish fringes to wing-feathers and belly. **Voice** Very varied; *call hoarse whep or wherp* alternating with single screeches or croaks; song medium-pitched series of separated clicks, whistles, clucks, often 2–3 notes repeated before next sequence, and whistled sequence alternating with clucks: *cluck-cluck... grickik... whit-whit... crick... whercluck-pit... crickik... clurk-clurk... whip... tikit...*, often as duet and incorporating imitations such as Frances's Sparrowhawk, Madagascar Coucal, Madagascar Bulbul, Common Newtonia and Common Myna; also lemurs and mechanical noises. **SHH** Endemic to Madagascar (nominate race) and Comoros; common except above 1,500 m; rare on C plateau; not limited to forests, but commoner within. Usually with mixed-species flocks, where perches conspicuously and takes flying insects flushed by other members; may follow terrestrial birds like couas.

Pied Crow *Corvus albus* — 45 cm
Only crow in Madagascar. From potential colonising House Crow (see Plate 83; recorded in Toamasina since 2013) by *white collar and belly, and lack of grey*. **Voice** Flat, loud, low-pitched *caah, caw* or *craaah*; also range of other bubbling or gurgling noises, and a low *cuh... cuh... cuh* from flocks. **SHH** Africa, including Comoros and other islands in Mozambique Channel and Aldabra group; in Madagascar common everywhere except on C plateau (where scarce except in towns); generally absent from primary rainforest.

Madagascar Starling *Hartlaubius auratus* — 20 cm
Dark *chocolate-brown* starling with *slim dark bill and white wing-flash*, conspicuous in flight; tail longer than Common Myna and lacks bare skin around eye; ♂ somewhat glossy purple on wings and tail; ♀ duller brown with streaked underparts. From Madagascar Bulbul by dark bill, short, slightly forked tail, and white in wings. **Voice** Not very vocal. Call, often given in flight, a slightly hollow, medium-pitched *plitt... plitt...* or *chlert... chlert... chlert...* or *tillit... tillit... tillit...* Song medium- to low-pitched, quiet, complex series *perodi... perodi... tillit... tillit... chititetet chert tsipserpserp trip trip trip...*; also quiet warbling subsong. **SHH** Endemic to Madagascar; fairly common in rainforest below c.1,000 m, less common in degraded areas and in plantations in E; scarce or absent on C plateau and local in CW, more common in NW. Often in forest canopy or at fruiting trees; usually in small family parties, rarely if ever on ground.

Common Myna *Acridotheres tristis* — 25 cm
Distinctive stocky brown starling; from all other birds by white wing-patches and outer tail tips, and *yellow facial skin and legs*. **Voice** Vocal and noisy. Fluty medium- to low-pitched *tuuu, chyuu* or *tweh* and similar, often repeated, sometimes with squawking quality, but also an amazing range of low- to high-pitched squealing, clicking, croaking, gurgling and whistling notes, often strung together in disjointed sequence as song; any of these components also used as calls; at roosts, many birds call together in cacophonous medley. Mimics other birds, occasionally causing confusion. **SHH** Native to S & SE Asia; widely introduced in Old World tropics including most of Malagasy region except coralline Seychelles. Widespread in Madagascar but commonest in E; less abundant at high altitudes and in S, mostly in towns and cultivation, not in forest; sometimes forms large roosts near towns.

PLATE 61: MADAGASCAR: SPARROW, WEAVERS AND MANNIKIN

House Sparrow *Passer domesticus*   16 cm
♂ distinctive. ♀ from Madagascar Fody by *lack of buffy-yellow tones* and larger size, and lacks ♀ Sakalava Weaver's bare skin around eye; smaller bill. **Voice** *A harsh chissk, short weesk or abrupt, lower-pitched chup, chup-up, or short, coarse trill, chukukukukuk.* Song a series of *chup* or *chip* notes; chorus of *chissk* notes given by many birds together, e.g. at roosts. Flight call a short clear *chep* or *cherp*. **SHH** Native to Eurasia and N Africa. Introduced almost worldwide, in Madagascar only CE coast, especially Toamasina, but can be expected to colonise other cities; seen in Antananarivo, Mahajanga and Antsiranana. Rare far from habitation.

Nelicourvi Weaver *Ploceus nelicourvi*   15 cm
Slim weaver of humid forest; *unstreaked olive upperparts and grey belly* distinctive; yellow head and neck with dark ear-coverts partly covered by black hood in breeding ♂. ♀ has black on head mainly replaced by yellow; olive patch on crown extends to darker ear-coverts. Juv has greener underparts, side of head uniform olive, yellowish restricted to chin and upper throat, *dull cinnamon undertail-coverts* diagnostic. Impressive hanging nests in forest, over rivers. **Voice** Distinctive, high-pitched *chizz chizz chizz*, also *sit* or *chet*; song, given while hanging from nest, *chiz... chizz chswiriisssiszz*. **SHH** Endemic to Madagascar; in rainforest to c.2,000 m, commoner below 1,200 m; scarcer in adjacent plantations, regrowth and even villages. Often in mixed-species flocks, searching for insects in canopy, in dead leaves and rotten wood. Sometimes given own genus, *Nelicurvius*.

Sakalava Weaver *Ploceus sakalava*   15 cm
Solid, thick-billed weaver of dry areas; from fodies and ♀ House Sparrow by *thicker bill, bare pink or red skin around eye*, three pale stripes across face; yellow head in ♂. **Voice** 'Swizzling' song given while hanging from nest, usually in chorus. Contact and flight call *short chert-chert or chep-chep*. **SHH** Endemic to Madagascar, common in dry areas of N, S & W outside forest and in clearings within, as well as cultivation and regrowth. Absent above c.1,000 m, from C plateau and E. Forms flocks feeding on ground. Nests in large messy colony, in large tree. May belong in own genus, *Saka*.

Madagascar Fody *Foudia madagascariensis*   13 cm
Breeding ♂ distinctive. ♀ and non-breeding ♂ from House Sparrow by *buffy ground colour*, lacks bare facial skin and has smaller bill than ♀ Sakalava Weaver. Rare flavistic birds more extensively yellow than ♂ Sakalava Weaver, with black eye-patch. See also Forest Fody. **Voice** A very high-pitched, slightly chirping *tsip* or *tsikip*; also a lower *chup* or *chutut*. Breeding ♂ has trill call, and buzzing *zzz-zzz-zzz*. Song lacks regular form: wheezes, chirrups, squeaks and *tsip* calls. **SHH** Probably originally endemic to Madagascar, but widely introduced to islands in region. Very common in towns, plantations and cultivation, also in degraded and secondary forest. Not in closed-canopy forest.

Forest Fody *Foudia omissa*   14 cm
♀ and non-breeding ♂ from Madagascar Fody (with difficulty) by *more olive-green* (less buffy) ground colour, and *longer and deeper bill*; hybrids probably are common. Extent of red on breeding ♂ of limited identification value: many Forest Fodies have odd red feathers on belly or back, like moulting Madagascar Fody, while moulting or variant ♂ Madagascar Fody often shows similar extent of red to ♂ Forest Fody. **Voice** Calls and song similar to Madagascar Fody but *higher-pitched and faster*. Song is long sequence of very high twittering and chirping notes: e.g. *tsisisisitisisititisisisitit... tsit tsit tsit tsit tsit tsit tsit... tsertsertser... tsisisisisisititititisisisitit* and variants, lasting 20–60 s. **SHH** Endemic to Madagascar. In rainforest to c.2,100 m; at least in SE, commoner at 1,000–1,500 m; also in adjacent degraded areas, plantations and C plateau forest fragments. In flocks or small groups, sometimes with Madagascar Fody; forages on ground and in canopy.

Madagascar Mannikin *Lepidopygia nana*   9 cm
Tiny; from all other species by *size and black bib*. Juv (usually with ad) lacks black lores and bib. **Voice** Distinctive *tsirrip* or *tsitup* or *piwisit* or *pisit*, often in tinkling chorus; sometimes a single *pit* or *tsit*. Song short, high-pitched rapid complex *tsirtsirriup*. **SHH** Endemic to Madagascar; common throughout, in cultivation and grasslands, in clearings in primary forest. Usually in small flocks.

PLATE 62: GRANITIC SEYCHELLES: LANDBIRDS I

Feral Pigeon *Columba livia*   33 cm
The *typical 'town pigeon'*, familiar worldwide, not particularly common in Seychelles but increasing in numbers. Plumage very variable, *usually blue-grey with purple and green sheen on neck, white rump and two black wingbars, some predominantly white, black or piebald*. Unlikely to be confused except at distance with Madagascar Turtle Dove, which has shorter wings and is more reddish-brown. **Voice** A long, moaning *o-o-orr*, repeated monotonously, or a hurried, rambling *coo-roo-ooo-ooo*. **SHH** Native to Eurasia and N Africa, but introduced worldwide including granitic Seychelles, on Mahé, Praslin, La Digue and Silhouette, extirpated Frégate which held the only recorded population in 1970s (likewise on Assumption, coralline Seychelles); also Madagascar, Comoros, Mauritius, Rodrigues, Reunion. Locally common, especially in urban areas, playing fields and sites providing food scraps or nesting ledges; may also feed in agricultural land, adjacent airstrips, grain warehouses and in other open areas. Flight powerful and fast on long, pointed wings.

Madagascar Turtle Dove *Nesoenas picturata*   28 cm
A sturdy, dark dove, *frequently seen on the ground*. Common, presumed introduced race *picturata* has *blue-grey head*, purple-pink breast, paler pink-buff belly, *brownish wings and grey-brown tail with ash-grey outer tail-feathers with slightly paler (not white) corners*. Race *rostrata*, endemic to granitic Seychelles, is *smaller with dark plum-purple head and variable blue-grey to pink-purple lower breast*. Juv lacks rufous, with darker brown upperparts, dull brown-grey breast and buff wash on lower breast. Dark reddish-brown on upperparts distinguishes it from Seychelles Blue Pigeon, which has mainly blue-black plumage, whitish bib and more rounded wings. **Voice** Low-pitched, double coo: *coo cooooo-uh*, the second *coo* louder, longer and rising at end; first *coo* is often not noticed. **SHH** Common on all larger islands; also Madagascar, coralline Seychelles, Comoros, Mauritius, Reunion and Agalega. Frequents wooded habitats, forest trails, mountain roadsides and open areas. Feeds mainly on ground from where often flushed with noisy wingbeats, but flight otherwise silent, very fast and usually low. Endemic *rostrata* could be treated as a separate species, but has hybridised with *picturata* so that pure forms are probably extinct; birds at least partially resembling it still occur.

Zebra Dove *Geopelia striata*   21 cm
The only small resident dove. A very distinctive, *tiny, greyish-brown dove, with extensive black barring*, otherwise pink or grey plumage and blue skin round eye. Juv paler, less smart, shorter-tailed and lacks blue eye-ring. In flight, long, wedge-shaped tail with narrow white tips. *Bounding flight on take-off* distinctive. At distance, when perched on wires or posts, shape and stance confusable with Seychelles Kestrel. **Voice** A medium- to high-pitched, *bubbling, quick cooruruwuh or wurruwawa*, first note cooing and last two sharper, last note sometimes repeated; in display, a harsher *caaw-caaw-caaw*. **SHH** Introduced (SE Asia via Mauritius) to all larger islands of granitic Seychelles; also coralline Seychelles, Mauritius, Rodrigues, Reunion, Glorieuses, Juan de Nova, Agalega. Common in open areas or habitation with scattered woody vegetation, including urban areas and tourist establishments. [Alt: Barred Ground Dove]

Seychelles Blue Pigeon *Alectroenas pulcherrimus*   24 cm
A sturdy, very short-tailed, broad-winged woodland pigeon. Ad has *deep blue upperparts, wings, tail and belly, contrasting sharply with pale greyish-white breast, neck, sides of face and nape*. Scarlet wattle (or caruncle) in front of eye extends to forecrown and around eye. Juv has dark grey head and bib, dark greenish-brown upperparts with pale yellow or buff fringes, blackish-grey primaries with bluish cast and narrow pale yellow outer edges, and dark pink wattle around eye. Almost never seen on ground, unlike Madagascar Turtle Dove, which is larger with reddish-brown not blue-black plumage and faster, more direct and usually lower flight on narrower, more pointed wings. **Voice** A low, strong but muffled coo: *kuk-kuk-kok-kok-kok-kok-kok-kok-kok-kok* of 15–30 notes rising slightly at first and ending abruptly. **SHH** Endemic to granitic Seychelles and fairly common on all larger islands and at all altitudes wherever there are trees. Solitary, in pairs or occasionally in small groups, especially when feeding. Flight strong and swift, in display may fly high above trees, before plummeting down at a steep angle, wings held rigidly forward and downward. Greatly increased after cessation of exploitation for food and, since 1980s, has naturally re-colonised Curieuse, Denis, Aride, North and Bird; also reintroduced Cousin.

PLATE 63: GRANITIC SEYCHELLES: LANDBIRDS II

Rose-ringed Parakeet *Psittacula krameri*   41cm (♂), 37 cm (♀)
The *only predominantly green parrot in Seychelles*. Long graduated pointed tail and red bill. ♂ has a rose-pink collar. ♀ has collar absent or indistinct, and shorter tail. Juv similar to ad ♀, with yellowish tinge to plumage, tail even shorter, and pink bill. **SHH** Tropical N Africa to India, SE China; Asian birds widely introduced including to Mahé but numbers reduced considerably by attempted eradication; also on Mauritius and probably Reunion. Mainly in secondary lowland forest. Gregarious, large flocks at roosts. **Voice** Most often a *raucous, far-carrying kee-ak kee-ak*; many other calls, varying in pitch and tempo, but most with similar strident tone. Most vocal at roost and in flight. Perched birds also utter various whistling and subdued chattering or twittering notes. [Alt: Ring-necked Parakeet]

Seychelles Black Parrot *Coracopsis barklyi*   32 cm
The *only grey-brown parrot in Seychelles*, darker above with blue or greyish-green reflections on primaries and primary coverts. Bill blackish-grey, becoming paler during breeding season. **Voice** A variety of *high-pitched whistles*, usually monosyllabic, with melodious trisyllabic whistles during breeding season. **SHH** Endemic to granitic Seychelles; restricted to Praslin and common in suitable habitat (population 520–900); rare but regular visitor to Curieuse; historically also Aride and Marianne. Inhabits woodland dominated by endemic palms, river valleys with tall trees and mixed forests, agricultural land and gardens. Typically seen foraging or flying above canopy. Fairly gregarious, usually in pairs or small flocks. Flight fast, powerful and often direct, with rapid and deep wingbeats. When perched, agile and acrobatic. Formerly considered a race of Lesser Vasa Parrot but vocal and morphological differences and DNA confirm separation. **VU** [Alt: Seychelles Parrot]

Barn Owl *Tyto alba*   35 cm
Unique in Seychelles, often detected by *distinctive screeching call*, or seen in headlights while driving at night. *Very pale plumage* but distinctly orange-brown on underparts, very different to darker and much smaller Seychelles Scops Owl. **Voice** Loud, toneless, grating screech *ksssccchh*. Hisses and 'snores' when at rest and makes loud bill-snaps when disturbed at or near nest. **SHH** Almost cosmopolitan; introduced granitic Seychelles: Mahé, Praslin, Curieuse, Silhouette and La Digue; regularly visits (may breed) North, Aride and other islands. Native on Madagascar, Comoros and Europa. Occurs at sea level in range of degraded habitats. Nocturnal, taking rodents and small terns (including Roseate and Fairy Terns), mainly on wing with low, buoyant flight.

Seychelles Scops Owl *Otus insularis*   20 cm
The *only small owl resident in Seychelles. Harsh 'sawing' call unique*. Short ear-tufts usually invisible. See vagrant Eurasian Scops Owl (Plate 114). **Voice** Territorial calls peak at dawn and dusk, a distinctive, low rasping *gork... gork... gork...* rhythmically repeated over 15 s to several min, or sometimes isolated single notes. Usually given by ♂, sometimes ♀ and pairs may duet, including frog- or duck-like quacks, and gurgling, frequently increasing in intensity and sometimes ending in copulation. **SHH** Endemic to granitic Seychelles; Mahé alone. Inhabits mixed mature damp forest usually above 100 m, more frequent above 400 m, especially around Morne Seychellois (905 m). Prey captured by both perch-and-pounce and active hunting. **EN**

Seychelles Swiftlet *Aerodramus elaphrus*   11 cm
The *only swift resident in Seychelles*, dark grey-brown above, slightly paler below with blunter, more rounded wings than the sickle-shaped vagrant *Apus* swifts (Plate 114). **Voice** Sometimes utters a soft twitter in flight. In breeding caves, uses echolocation to navigate in darkness, call a sharp metallic click *nik-nik-nik-nik*. **SHH** Endemic to granitic Seychelles; resident Mahé, Praslin and La Digue, occasionally recorded elsewhere in granitic islands. Occurs from sea level to mountains, hawking for insects 5–20 m above the ground. Rapid, twisting flight. **VU**

Barn Swallow *Hirundo rustica*   18 cm
The only migrant swallow recorded annually, from vagrant hirundines by *blue-black upperparts, dark throat* and long tail-streamers. Imm duller, browner, has pale buff on chin and throat, whiter underparts and shorter tail-streamers. **Voice** Contact call *witt-witt*, sometimes run together as series of *tswitt* notes. Song, a more prolonged bubbling twitter, not recorded in Malagasy region. **SHH** Breeds Eurasia and North America, spending boreal winter in sub-Saharan Africa, Asia and South America. Rare migrant across much of Malagasy region; regular Seychelles, mainly on passage Nov and Mar–Apr. A typical hirundine, attracted to concentrations of flying insects, for which it forages aerially, in very agile flight.

PLATE 64: GRANITIC SEYCHELLES: LANDBIRDS III

Seychelles Bulbul *Hypsipetes crassirostris*   30 cm
Sturdy, noisy woodland species. Mainly *grey-brown above with olive-green hue, slightly paler below, shaggy black crest and bright orange bill and legs*. Juv has chestnut flight-feathers, blackish bill and dark brown legs. No similar species. **Voice** Loud and highly vocal, including a frequently repeated cluck, a raucous *nyeeeer*, a rapid chatter *ak-ak-ak-ok-ok-ok-ak* and a song comprising a series of clucks, squawks, whistles and chatters, given by one member of pair, usually from prominent perch, with wings drooped. **SHH** Endemic to granitic Seychelles; common on Mahé, Praslin, La Digue, Silhouette and Cerf. Frequents woodland, less common in lowlands. Gregarious; small parties noisily chase each other between trees in rapid, direct flight.

Seychelles Magpie-robin *Copsychus sechellarum*   23 cm
Unmistakable, tame, *ground-feeding bird, entirely black except white wing-patches*. Ad has violet-blue sheen in sunlight. Juv lacks gloss, white inner wing-coverts tinged tawny-brown with narrow blackish fringes. No similar species. **Voice** ♂ song is a series of repeated, melodious, fluty warbles and harsher phrases, sometimes including mimicry. ♀ sometimes produces a simpler song, occasionally in duet with ♂. **SHH** Endemic to granitic Seychelles; restricted to Frégate, Cousin, Cousine, Aride and Denis. Formerly present throughout larger granitics, range reduced to Frégate (population 12–15 in 1965 and still only 21 in 1990) until translocation to other islands (total population *c*.250 in 2014). A territorial and terrestrial feeder in woodland or cultivation. **EN**

Seychelles Paradise Flycatcher *Terpsiphone corvina*   18 cm + 16 cm tail-streamers (♂)
Unmistakable ♂ *entirely black with long tail-streamers*, showing a deep purple sheen in sunlight. ♀ *chestnut-brown on upperparts*, wings, rump and tail, with *creamy white underparts and an all-black head*. Juv resembles ♀, but duller, browner and less smart. **Voice** Song a medium-high clear warble, *tui... titertitaytiteri* and variants, slow and measured, lasting 2–4 s, repeated after 2–4 s. Call a strong but thin, high, creaky *crikikikerkikerki*, with many variants, between groups of birds; also single *cri... cri... cri* and *tui... tui*. Alarm a harsh *zweet*. **SHH** Endemic to granitic Seychelles; restricted to La Digue and Denis. Formerly found elsewhere; range reduced to La Digue until translocation to Denis. Occasionally reported on Marianne, Praslin and Félicité. Frequents mature woodland often adjacent to marshes. Darts between trees in undulating flight, picking insects from leaves or flycatching. **CR**

Seychelles Warbler *Acrocephalus sechellensis*   14 cm
Only resident warbler of granitic Seychelles. Ad has *dull greenish-brown upperparts, dingy white underparts* and an indistinct pale supercilium. Long, slender bill with yellowish-flesh base, and red-brown eye. Juv darker with some speckling on breast. **Voice** Clear, whistling, rich and melodious song. Alarm call a harsh chatter. Juv makes rasping *zhzh-zhzh-zhzh* notes. **SHH** Endemic to granitic Seychelles; restricted to Cousin, Aride, Cousine, Denis and Frégate. Formerly found elsewhere; range reduced to Cousin (population <30 in 1960s) until translocation to four other islands (total population *c*.3,000 by 2012). A woodland species, solitary, in pairs or small family parties. Tame and approachable, but often unobtrusive. Flitting flights between feeding stations. **VU**

Seychelles Sunbird *Nectarinia dussumieri*   12 cm
Only sunbird on granitic Seychelles. Breeding ♂ mostly *dull slate-grey* with orange or yellow pectoral tufts (not always visible) and *dark blue iridescence on head and throat*. ♀ duller, slightly smaller and lacks pectoral tufts. Non-breeding ♂ and ♀ lack iridescence. **Voice** ♂ song given throughout day, though mainly in morning, is a series of enthusiastic, harsh, high-pitched notes interspersed by occasional rasping calls and a quieter subsong. ♀ may also sing. Call a high-pitched, squeaky, repeated *crisp*, sometimes in long sequence with *c*.0.5 s gaps between them as a song; also high *trilling trrrp... trrrp... trrrp*. **SHH** Endemic to granitic Seychelles; resident and common on all larger islands; translocated to Bird. Frequents gardens and open woodland. Very active.

Seychelles White-eye *Zosterops modestus*   11 cm
Small grey-brown bird with broad white eye-ring. Head, upperparts and tail dark olive-grey to brown, paler and tinged yellowish on rump. Underparts pale grey, dull pale mustard-yellow on throat, the belly having a faint yellowish tinge, and brownish on flanks. **Voice** A low, soft, nasal contact call, *cheer-cheer*. Alarm a loud *chewick* and a chattering trill. Song is varied and pleasant, most frequently heard at dawn. **SHH** Endemic to granitic Seychelles; known only from Mahé and Conception until translocated to Frégate, Cousine and North. Frequents open woodland and gardens. Weak, undulating flight. **EN**

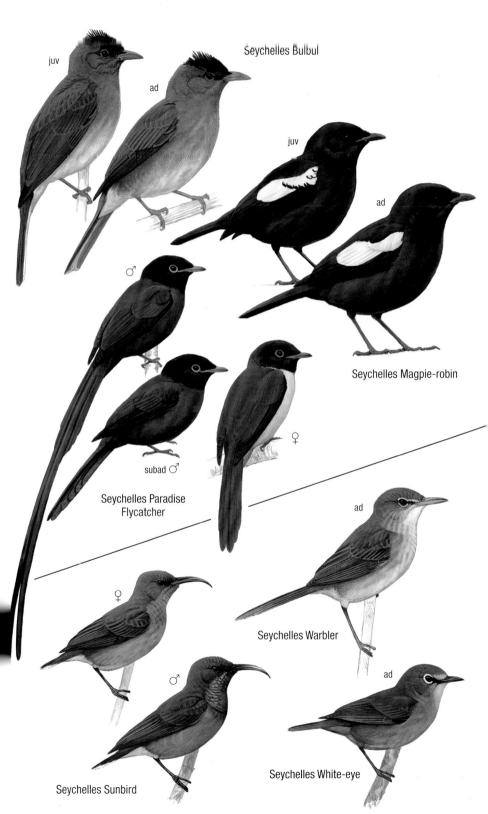

PLATE 65: GRANITIC SEYCHELLES: LANDBIRDS IV

Common Myna *Acridotheres tristis* — 25 cm
Distinctive, large passerine, from all other birds by *brown plumage with white wing-patches and tail tip (prominent in flight); yellow legs, bill and skin round eye*. **Voice** Vocal and noisy. Fluty medium- to low-pitched *tuuu*, *chyuu* or *tweh* and similar, often repeated, sometimes with squawking quality, but also an amazing range of low- to high-pitched squealing, clicking, croaking, gurgling and whistling notes, often strung together in disjointed sequence as song; any of these components also used as calls; at roosts, many birds call together in cacophonous medley. Mimics other birds, occasionally causing confusion. **SHH** Introduced (S Asia via Mauritius) throughout granitic Seychelles, Bird and Denis, common at all altitudes and virtually in all habitats including urban areas, primary forest, agricultural land, grassland, mangroves and open coasts; also Madagascar, Comoros, Mauritius, Rodrigues, Reunion, Agalega. Strident voice and bold, commensal habits make this a familiar species.

Seychelles Fody *Foudia sechellarum* — 12 cm
Small drab, dark brown, dumpy weaver, darker on back and wings. Breeding ♂ acquires *yellow crown, face and chin*, merging to sooty-brown then dark brown on throat and rear crown. Often has a small white wing-patch. Non-breeding ad and juv are dark brown above and below, darker on back and wings. *Dark brown plumage and dumpy shape* readily distinguish from Madagascar Fody. **Voice** Both sexes give *characteristic tok-tok-tok*. In distress or alarm, *tok* calls rapidly increase in intensity and frequency, also given singly or with long pauses between notes, while feeding. Song given by both sexes, especially ♂, single high-pitched notes with pauses, which become shorter so that high notes become almost continuous. A different alarm call, perhaps higher intensity, *tchr*. **SHH** Endemic to granitic Seychelles; restricted to Cousin, Cousine, Frégate, Aride, Denis and also D'Arros (coralline Seychelles, outside native range). Formerly found elsewhere; range reduced to first three named islands until translocations to others (population <500 individuals in 1959, now >8,000); also visitor to Bird. Frequents mainly woodland and edges, but has few habitat constraints. Social, often in groups of >10, but less gregarious than Madagascar Fody (family units rather than large loose flocks). Flight is direct, compactness obvious and appears almost tailless. Constantly flicks wings. Little overlap with Madagascar Fody in feeding, being mainly insectivorous. **NT**

Madagascar Fody *Foudia madagascariensis* — 13 cm
Breeding ♂ distinctive with *entirely red underparts*. ♂ moults into ♀-like non-breeding plumage (Jun–Oct). Flavistic birds regular but rare, and unmistakable on Seychelles. Gregarious habits and preference for open areas, where it feeds mainly on seeds, often on ground. *From House Sparrow (recorded occasionally) by smaller size and more greenish-brown or buffish-yellow plumage.* Sibilant *tsip* call not given by any confusion species. **Voice** A very *high-pitched, slightly chirping tsip or tsikip*; also a lower *chup* or *chutut*. Breeding ♂ has trill call, or buzzing *zzz-zzz-zzz*. Song lacks regular form: wheezes, chirrups, squeaks and *tsip* calls, with up to c.1 s gap between phrases: *tsitsitsitsitstit tiptiptip titititit chipchipchip*. **SHH** Endemic to Malagasy region; introduced throughout granitic Seychelles, Bird and Denis, and to coralline Seychelles, Comoros, Mauritius, Reunion, Rodrigues, Glorieuses, Juan de Nova, Agalega; native to Madagascar. Highly gregarious seed-eater, flocks sometimes in hundreds; also takes insects. Very common in urban areas, cultivation and secondary forest, but absent from high-mountain forest except in clearings.

Common Waxbill *Estrilda astrild* — 12 cm
Very small size and long tail, combined with *brown upperparts*, paler underparts, *red eye-patch and red bill* unique in Seychelles. Juv has black bill, black line below red lores and indistinct barring. **Voice** Medium- to high-pitched *tewk* and hesitant *tetetete* or *tewtewtew*, in alarm a nasal *jaa* or longer *weeez*. In groups, weak twittering, sparrow-like chirping and shrill *tseep*. Song ends with a long, buzzy, upslurred note, *ji-ji-kweeezz*. **SHH** Introduced (Africa) to Malagasy region. On granitic Seychelles restricted to Mahé and La Digue; also Alphonse (coralline Seychelles), Mauritius, Reunion, Rodrigues, Juan de Nova, possibly Madagascar. Not very common; in grassy lowland clearings, gardens and roadsides. Gregarious except when breeding, in tight flocks of 5–20 birds, which fly off and move *en masse*.

PLATE 66: CORALLINE SEYCHELLES: LANDBIRDS I

Madagascar Turtle Dove *Nesoenas picturata* 28 cm

Sturdy, dark dove, *frequently seen on the ground*. Nominate race (introduced to Amirantes) has *blue-grey head*, purple-pink breast, paler pink-buff belly, *brownish wings and grey-brown tail with ash-grey outer tail-feathers tipped white*. Race *coppingeri* (native to Aldabra group) has *dark purple-mauve head* with black necklace, dark pink-purple breast merging to greyish belly, white vent, mauve-brown upperparts and rump. Dark reddish-brown upperparts distinguish from Comoro Blue Pigeon, which has mainly blue-black plumage, whitish bib and more rounded wings. **Voice** Low-pitched, double coo: *coo cooooo-uh*, the second *coo* louder, longer and rising at end; first *coo* often not noticed. **SHH** Endemic to Malagasy region; on coralline Seychelles, race *coppingeri* resident Aldabra and possibly this form Cosmoledo; nominate race breeds D'Arros, St Joseph's Atoll and Rémire, recorded Desroches, Platte; nominate and other races breed granitic Seychelles, Madagascar, Mauritius, Comoros, Reunion. Some races are fairly distinct and might merit species status, but further study needed as *coppingeri* is rather similar to *comorensis* (Comoros). Another race endemic to Amirantes is extinct. Common in wooded habitats and open areas. Feeds mainly on ground from where often flushed with noisy wingbeats, but flight otherwise silent, very fast and usually low.

Zebra Dove *Geopelia striata* 21 cm

The only small resident dove. A very distinctive, *tiny, greyish-brown dove, with extensive black barring*, otherwise pink or grey plumage and blue skin round eye. Juv paler, less smart, shorter-tailed, and lacks blue eye-ring. In flight, long, wedge-shaped tail with narrow white tips. *Bounding flight on take-off* distinctive. **Voice** A medium- to high-pitched, bubbling, quick *cooruruwuh* or *wurruwawa*, first note cooing and last two sharper, final note sometimes repeated; in display, a harsher *caaw-caaw-caaw*. **SHH** Introduced (SE Asia via Mauritius) to D'Arros, St Joseph's Atoll, Desroches, Coëtivy and Farquhar; also granitic Seychelles, Mauritius, Rodrigues, Reunion, Glorieuses, Juan de Nova, Agalega. Common in open areas or habitation with scattered woody vegetation, including around houses and hotels. [Alt: Barred Ground Dove]

Comoro Blue Pigeon *Alectroenas sganzini* 27 cm

Plumage unique in its range: *deep blue upperparts, belly and tail and a silvery-white head, neck and breast*. Juv less attractively patterned than ad, but still distinctive. Madagascar Turtle Dove is all dark. **Voice** A *series of how or hrrowr notes* (often 3–5, sometimes ten; one note per 1–2 s), harsher and more rapid than coos of Madagascar Turtle Dove, level in pitch or with upward inflection; sometimes a final drop in pitch. Pitch, speed and quality vary: sometimes hoarse, with overall effect varying from whooping to almost groaning. **SHH** Endemic to Malagasy region: race *minor* on Aldabra, formerly throughout Aldabra group and recorded again on Assumption since 2012; larger, nominate race on Comoros. Mainly in native evergreen forest (including partially degraded areas), dry forest, secondary forest, plantations and cultivated areas with trees; rarely mangroves. A gregarious, arboreal, fruit-eating pigeon, often using prominent treetop perches. Flight fast, with frequent wing-claps.

Madagascar Coucal *Centropus toulou* 40 cm (♂), 43 cm (♀)

Ad breeding mainly *black except chestnut wings*. Non-breeder duller, dark greyish-brown, browner on wings, streaked and spotted whitish on throat and breast. ♀ is significantly larger than ♂. Juv has blackish upperparts, speckled cream on mantle and crown, cinnamon-rufous wings barred dusky-brown, dark brown throat, dusky-brown underparts striated on upper breast, and finely barred rump and tail. **Voice** A repeated hollow *pop... pop... pop...*, a loud, toneless rattle *kakakakakakekekek*, and single, toneless *kisschheerrrrk* or *tish* dropping at end. Song, a distinctive, far-carrying, rapid, low popping, *popopopopopopopopopopopopopopewpew*, falling and slowing at end. Usually a duet in overlapping sequence at lower or higher registers. **SHH** Endemic to Malagasy region including race *insularis* on Aldabra; nominate race (slightly more brightly coloured) on Madagascar. Extinct Cosmoledo; also extinct Assumption since c.1909 but present 2013. Frequents woodland, moving through dense vegetation with agility. Weak flight, often only moving short distances. Generally alone or in pairs, skulking, but fairly tame.

PLATE 67: CORALLINE SEYCHELLES: LANDBIRDS II

Madagascar Nightjar *Caprimulgus madagascariensis* — 22 cm
Grey-brown upperparts mottled black, finely barred underparts, paler on belly and undertail-coverts. ♂ has *broad white wingbar and tips of two outer pairs of tail-feathers also white*. ♀ has wingbar less extensive and buff spots on outer tail-feathers. Juv similar to ♀, more cryptically coloured. **Voice** Song, heard very frequently, *loud single note followed by somewhat lower, short trill tip-tttttttrrrr*. Call *pow-pit, wow-up, wow-wow, tpit-woau, wau, wa-pit,* or *ku-wuh*, often several birds calling together, in flight or perched. Also, a long, low, glugging or clucking trill lasting 10–30 s, moving up and down scale and becoming faster and slower. **SHH** Endemic to Aldabra (race *aldabrensis*) and Madagascar (nominate race, slightly darker). Feeds in flight, on Aldabra often over open areas including around settlement.

Broad-billed Roller *Eurystomus glaucurus* — 29 cm
Stocky, *large-headed, long-winged roller, mainly cinnamon above and deep lilac below with short broad yellow bill*. Cinnamon forewing, rest of wing dark blue, rump blue and uppertail-coverts and undertail-coverts greyish-blue (nominate) or clear blue (other races). Juv duller, underparts below breast dull blue and mottled greyish, upperparts rufous-brown, forewing rufous-brown and rest of wing dull greenish-blue. **Voice** Silent in Seychelles. **SHH** Sub-Saharan Africa and Madagascar; nominate race breeds in Madagascar and migrates to Africa, a few passing through Aldabra group (and also Comoros), mainly Oct–Nov; vagrant elsewhere in region. Usually in areas with scattered tall trees, perching prominently on branches.

Madagascar Bulbul *Hypsipetes madagascariensis* — 24 cm
Medium-sized bulbul, mainly *grey-brown above, paler below with a shaggy crest and bright slender orange bill*. Juv duller, underparts greyer washed strongly brown, chestnut margins to tail- and flight-feathers, and duller bill. No similar species on Aldabra. **Voice** Very vocal; a squeaky *kee kee keer*, whining *keeeeer* or clucking *tut tut tut tutt* interspersed by short dropping *tchiliyu*, or given together *chup-tchilyu*. In song, an extended *chip-er-chilyu-chip-chip-chilyu-chup-eee-chup-ee-tsiliyup*. **SHH** Endemic to Malagasy region; brownish race *rostratus* restricted to Aldabra and Glorieuses, extinct Astove and Cosmoledo; greyer nominate race on Madagascar and Comoros. A common, conspicuous, noisy, aggressive and curious inhabitant of any habitat with trees or shrubs.

Aldabra Drongo *Dicrurus aldabranus* — 25 cm
All black with long forked tail and heavy black bill. Long nasal plumes visible at close range. Deep red eye, heavy, hooked, black bill and black legs. Juv mainly greyish-brown above and (most unusually among drongos worldwide) very pale brown below with darker streaking, shorter tail fork and brown eye. **Voice** Very vocal and noisy. Call a harsh chuckle. When chasing other birds, a nasal *titi-po fa fa* often repeated; after repelling an intruder may land on and call *fa-wip* a few times. **SHH** Endemic to Aldabra, where common. Arboreal with a marked preference for *Casuarina* woodland, mangroves and dense scrub. Follows Madagascar Coucal, using agility to steal prey, or Aldabra Giant Tortoises *Aldabrachelys gigantea*, catching flying insects. Territorial and aggressive, especially when nesting, mobbing intruders. **NT**

Pied Crow *Corvus albus* — 45 cm
Only crow in Seychelles. May soar like a raptor but easily distinguished by *black-and-white plumage* and typical corvid silhouette with relatively short wings and stout bill. **Voice** Flat, loud, low-pitched *caah, caw* or *craaah*; also a low *cuh... cuh... cuh* from flocks and a range of other bubbling or gurgling noises. **SHH** Sub-Saharan Africa and Malagasy region, including Aldabra group where uncommon except around settlement on Picard (Aldabra); also Madagascar, Comoros, Glorieuses, Juan de Nova, Europa. Flight powerful, frequently seen flying over forest or patrolling beach for turtle hatchlings. Predates nests or kills birds as large as Madagascar Turtle Dove.

PLATE 68: CORALLINE SEYCHELLES: LANDBIRDS III

Barn Swallow *Hirundo rustica*   18 cm
Only swallow recorded annually throughout Seychelles, distinguished from vagrant Mascarene, Sand and House Martins by *blue-black upperparts, dark throat and long tail-streamers*. Imm duller, browner, has pale buff on chin and throat, whiter underparts and shorter tail-streamers. See vagrant swallows, especially Wire-tailed (Plate 116). **Voice** Contact call *witt-witt*, given at any time, sometimes run together as series of *tswitt* notes. Song a more prolonged, bubbling twitter, not recorded in Malagasy region. **SHH** Breeds Eurasia and North America, spending boreal winter in sub-Saharan Africa and South America. Rare migrant across much of Malagasy region; regular Seychelles mainly on passage Nov and Mar–Apr.

Madagascar Cisticola *Cisticola cherina*   11 cm
Tiny passerine. Tawny-brown upperparts and wings with heavy blackish-brown streaks on back and pale brown unstreaked rump. Underparts whitish washed pale brown on flanks, throat and chin white. White supercilium gives capped appearance. *Graduated tail is tipped white with blackish subterminal bars* and brown central rectrices. Juv has rusty upperparts with heavy brown streaking, a yellow wash to chin, throat and upper breast, suffused rusty on flanks. From vagrant warblers by tail shape and pattern. **Voice** Song a very high-pitched, metallic *tint... tint... tint* or *pik... pik... pik... pik...*, repeated at 0.5–1.0 s intervals; also a high-pitched, metallic, repeated *triit* or *chiit* or *chreep*, slightly trilled and modulated. Both given in song-flight, and in duet, when may be delivered more rapidly. Alarm note *tic*. **SHH** Endemic to Malagasy region including Cosmoledo and Astove; also Madagascar and Glorieuses. Common in low dense vegetation; frequently cocks tail.

Aldabra Brush Warbler *Nesillas aldabrana*   19 cm
Only resident warbler on Aldabra, though Marsh and Sedge Warblers have been recorded as vagrants to Seychelles and other *Acrocephalus* species could occur. *Dingy greyish-brown, lacking rufous tones, with long bill and tail* (longest in the genus). **Voice** A very brief *chak* sometimes given as loud series, beginning slowly and increasing in tempo, with a few slower calls at end. Also a loud nasal *chir*, always given after 1–2 *chak* calls. *Chak-chir* phrases given by both sexes but especially ♀. **SHH** Endemic to Aldabra, confined to W end of Malabar but now almost certainly extinct. Discovered 1967, last recorded 1983. Last populations found in dense mixed scrub of closed-canopy vegetation with leaf-litter and soil below. **EX**

Souimanga Sunbird *Nectarinia souimanga*   11 cm
Only sunbird in the Aldabra group. ♂ has *head and neck dark metallic green* with a well-defined maroon chest-band (broader in *buchenorum*), dark brown upperparts and wings with yellow pectoral tufts. *Lower abdomen is dull white to yellowish white* (*aldabrensis*), *dark brown* (*abbotti*) or *blackish* (*buchenorum*). Thus *aldabrensis* most like Madagascar races, but with broader black and red breast-bands. Non-breeding ♂ lacks metallic sheen. ♀ is dark grey-brown above, varying among races in extent and mottling of black on throat and breast. Juv similar to ♀, sometimes with black chin and throat, and stronger olive wash to upperparts. **Voice** Call a single *plit* or *chit* or *chip* in flight; also high-pitched, drawn-out, peevish (or mewing) *chweee*. Song high-pitched twittering *plit... chit... chirrit... chirrititiwitittirrit*; last phrase very complex and sometimes lasting 1.0–1.5 s. **SHH** Endemic to Malagasy region including throughout Aldabra group, with three endemic races: *aldabrensis* on Aldabra, *abbotti* on Assumption and *buchenorum* on Astove and Cosmoledo; nominate race resident Madagascar and Glorieuses. DNA suggests *buchenorum* is more closely related to nominate race (Madagascar) than to *abbotti* or *aldabrensis*, thereby not supporting the separation of 'Abbott's Sunbird' (races *abbotti* and *buchenorum*) as a separate species. More logical to lump all in one species, or split into four species.

Madagascar White-eye *Zosterops maderaspatanus*   11 cm
Only white-eye on the coralline Seychelles. Readily identified by white eye-ring and mainly green-and-yellow plumage. **Voice** Flocks emit varied nasal *tew*, *ter* or *ter-tew* notes, and peevish rising *wrrrri* or *trrrrri*. Song a medium-pitched rapid warble *tewteerweerteerteerpiwitewteerteertewtewpiwi*, often given around dawn. **SHH** Endemic to Malagasy region including Aldabra group, where endemic race *aldabrensis* (yellower above than nominate from Madagascar, with broader yellow band above lores) occurs on Aldabra; birds on Cosmoledo (Menai I alone) and Astove resemble nominate, although former proposed as race *menaiensis*; also Madagascar, Comoros, Glorieuses, Europa. Locally common in wooded habitats, gregarious and highly active. Race *aldabrensis* is genetically distinctive and could be treated as a species; DNA from Cosmoledo and Astove birds not studied.

PLATE 69: CORALLINE SEYCHELLES: LANDBIRDS IV

House Sparrow *Passer domesticus*   16 cm
Breeding ♂ distinctive. ♀ resembles ♀ or non-breeding ♂ Madagascar Fody, but larger with proportionately smaller bill and a *greyish rather than yellowish-buff hue to upperparts*. ♂ has bright *white cheeks, almost white underparts, chestnut back* (paler than European birds) heightening contrast with black streaks. **Voice** A harsh *chissk*, short *weesk* or abrupt, lower-pitched *chup, chup-up*, or short, coarse trill, *chukukukukuk*. Song a series of *chup* or *chip* notes; chorus of *chissk* notes given by many birds together, e.g. at roosts. **SHH** Introduced and commensal worldwide, to Amirantes on all islands from Rémire S to St François; also locally Madagascar, Comoros, Mauritius, Rodrigues, Reunion. Highly social and gregarious. Flight rapid and direct, on whirring wings.

Aldabra Fody *Foudia aldabrana*   14 cm
No other fody shows breeding ♂'s combination of *red on head, breast and rump, and yellow on underparts*; bill is stoutest of any fody. ♀ has notably *yellowish-brown head, upperparts and rump*. Non-breeding ♂ loses bright red plumage typically around Jul, resembling ♀. **Voice** A high-pitched metallic *two-eet* or *twee* call, with lower or more staccato notes interspersed. In alarm ♂ gives an excited trill in which *two-eet* predominates, ♀ a continuous, loud scolding *tic-tic tzip tic tzip tzip tic-tic*. **SHH** Endemic to Aldabra. Extinct Astove, Cosmoledo and Assumption. Frequents woodland. Not closely related to Comoro Fody as formerly suggested. Not currently recognised by BirdLife; may be threatened or Near Threatened due to restricted range and small population.

Seychelles Fody *Foudia sechellarum*   12 cm
Small drab, dumpy weaver, darker on back and wings. Breeding ♂ acquires *yellow crown, face and chin*, merging to dark brown on throat and rear crown. Often has a small white wing-patch. Non-breeding ad and juv mainly *dark brown, darker and rounder than Madagascar Fody*. The two have hybridised on D'Arros: hybrids variable, sometimes resembling Seychelles Fody but with bright red, orange or yellow head, sometimes with black band through eye. **Voice** Both sexes give *characteristic tok-tok-tok*. In distress or alarm, *tok* calls rapidly increase in intensity and frequency, also given singly or with long pauses between notes, while feeding. Song given by both sexes, especially ♂, single high-pitched notes alternating with pauses, which become shorter thus high notes become almost continuous. A different alarm call, perhaps higher intensity, *tchr*. **SHH** Endemic to granitic Seychelles but introduced to D'Arros in Amirantes (1965, not mapped). Frequents mainly woodland and edges, but has few habitat constraints. Social, often in groups, but less gregarious than Madagascar Fody. Flight is direct, compactness obvious, appearing almost tailless. Constantly flicks wings. **NT**

Madagascar Fody *Foudia madagascariensis*   13 cm
Breeding ♂ usually distinctive with all-red underparts, but red may be patchy; moults into ♀-like non-breeding plumage (Jun–Oct). Gregarious habits and preference for open areas, where it feeds mainly on seeds, often on ground. *From House Sparrow by smaller size and more greenish-brown or buffish-yellow plumage*. Sibilant *tsip* call not given by any confusion species. **Voice** A very high-pitched, slightly chirping *tsip* or *tsikip*; also a lower *chup* or *chutut*. Breeding ♂ has trill call, or buzzing *zzz-zzz-zzz*. Song lacks regular form: wheezes, chirrups, squeaks and *tsip* calls, with up to *c*.1 s gap between phrases: *tsitsitsitsitstit tiptiptip titititit chipchipchip*. **SHH** Endemic to Malagasy region; introduced to Platte, Amirantes (Rémire, D'Arros, St Joseph's Atoll, Desroches), entire Farquhar group and Assumption, as well as to granitic Seychelles, Comoros, Mauritius, Reunion, Rodrigues, Glorieuses, Juan de Nova, Agalega; native to Madagascar. Has reached Aldabra (Grande Terre), probably in early 21st century, and hybridised with Aldabra Fody. Frequents woodland, plantations and settlement areas.

Common Waxbill *Estrilda astrild*   12 cm
Very small size and long tail, combined with *brown upperparts*, paler underparts, *red eye-patch and red bill* unique in Seychelles. Juv has black bill, black line below red lores and indistinct barring. **Voice** Medium- to high-pitched *tewk* and hesitant *tetetete* or *tewtewtew*, in alarm a nasal *jaa* or longer *weeez*. In groups, weak twittering, sparrow-like chirping and shrill *tseep*. Song ends with a long buzzy upslurred note, *ji-ji-kweeezz*. **SHH** Introduced (Africa) to Alphonse; also granitic Seychelles, Mauritius, Rodrigues, Reunion, Juan de Nova, possibly Madagascar. Found in grassy clearings, gardens and roadsides. Gregarious except when breeding, in tight flocks that fly off and move *en masse*.

PLATE 70: COMOROS: LANDBIRDS I

Feral Pigeon *Columba livia* — 33 cm
Typical *'town pigeon'*; tight flocks of pigeons in urban areas or flying high likely to be this species. Plumage very variable but never known to resemble other large pigeons on Comoros, which are not town-dwelling. **Voice** A long, moaning *o-o-orr*, repeated monotonously, or a hurried, rambling *coo-roo-ooo-ooo*. **SHH** Native to Eurasia and N Africa, but introduced worldwide. Status on Comoros uncertain: domesticated in many towns and villages on all four islands; commonest in and around Moroni (Grande Comore) but fewer on Mohéli and Anjouan. Also Madagascar, Seychelles, Mauritius, Rodrigues, Reunion. Flight powerful and fast, on long, pointed wings, often in tight flocks.

Comoro Olive Pigeon *Columba pollenii* — 38 cm
Large size and dark-looking plumage at all ages, with *bright yellow bill* in ad (dull greenish-brown in imm), unique. Feral Pigeon is confined to urban areas and usually has variegated plumage. Comoro Blue Pigeon is 25% smaller; ad has whitish head and neck; darker-headed juv never has plain, greenish look of Comoro Olive Pigeon. Madagascar Turtle Dove also much smaller, with reddish-brown upperparts, paler tip to tail and more terrestrial habits. **Voice** A *very low-pitched*, moaning *guk-ohoooo hoo hooo*, usually repeated 2–3 times, the first note low and guttural, the second lower, slightly disyllabic and tapering off; sometimes heard as a single *coo*, very deep. **SHH** Endemic to Comoros, all four islands, mainly in forested uplands, but also tree plantations and degraded mosaics of cultivated and woody land including in lowlands. Fairly common on Mt Karthala, but highest densities on Mayotte. Arboreal, canopy-living pigeon, sometimes fairly tame. **NT**

Madagascar Turtle Dove *Nesoenas picturata* — 28 cm
Only sturdy, short-tailed, dark dove on Comoros. Dark reddish-brown head and upper body, and *tail with only slightly paler (not white) corners*. Comoro Olive Pigeon is larger and more uniformly dark with yellow bill and legs. In flight any hint of coloration should distinguish from Comoro Blue and Madagascar Green Pigeons. **Voice** Low-pitched, double cooing: *coo cooooo-uh*, the second *coo* louder, longer and rising at the end; first *coo* is often not noticed. **SHH** Endemic to Malagasy region; also Madagascar, Seychelles, Mauritius, Reunion, Agalega, with red-headed endemic race *comorensis* on Comoros, all four islands, in wooded habitats and nearby open areas; common except on Anjouan and Mt Karthala summit zone. Some races are rather distinct and might merit species status, but further study needed as *comorensis* is rather similar to *coppingeri* (Aldabra group). Feeds mainly on ground from where often flushed in forest with noisy wingbeats, but flight otherwise silent, very fast and usually low.

Madagascar Green Pigeon *Treron australis* — 32 cm
Only *'green'* pigeon on the Comoros, although in fact yellow below and greyish-green above. In poor light or in flight, green colour not always obvious but *pale yellow wingbars* striking; tail shows dark base with bluish-grey terminal band. Juv probably as ad. **Voice** Remarkable song like other *Treron* pigeons, a *fast medley of whoops, whistles, trills and yaps, tiwiorrr... whip... towiorrr... whip-por-whip koorr*, and variants. **SHH** Endemic to Malagasy region; also Madagascar, with poorly known endemic race *griveaudi* on Comoros, restricted to Mohéli, where uncommon within its tiny range, in canopy of evergreen forest, secondary forest and coconut plantations. An arboreal fruit-eating pigeon, seen singly or in small flocks. Differs significantly from Madagascar birds, with grey (not red) cere and reddish-purple (not yellow-orange) feet, as well as plumage difference; may represent a species (Comoro Green Pigeon *T. griveaudi*), as recently recognised by BirdLife and treated as **EN**.

Comoro Blue Pigeon *Alectroenas sganzini* — 27 cm
Plumage unique in its range: *deep blue upperparts, belly and tail and silvery-white head, neck and breast*. Other medium-large pigeons are the all-dark Madagascar Turtle Dove and Comoro Olive Pigeon throughout the Comoros, and green-and-yellow Madagascar Green Pigeon on Mohéli alone. Juv less attractively patterned than ad, but still distinctive. **Voice** A *series of how or hrrowr notes* (often 3–5, sometimes ten; one note per 1–2 s), harsher and more rapid than cooing of Madagascar Turtle Dove, level in pitch or with upward inflection; sometimes a final drop in pitch. Pitch, speed and quality vary: sometimes hoarse, with overall effect varying from whooping to almost groaning. **SHH** Endemic to Malagasy region, nominate race on Comoros; also coralline Seychelles (Aldabra). Common in suitable habitat on all four islands. Mainly in native evergreen forest including partly degraded areas, dry forest, secondary forest, plantations and cultivated areas with trees; rarely mangroves. A gregarious, arboreal, fruit-eating pigeon, often using prominent treetop perches. Flight fast, with frequent wing-claps.

PLATE 71: COMOROS: LANDBIRDS II

Ring-necked Dove *Streptopelia capicola*   25 cm
Only small, greyish-brown dove on Comoros, with characteristic *far-carrying, monotonous song. Solid black neck-patch and, especially in flight, white tail-corners* diagnostic. Madagascar Turtle Dove is much larger and darker reddish-brown; Tambourine Dove smaller and shorter-tailed, dark brown above, with striking white face markings and conspicuous chestnut wing-patches in flight. **Voice** As in Africa, an endless rolling medium-high tut-tooor-tut-tut-tooor-tut-tut-tooor-tut..., the middle coo note vibrato and shorter notes popping, rendered 'how's father' or 'cure your cough'. **SHH** Africa and Comoros, where common in lowlands of all four islands: open country, especially cultivation, forest clearings and dry coastal scrub, rarely above 500 m; mainly terrestrial but displays from elevated perches. Madagascar Turtle Dove prefers thicker cover. [Alt: Cape Turtle Dove]

Tambourine Dove *Turtur tympanistria*   23 cm
Only small, short-tailed dove on the Comoros, also distinguished from Ring-necked Dove by dark brown upperparts, largely white underparts, face and supercilium (grey in imm). Often seen in fast flight, when flushed or moving between patches of cover, showing very conspicuous chestnut wing-patches above and below, and short, all-brown tail without white corners. **Voice** Song distinctive, *a slow, medium tooting too... too... too... tootootootootootoo*, hesitant at first, and falling slightly. **SHH** Africa and Comoros, where common on all four islands, but secretive so often overlooked; avoids forest, preferring gardens, overgrown cultivation and other areas of shrubbery in lowlands, with small thickets and often mango trees; also mangroves. Occurs singly or in pairs, feeding on ground; wings clatter when flushed but flight is otherwise silent.

Grey-headed Lovebird *Agapornis canus*   15 cm
Only lovebird on the Comoros (and in region). Easily identified when perched; juv similar to ♀ but has darker bill, grey hood of ♂ suffused green. *Compact shape and distinctive bat-like flight,* often chirping. **Voice** High-pitched, single *chik* or *chilp*, and more complex, slightly lower *tswik* or *tsiwik*, singly or many calling together. Song a pleasant twitter of such notes, *chickitwick-psick-tsiwick-tsikichick*; occasionally also squawks or whistles. **SHH** Endemic to Malagasy region; probably introduced to Comoros from Madagascar. On all four islands, avoids extensive forest in favour of open or bushy areas, wooded savanna, woodland or forest patches with grassland and cultivation. Perches in trees, but feeds on ground. Gregarious, flocks sometimes numbering dozens.

Greater Vasa Parrot *Coracopsis vasa*   50 cm
Sooty-brown plumage very similar to Lesser Vasa Parrot, but on Comoros (unlike Madagascar) identification is normally simple as *local race of Lesser is very small, such that Greater is 50% larger* in linear measurements. In Madagascar, incubating or chick-rearing ♀ develops bald or partially bald head with orange or yellow skin; not confirmed, but likely, in Comoros. **Voice** Wide variety of mono- or disyllabic calls when feeding, with a loud screeching given in alarm; squawks and whistles including loud *kwa... kwa... kwa...*, single *kark* or *kia-kark*, series of *karr krii kaarrrr grrk*; in flight, a high *weet*. Song of breeding ♀, if like that in Madagascar, a long sequence of squawks and whistles mixed with lower rasps. **SHH** Endemic to Malagasy region, with endemic race *comorensis* on Comoros (absent on Mayotte), paler than Madagascar birds with less blue-grey fringing on wings. Found in primary and secondary forest, plantations, other wooded areas and farmland, up to edge of tree-heath on Mt Karthala. Fairly common on Grande Comore, less so on Mohéli and Anjouan. Often in groups of 4–10; roost flights may number hundreds. Active and vocal at night, especially if moonlit. Flight strong, recalling a raptor. Not known to associate with Lesser Vasa Parrot.

Lesser Vasa Parrot *Coracopsis nigra*   32 cm
Confusable only with Greater Vasa Parrot, but on Comoros is *much smaller*. **Voice** Long sequence of clear, squeaky notes in rapid sequence, *ki-ki-ki-ke-ke-ke-ker-ker-ker-kuk*, last note lower and truncated; also similar sequence of more modulated notes, *kiu-kiu-kiu-kiu-kier-kier-kier-kier*; usually markedly drop, but sometimes all on same pitch; also single *ki*, *keer*, *kiwik* or *kowik* notes. Poorly known on Comoros; unclear whether voice similar to Madagascar populations. **SHH** Endemic to Malagasy region; endemic Comoros race *sibilans* restricted to Grande Comore and Anjouan; also on Madagascar. Scarce: restricted to mid-altitude evergreen forest on Mt Karthala, Grande Comore, and remnant upland forest on Anjouan. Arboreal on Comoros, and unobtrusive compared to Greater Vasa. Much smaller than Madagascar races and not closely related to very similar-looking Seychelles Black Parrot; may be a species (Comoro Black Parrot, or Comoro Parrot), as recently recognised by BirdLife and treated as **NT**.

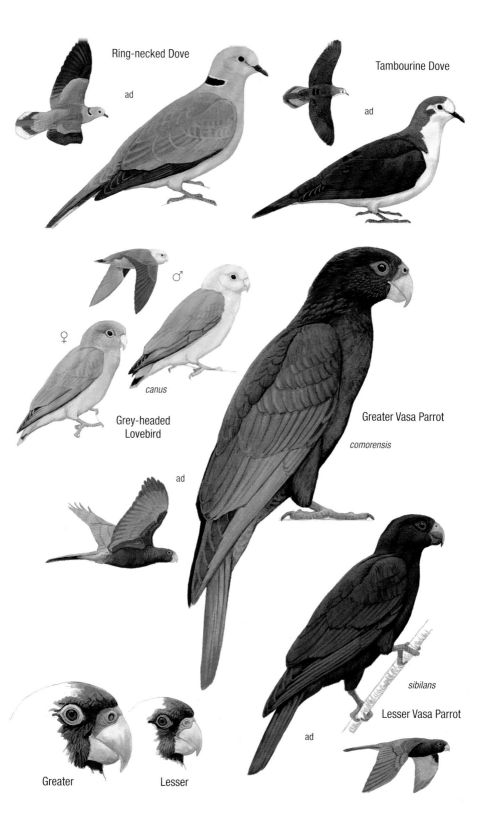

PLATE 72: COMOROS: LANDBIRDS III

Barn Owl *Tyto alba*   35 cm
Unique on Comoros, where often detected by distinctive **screech call**, familiar in much of world. Barn Owls on Comoros may be obviously orange-brown on underparts, but **very pale plumage** always very distinct from other local owls, which are all much smaller, darker scops owls; however, note unconfirmed reports of a larger owl, perhaps Marsh Owl (see Plate 42), on Mayotte. **Voice** Loud, toneless, grating screech *ksscchh*. Hisses and 'snores' at rest and makes loud bill-snaps when disturbed at or near nest. **SHH** Almost cosmopolitan; throughout Comoros including islets near Mohéli and Mayotte; on Grande Comore, up to 1,000 m, occasionally 1,800 m on Mt Karthala. Also Madagascar, granitic Seychelles, Europa. Found in towns, villages, forests (perhaps mainly edge) and a range of degraded habitats. Nocturnal hunter, taking many rats, mainly on wing with low, buoyant flight.

Karthala Scops Owl *Otus pauliani*   19 cm
Only small owl on Grande Comore; just one morph known. Ear-tufts not visible in field, and iris can be dark or yellow. Only other owl on Grande Comore is Barn Owl, which could overlap in distribution but is much larger with completely different plumage and voice. Vasa parrots also active and vocal at night, but have very different whistling or screeching calls. **Voice** *Long sequence of simple, repeated short notes*, *cho* or slightly modulated *kyu*, repeated over long periods from perch: 10–30 or many more notes, sometimes >1,000 and lasting tens of minutes; calls year-round. Also a slower, medium-pitched *kuiiip... kuiiip... kuiiip...*. **SHH** Endemic to Grande Comore, restricted to primary and degraded forest, and partly forested mosaics (but not pure tree-heath) on upper slopes of Mt Karthala, 800–1,900 m. Local densities, e.g. around La Convalescence, high, but overall population low and area of habitat declining. **CR**

Anjouan Scops Owl *Otus capnodes*   22 cm
Only small owl on Anjouan. Ear-tufts rarely seen in field, as obscured by long crown feathers. Pale and dark morphs both common. Only other owl on Anjouan is Barn Owl, which occurs in same habitat but is much larger and completely different in colour. Whistling call unique; vasa parrots often give whistled calls at night but these differ in tone, structure and inflection. **Voice** *Sequence of beautiful, modulated, high-pitched whistles*: plaintive di- or trisyllabic *peeoo* or *peeooee*, not all notes in one sequence identical; often in duet, and reminiscent of truncated Grey Plover call. Possibly also a screech call similar to Barn Owl. Calls all night and apparently all year, including midwinter. **SHH** Endemic to Anjouan, where occurs in a range of forest types, including degraded forest and agroforestry areas, but not exotic plantations below 300 m; especially common in native forest remnants above 800 m. Calls from perch and in flight. **CR**

Mohéli Scops Owl *Otus moheliensis*   22 cm
Only small owl on Mohéli. Ear-tufts often visible but not large as Mayotte Scops Owl. Brown and rufous morphs known. Voice distinctive, although can be confused with Mongoose Lemur *Eulemur mongoz*; other vocal nocturnal species on Mohéli include Greater Vasa Parrot, Tropical Shearwater and Barn Owl, but none known to give a call similar to Mohéli Scops Owl. **Voice** *A low, coarse hissing* keooroooh *or* kroooh *lasting 1–5 s* (territorial call); also a screech. Sings mainly at night, but also in late afternoon. **SHH** Endemic to Mohéli, where restricted to dense evergreen forest and edges of W & C, mainly above 450 m but occasionally lower. Prefers intact forest but also in degraded forest; very local but not rare in tiny range. **CR**

Mayotte Scops Owl *Otus mayottensis*   23 cm
Only small owl on Mayotte. A large scops owl, with grey-brown and (rarer) rufous morphs, and prominent ear-tufts (like Madagascar Scops Owl). Only other nocturnal species are Barn Owl and, possibly, a second, unidentified large owl species, perhaps Marsh Owl. **Voice** *Slow sequence of 3–10 (usually 3–5) simple medium-low hoots*, *poo poo poo poo*, each lasting c.1 s. Sings from midstorey, mostly at night but also 1–3 hr before dusk. **SHH** Endemic to Mayotte, where found throughout Grande Terre, from woodland on coast, even in towns (Mamoudzou) to forested peaks of C; more widespread than other Comorian scops owls on their respective islands.

PLATE 73: COMOROS: LANDBIRDS IV

Madagascar Spinetail *Zoonavena grandidieri*   12 cm
Only small, short-tailed swift on the Comoros. May occur with Palm Swifts, which are paler, with longer, narrower, backswept wings and much longer, deeply forked tail often appearing pointed; Palm Swifts with broken tail may suggest spinetail. African Black Swift much larger. Race *mariae* has more blackish-brown upperparts than nominate, underparts more uniform, lacking paler throat. **Voice** Rarely heard, and vocalisations quiet: a 'ticking trill' noted on Comoros; in Madagascar, a short *chip... chip... chip...* or *zip... zip... zip...*, often in chorus or in short trill *zrizrizri*. **SHH** Endemic to Malagasy region. Race *mariae* endemic to Grande Comore, mostly on Mt Karthala at 500–1,470 m, but often higher; occasional elsewhere down to coast. Absent elsewhere on Comoros, but nominate race in Madagascar. Mainly forest, both closed-canopy and degraded areas, clearings and along roads; occasionally visits non-forest areas.

African Palm Swift *Cypsiurus parvus*   16 cm
Small size, brown colour and long tail characteristically held closed thus appearing pointed make this a distinctive swift; long, narrow and typically backswept wings also unique. Madagascar Spinetail is darker and stockier with short, square tail; African Black Swift is larger and darker, with shallowly forked tail. Comoros race *griveaudi* differs from *gracilis* (Madagascar) in having less dark upperparts and paler underparts, and streaked whitish area extending to sides of chin and throat. **Voice** Short, vowel-less, scratchy *tchh tchh tchh* sometimes in long sequence. **SHH** Africa and Malagasy region, where also in Madagascar. Race *griveaudi* endemic to Comoros, ubiquitous in lowlands on all four islands, often associated with coconut and other palms; wanders up to 2,000 m on Mt Karthala. Gregarious, also mixing with Madagascar Spinetail.

African Black Swift *Apus barbatus*   16 cm
Obviously *larger than other resident swifts on Comoros*: resembles the widespread *Apus* species such as Common Swift (vagrant to region: Plate 114) and its close relatives over much of the Old World. Palm Swift is slimmer and paler, with long tail and narrow, backswept wings. Madagascar Spinetail smaller, with short, square tail. Comoros race *mayottensis* differs from *balstoni* (Madagascar) in paler forehead, slightly greyer malar area, and pale patch on chin and upper throat more restricted, sometimes obscure. **Voice** High screaming trill, *zzzzziiieeewwww*, dropping at end, often in chorus. **SHH** Africa and Malagasy region, where also in Madagascar. Race *mayottensis* endemic to Comoros, throughout all four main islands and islets although rarely abundant. Typical all-dark *Apus* swift, most often seen in screaming flocks of 20–100, often higher than other swifts on Comoros, but sometimes joins them in lower airspace. Taxonomy uncertain; Malagasy region birds may form a separate species, thus Madagascar Black Swift *Apus balstoni*.

Olive Bee-eater *Merops superciliosus*   30 cm
Only resident bee-eater on the Comoros; very distinctive at all ages, with *olive-green plumage and contrasting brown crown*. Juv lacks tail-streamers and is duller. See vagrant Blue-cheeked Bee-eater (Plate 115). **Voice** *Sequence of rapid, hollow and rolling medium-pitched notes*, *krioor krioor krioor* or *kriew kreiw kriew*, often repeated and in chorus from flocks; also single *prioop*. **SHH** Africa and Malagasy region, where also on Madagascar and Juan de Nova. Common breeder, present year-round, on all four Comoro islands (scarcer on Anjouan), mainly in lowlands but locally inland and some non-breeders possibly move to E Africa; vagrant Aldabra Aug–Oct. Prefers open country with or without scattered trees, beaches, towns and villages; also hunts over the sea, close to shore. [Alt: Madagascar Bee-eater]

Barn Swallow *Hirundo rustica*   18 cm
Only swallow regularly recorded on Comoros, distinguished from vagrant Mascarene Martin (Plate 48) by *blue-black upperparts, white underparts and underwing-coverts, and long tail-streamers*. Imm duller, browner, has pale buff on chin and throat, whiter underparts and shorter tail-streamers. **Voice** Contact call *witt-witt*, given at any time, sometimes run together as series of *tswitt* notes. Song, a more prolonged bubbling twitter, not recorded in Malagasy region. **SHH** Breeds Eurasia and North America, spending boreal winter in sub-Saharan Africa, Asia and South America. Rare migrant across much of Malagasy region; irregular, occasional migrant to Comoros, where recorded on Grande Comore, Mohéli and Mayotte. A typical hirundine, attracted to concentrations of flying insects, for which it forages aerially, in very agile flight, appearing more flexible and less direct than Mascarene Martin.

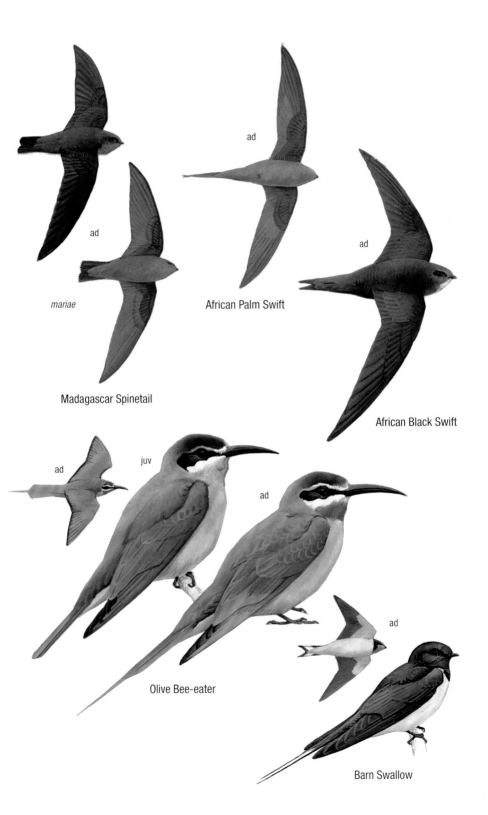

PLATE 74: COMOROS: LANDBIRDS V

Madagascar Malachite Kingfisher *Corythornis vintsioides*   13 cm
Only kingfisher on the Comoros, where unique although, despite *brilliant colours*, sometimes inconspicuous and *attention often drawn only by calls*. Race *johannae* endemic to Comoros, more turquoise than nominate, upperparts and especially rump paler. **Voice** High-pitched *tsip* or *tsrip*, often repeated, *tsip-ip* or *tsrip-ip*; also flatter *chipchipchip...chipchip*. **SHH** Endemic to Malagasy region; also in Madagascar and vagrant (race unknown) to Aldabra. Tied to wetlands, especially mangroves but also rocky shores, jetties, streams, lakes, pools. Common on all four islands except on Grande Comore, which lacks fresh water, leaving kingfishers restricted to within 500m of coast and Mitsamiouli crater lake.

Broad-billed Roller *Eurystomus glaucurus*   29 cm
Only roller on Comoros. Very distinctive at all ages: *dark plumage highlights broad, bright yellow bill and, in flight, largely turquoise tail*. *Rattling calls* also typical. **Voice** A distinctive, *loud, croaking or rattling gagagagaggrikgrikagrika... grik... grik* or *karr kriik kriik karr krikrikrikrikrikrikukrikukrikukuk... uk... uk... uk.* Many variants, longer and shorter, with single notes given when mobbing intruders. However, rarely if ever heard outside breeding range in Madagascar. **SHH** Africa and Malagasy region; endemic race *glaucurus* breeds in Madagascar and migrates to Africa, some passing through all four Comoro islands in Mar–May and Oct–Nov; also annual Aldabra group, vagrant rest of Seychelles and elsewhere in region. Usually in areas with scattered tall trees often including riparian vegetation. Noisy and conspicuous, spending long periods perched prominently in treetop (often dead) or other exposed perch.

Cuckoo-roller *Leptosomus discolor*
    42 cm (*gracilis*), 45 cm (*intermedius*), 50 cm (nominate)

Highly distinctive; *recalls a roller or raptor when perched*, but shape of head and bill prevent anything but momentary confusion; no raptor on Comoros comes close. Race *gracilis* (Mt Karthala, Grande Comore) strikingly smaller than nominate (Mohéli and Mayotte; also Madagascar), ♂ has better defined blue-grey face and breast contrasting with white underparts; ♀ paler than nominate, more heavily spotted below. Race *intermedius* (Anjouan) ♂ similar to nominate; ♀ deeper cinnamon-buff below and has extensive gloss like ♂. In flight, *long, broad wings also give raptor-like silhouette* but distinguished by *sweeping wingbeats* quite unlike a raptor. **Voice** Aerial display a characteristic sight and sound of Comoros, as in Madagascar: repeated glides and swoops by ♂♂, sometimes joined by ♀♀, wings flapped only during ascents and held motionless during descents, tips slightly below level of body, accompanied by *spectacular, high-pitched, fluid, rolling calls, pee-eww pew pew*; higher-pitched on Grande Comore, lowest on Mohéli and Mayotte. Also, a simple, ascending series of 5–7 short whistles, easily mistaken for Greater Vasa Parrot. Wide repertoire of other whinnying, whistling or trilled vocalisations in Madagascar, some no doubt given on Comoros. **SHH** Endemic to Malagasy region, where also in Madagascar. All occur in primary and secondary forest, on Mohéli and Mayotte also in more human-altered vegetation along coasts, thus commoner than endemic forms on Grande Comore (Mt Karthala) and Anjouan, where largely restricted to 400–1,200 m (rarely sea level to 1,500 m). Not closely related to any other bird, with many extraordinary anatomical and behavioural features, yet poorly known. Races on Anjouan and especially Grande Comore are distinctive and could be treated as two species (or one species with two races); they are also rare.

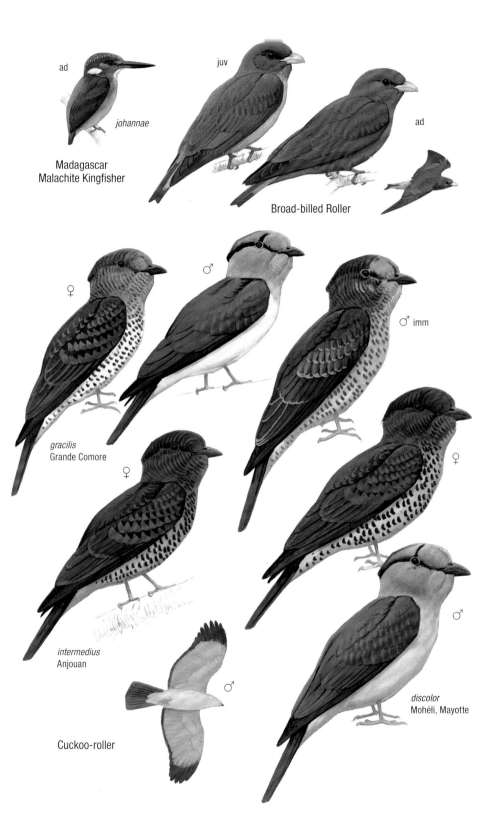

PLATE 75: COMOROS: LANDBIRDS VI

Madagascar Cuckoo-shrike *Coracina cinerea* 22 cm
Exists in *olive and grey morphs*. Confusable with *Hypsipetes* bulbuls, but *shorter-tailed, with black (not pinkish orange) bill, and gives more whistled calls, unlike the bulbuls' distinctive chattering or wheezing calls*. Grey-morph race *cucullata* (Grande Comore) darker than *moheliensis* (Mohéli). Juv and subad distinctively scaled and streaked, equally unique on Comoros. **Voice** A rapid, high-pitched *piitikakewkewkew-kew-kew*, less explosive and quieter than Madagascar races; also (Grande Comore) a high *whi-whew*, a penetrating *pew* or *pew-pew*, and (Mohéli) a short, sharp *chi-we* or *che-wi*. **SHH** Endemic to Malagasy region, where also in Madagascar. On Grande Comore, at 400–1,800 m on Mt Karthala, fairly common; also on forested main ridge of Mohéli, where scarce; all in evergreen forest, occasionally adjacent heath-forest on Mt Karthala. Keeps to forest canopy and midstorey, perch-hunting. Small Comoros races rather distinct from those in Madagascar, where no olive morphs occur, and could be treated as two separate species, or one species with two races.

Madagascar Bulbul *Hypsipetes madagascariensis* 24 cm
On Anjouan and Mayotte, the only medium-large grey passerine. On Grande Comore and Mohéli, must be distinguished from Madagascar Cuckoo-shrike and Grande Comore and Mohéli Bulbuls; see those species. **Voice** Very vocal. Calls include a squeaky *kee kee kee kee*; *'cat-call', a high-pitched, whining keeeeer or chair*; a clucking *tut tut tut tut tut* or *chup-chup-chup-chup* interspersed by dropping *tisilyu*. Gatherings produce a **continual hubbub of chattering**. **SHH** Endemic to Malagasy region. On Comoros, nominate race (also Madagascar and Glorieuses; browner race *rostratus* on Aldabra) *abundant outside highest parts of Grande Comore and Mohéli*, where replaced by endemics: on Grande Comore, up to 700 m, occasionally 1,000 m; on Mohéli, up to 400 m, occasionally 700 m; on Anjouan and Mayotte, throughout including some islets. A conspicuous, noisy, aggressive and curious inhabitant of any habitat with trees or shrubs, including in towns.

Mohéli Bulbul *Hypsipetes moheliensis* 27 cm
Restricted to uplands of Mohéli; similar Madagascar Bulbul occurs in lowlands. Birds deep in forest on island's main ridge most likely to be Mohéli Bulbul, and those in lowlands Madagascar Bulbul, with overlap mainly at 400–550 m, occasionally 200–700 m. Distinguished by *upperparts washed olivaceous-green, and underparts washed yellow*. **Voice** Varied vocalisations typical of *Hypsipetes* bulbuls, but rather deep, including a wooden, rattling, monotonous *cuk-tuk-cuk-cuk-tuk-cuk-cuk...* and variants. **SHH** Endemic to Mohéli, where restricted to humid forest, clearings and degraded areas; areas of overlap with Madagascar Bulbul often mosaics of degraded forest and regrowth. Habits as Madagascar Bulbul. Not currently recognised by BirdLife.

Grande Comore Bulbul *Hypsipetes parvirostris* 26 cm
Restricted to uplands on Grande Comore; similar Madagascar Bulbul occurs in lowlands. Birds deep in forest on Mt Karthala most likely to be Grande Comore Bulbul, and those in lowlands to be Madagascar Bulbul. Grande Comore Bulbul has *upperparts washed olivaceous-green, and extensive white on abdomen with a yellow wash*, becoming almost plain white on undertail-coverts; Madagascar Bulbul has dark grey upperparts without green wash, and pale grey underparts usually with little white and never a yellow wash. **Voice** Varied vocalisations typical of *Hypsipetes* bulbuls; 'cat-call' deeper, less whining yet more strident and wheezing than Madagascar Bulbul, and clucking notes lower-pitched and more muffled. **SHH** Endemic to Grande Comore, in humid forest (rarely tree-heath), clearings and adjacent degraded areas from 500 m to upper limit of forest. Sympatric with Madagascar Bulbul in mosaics of forest, degraded areas and plantations: on Mt Karthala at 500–750 m (exceptionally 350–1,000 m) and on La Grille (in N) at 800–900 m. Habits as Madagascar Bulbul.

Comoro Thrush *Turdus bewsheri* 23 cm
Only true thrush (genus *Turdus*) in the region. *Shorter-tailed than Hypsipetes bulbuls, which are never terrestrial*; brownish upperparts and (less strident) voice of thrush also very different. **Voice** Variable; on all islands, a high-pitched call, *tssiet, siiw* or *plik*, sometimes ending in a trill. Other notes include a sharp, popping *ka*, or *kakak*, and a sharp *krrk* or *brrp*, often heard from many individuals at dusk (Anjouan). Typical thrush song: prolonged, often beautiful, rich and varied warbling, with structure varying between islands. **SHH** Endemic to Comoros, restricted to Grande Comore (Mt Karthala and La Grille), Mohéli and Anjouan, mostly above 400 m, but occasionally to sea level on Mohéli. Inhabits humid forest, including underplanted areas, and dense tree-heath. Seen singly or in groups of up to four. Forages on or near ground but perches in trees. Strong inter-island variation in plumage and voice; may be 2–3 species, with Anjouan population most distinct.

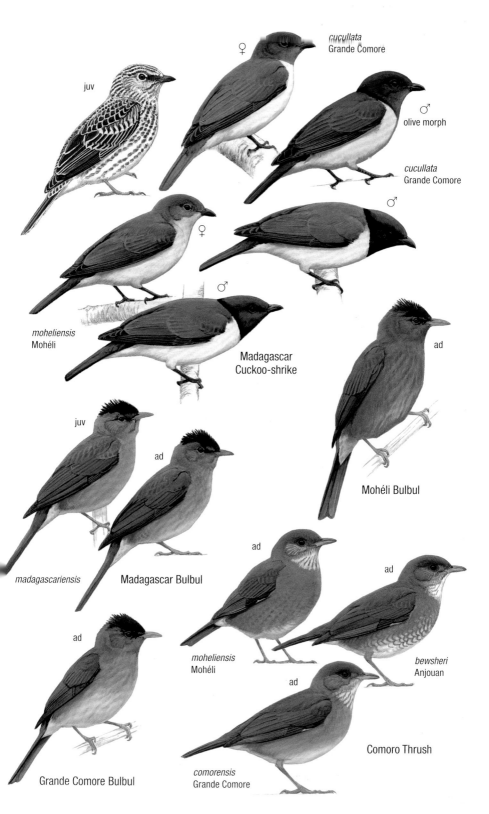

PLATE 76: COMOROS: LANDBIRDS VII

African Stonechat *Saxicola torquatus* 13 cm
Persistent prominent perching and calls unique on Comoros. Juv confusable with juv Anjouan Thrush, which is smaller, thicker-billed, found in forest interior and does not perch prominently. **Voice** A short *hii hii*. Song a long, varied jumble of repeated *titititititit* phrases, grating *krrr* notes and short, high-pitched warbles. **SHH** Africa and Malagasy region, where also in Madagascar. Race *voeltzkowi* (♂ with less white on rump and wing than *sibilla* of Madagascar; often has white mark over eye) endemic to Grande Comore, in open country above 500 m including clearings, tree-heath and crater of Mt Karthala. Perches conspicuously with upright stance, flicking wings and tail.

Humblot's Flycatcher *Humblotia flavirostris* 14 cm
Reminiscent of African and Eurasian flycatchers but no other such species is resident in Malagasy region (see vagrant Spotted Flycatcher, Plate 119). **Voice** Distinctive: *high-pitched, slightly rasping, 'shivering' or trilled siiiiii, or tseeeeuuu*; also a loud, metallic rattle *pttttttttr*, falling at end. No song described. **SHH** Endemic to Grande Comore, where restricted to forest and, marginally, tree-heath of Mt Karthala at *c*.800–1,850 m. Mainly frequents forest interior, midstorey or below, perching motionless apart from wing-flicks and sallies for prey; may emerge into adjacent open areas at dusk. The only genus endemic to the Comoros, although probably closely related to African *Muscicapa* flycatchers, which have not otherwise colonised the Malagasy region. **EN**

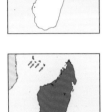

Madagascar Paradise Flycatcher *Terpsiphone mutata* 17 cm + 15 cm tail-streamers (♂)
Despite complex geographical and age-related plumage variation, *coloration distinctive at all ages and in all plumages*. **Voice** Variable, *harsh scolding calls*, e.g. *kecharcharcharchar kitchiwi*, often higher-pitched at end. Song a series of melodious bubbling whistled notes combined with chattering and scolding calls. **SHH** Endemic to Malagasy region, where also in Madagascar. Separate endemic races on each island of Comoros: *voeltzkowiana* (Mohéli) and *pretiosa* (Mayotte) widespread at all altitudes, *vulpina* (Anjouan) mainly above 400 m but locally to sea level, and *comorensis* (Grande Comore) in forests of Mt Karthala and La Grille up to 1,900 m. An active, restless bird of the midstorey. Could be treated as three species, of which two endemic to Comoros: Mayotte race similar to Madagascar birds; Mohéli and Anjouan birds similar to each other (short tail, ♀ with white in wing); and Grande Comore race unique with velvet, not glossy, crown.

Madagascar Brush Warbler *Nesillas typica* 18 cm
Only plain brown insectivorous passerine on Anjouan, but on Mohéli rather similar to Mohéli Brush Warbler, which see. **Voice** Calls as in Madagascar but may be higher-pitched: call *medium-low trkkk or chek or chekek*, repeated slowly or more rapidly as repeated rhythmic *tk-tk-teketek* or *teketeketek*, or other rhythms. Also long rattle on one note. Comoros populations similar to each other, although Anjouan birds higher-pitched. **SHH** Endemic to Malagasy region, where also in Madagascar. Race *moheliensis on Mohéli in forest undergrowth* on main ridge above 400 m; *longicaudata* throughout Anjouan in any habitat with dense undergrowth. Forages nimbly in low vegetation, sometimes descending to ground but usually not up trees. Anjouan birds relatively distinct from other races, and may be a separate species; on Mohéli, very like nominate (Madagascar).

Grande Comore Brush Warbler *Nesillas brevicaudata* 16 cm
Only warbler on Grande Comore. **Voice** A short, harsh, medium-pitched *brek... chip... trkk... chirk... whek*, repeated in rattling sequence, often dropping at end; also a low, nasal *peut*. Song a series of varied, harsh and more musical notes, *whek whek chipuril chep whek chipurilychip*, or variants. **SHH** Endemic to Grande Comore, on Mt Karthala and La Grille above 500 m in tall forest, tree-heath, clearings, and thickets including exotic Strawberry Guava. Forages from undergrowth to high in trees, typically in shrub layer.

Mohéli Brush Warbler *Nesillas mariae* 15 cm
Occurs alongside Madagascar Brush Warbler but *smaller, shorter-tailed, paler, greener (less brown), without an obvious supercilium* (well marked in *moheliensis*), recalling Grande Comore species; reddish eye-ring unique but inconspicuous in field. Song distinct, with complex phrases. **Voice** Calls include irregular series of harsh *chik... chip... trkk... chirk... chark...* notes, higher-pitched and more uniform in rhythm than congeners. Song a sequence of harsh and high-pitched notes in an irregular sequence: *chip trrk tsssk tsip tsirp tut tut tsirp chip trrrk*, notably *prolonged and well structured* compared to congeners. **SHH** Endemic to Mohéli, from forested ridge down to 150 m in secondary and exotic habitats; forages *in bushes, shrubs and lower parts of trees, usually at 7–8 m*, thus typically well above levels used by Madagascar Brush Warbler (1–3 m).

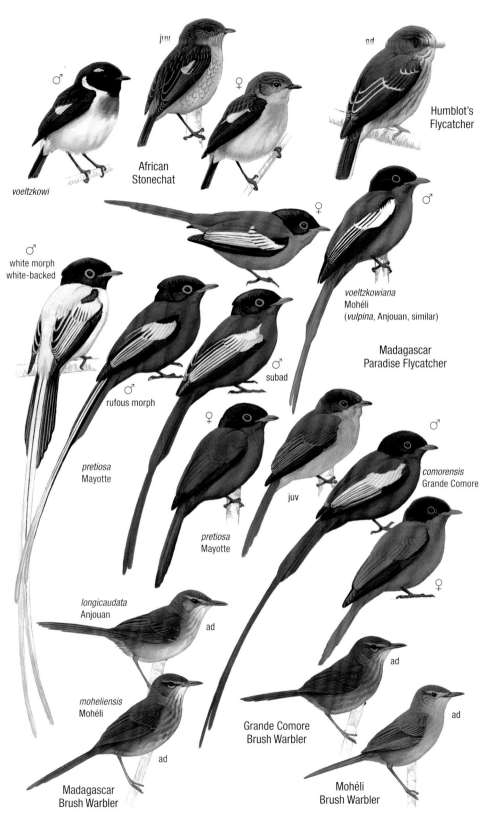

PLATE 77: COMOROS: LANDBIRDS VIII

Madagascar Green Sunbird *Nectarinia notata* — 15 cm
As well as by plumage, differs from Humblot's Sunbird by *much larger size, longer bill and deeper voice*. No non-breeding plumage known in ♂ but subad can have pale belly. **Voice** A medium-pitched *tew tup-tup* or *cher-chip-chip* and chacking *chuchuchuchuchu*. Song a prolonged warble with complex and fluid segments interspersed with halting phrases, and varied pitch and tone, reminiscent of a paradise flycatcher. **SHH** Endemic to Malagasy region, where also in Madagascar. Races *moebii* and *voeltzkowii* endemic to Grande Comore and Mohéli, respectively, in all habitats with trees or shrubs on both islands. Typical large sunbird, with strong flight. Comoros races are each other's closest relatives and could be treated as an (or two) endemic species.

Humblot's Sunbird *Nectarinia humbloti* — 11 cm
Only small sunbird in range; for distinctions from the other sunbird on these two islands, see Madagascar Green Sunbird. No record of non-breeding plumage in ♂. **Voice** High-pitched, peevish *cwee* or *creee*, metallic and slightly descending, repeated in long sequences; also a lower-pitched short *clik*. Song a high-pitched, rapid, jumbled sequence of metallic notes. **SHH** Endemic to Comoros, with races *humbloti* and *mohelica* on Grande Comore and Mohéli, respectively; abundant throughout both islands in all habitats with trees or shrubs. A typical small sunbird, often in mixed-species flocks; takes insects by gleaning and hover-gleaning, and feeds on nectar of many plant species.

Mayotte Sunbird *Nectarinia coquerellii* — 10 cm
Only sunbird on Mayotte; ♂ *distinctive and beautifully marked*, with no record of non-breeding plumage. **Voice** A peevish, mewing *cheer*, given in isolation, repeated or interspersed within song phrases; also a nasal *chit*, often rapidly repeated. Song frequent, a jumble of fluted rapid notes, *tsivitisriticherrrrtsitterr*, sometimes prefaced by soft *pi* notes. **SHH** Endemic to Mayotte, where found throughout main island and all major satellites. Common but not abundant in all habitats with any woody vegetation. Typical small sunbird.

Anjouan Sunbird *Nectarinia comorensis* — 10 cm
Only sunbird on Anjouan. No non-breeding plumage recorded in ♂. **Voice** *A short, high-pitched chip*, very frequently uttered, sometimes repeated in rapid sequence; also gives long, rapid-fire sequences of peevish *chew* notes. Song a jumble of rapid, high-pitched squeaks and chips. **SHH** Endemic to Anjouan, where abundant at all altitudes and in all habitats with trees or shrubs. Typical small sunbird.

Karthala White-eye *Zosterops mouroniensis* — 13 cm
Dominant white-eye in its tiny range, but overlaps with similar Madagascar White-eye of yellow-bellied race *kirki* around heath–forest ecotone. Karthala White-eye is *larger (20% longer-winged and -tailed), darker yellow above and duller yellow below, with a narrower white eye-ring and black (not grey) legs*. **Voice** Calls typical of a white-eye but rather slow and low-pitched, hoarser and more powerful than Madagascar White-eye, with occasional *sharp, louder and harsh calls*: typical calls *peer*, a loud trilling chatter and harsh churrs, sometimes combined *pew pweerwhititup*. Song reported to be a warbling *fi-fi-fi-pi-pi-pyo-pyo*. **SHH** Endemic to Grande Comore, where restricted to summit of Mt Karthala, using all forms of heath vegetation, typically above 1,700 m but also locally down to c.1,500 m. Overlaps with Madagascar White-eye at least across 1,500–1,900 m. **VU**

Madagascar White-eye *Zosterops maderaspatanus* — 11 cm
Only white-eye on the Comoros except high on Mt Karthala, Grande Comore, where Karthala White-eye occurs; see that species for separation. Readily identified by *white eye-ring and mainly green-and-yellow plumage*. **Voice** Flocks emit a medley of nasal *tew, ter* or *ter-tew* notes, or a distinctive, querulous, rising *wrrrri* or *trrrrri*. Song a medium-pitched warble, running slightly up and down scale, e.g. *pita weer weer whit weer weer witwit weer*, recalling a canary. Some inter-island variation in pitch and tone ('hoarseness'). **SHH** Endemic to Malagasy region, where also on Madagascar, coralline Seychelles, Glorieuses, Europa. Races *kirki, comorensis, anjuanensis* and *mayottensis* endemic to Grande Comore, Mohéli, Anjouan and Mayotte, respectively, all abundant throughout their islands, including numerous islets around Mayotte; absent only above c.1,900 m on Mt Karthala, Grande Comore, where replaced by Karthala White-eye, with overlap at 1,500–1,900 m. Underparts colour has led to suggested splitting of (only) Grande Comore and Mayotte populations as species, but this is refuted by DNA analysis indicating Anjouan birds to be the best candidate; all four taxa (and Aldabran birds) could be treated as species, or all as races of Madagascar White-eye.

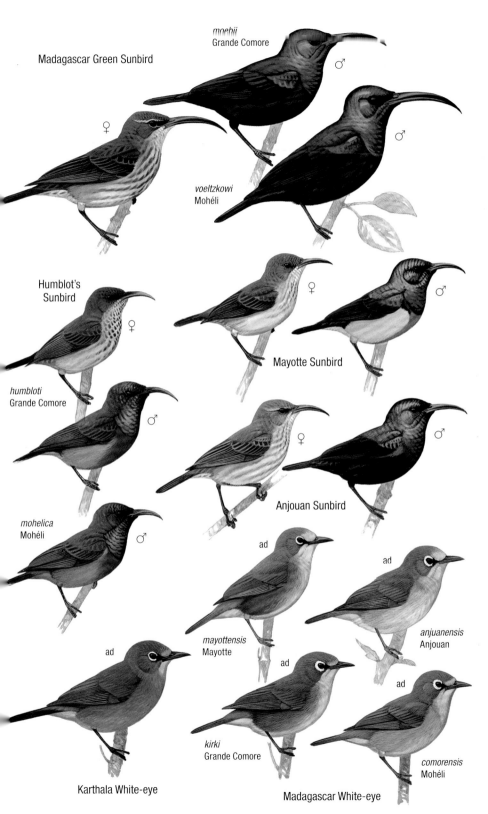

PLATE 78: COMOROS: LANDBIRDS IX

Blue Vanga *Cyanolanius madagascarinus*   19 cm
Only vanga on Comoros, easily recognised by active, acrobatic foraging behaviour, simple two-toned pattern, and ♂'s *spectacular blue upperparts*. **Voice** Limited vocabulary: harsh *craak crkkcrkk*; apparent song a grating *erch-chhh-crkkk-chh-chhh-chhh-crrk* or a slightly less grating *crew-crew-crew*. **SHH** Endemic to Malagasy region, where also in Madagascar. Race *comorensis* common on Mohéli above 300 m; race *bensoni* on Grande Comore extremely rare, restricted to mid-elevations on Mt Karthala. Confined to evergreen forest, characteristically foraging among outer branches and foliage. Comoros birds large and fairly distinct in plumage (especially ♀); perhaps a separate species.

Grande Comore Drongo *Dicrurus fuscipennis*   28 cm
Only drongo, or large black passerine, on Grande Comore. **Voice** Varied selection of clucking and whistling notes. Song a medley of musical, medium-pitched, bubbling whistles, squawks and chirps, often delivered as a cacophonous duet; also a shorter, very rapid babbling sequence. **SHH** Endemic to Grande Comore: restricted to Mt Karthala, where rare and known from few localities at 100–1,120 m, mostly at 500–900 m. Distribution apparently fragmented and not limited to primary forest: uses degraded forest with broken canopy, clearings in intact forest, cultivated areas with scattered trees and plantations including coconut and cacao. Behaviour typical of a drongo. **EN**

Mayotte Drongo *Dicrurus waldenii*   37 cm
Only drongo, or large black passerine, on Mayotte, with *strikingly long, deeply forked tail*. **Voice** Extensive repertoire of buzzy and scratchy calls, including a loud *tyok* repeated several times, *squa-aa-kuchuk*; quiet *quit* calls given when sallying for insects. Song a harsh repetitive medley of chirps and squeaks, with varied whistles, cackles, squawks and nasal grinding notes; also a single, melodious note. Pairs duet. **SHH** Endemic to Mayotte, where restricted to Grande Terre (main island); common in mountains with large expanses of semi-natural humid forests, with smaller numbers around or between these strongholds and in mangroves, but absent from S peninsula. Behaviour typical of a drongo. **VU**

Crested Drongo *Dicrurus forficatus*   29 cm
Only drongo, or large black passerine, on Anjouan. **Voice** Song varied, a series of separated clicks, whistles, clucks, often 2–3 notes repeated before next sequence, and whistled sequence alternating with clucks, often with a very clear rhythm; new elements introduced as song progresses; often delivered as duet. Calls often components of song, with mimicry; also a slightly hoarse *whep* or *wherp* alternating with single screeches or croaks. **SHH** Endemic to Malagasy region, where also in Madagascar. Large race *potior* endemic to Anjouan, where uncommon and poorly known, with records widely scattered in forest, woodland, cultivated areas with large trees and coconut plantations, especially above 700 m but occasionally to sea level. Behaviour typical of a drongo.

Pied Crow *Corvus albus*   45 cm
Only crow on the Comoros. Often soars, when confusable with raptor, but distinguished by *black-and-white plumage* and *typical corvid silhouette* with relatively short wings and stout bill. **Voice** Flat, loud, low-pitched *caah, caw* or *craaah*; also a low *cuh... cuh... cuh* from flocks and a range of other bubbling or gurgling noises. **SHH** Sub-Saharan Africa and Malagasy region, including lowlands of all four Comoro islands, and islets of Mohéli and Mayotte; usually absent above 700 m but recorded at 1,800 m on Mt Karthala. Inhabits almost all non-forest habitats; often commensal in town centres, open ground near villages and rubbish dumps, but also degraded forest and mosaics of forest and open country. Flocks of up to 25 occur on Comoros. Observers on Comoros should look out for House Crow (Plate 83), a potential and unwelcome colonist.

Common Myna *Acridotheres tristis*   25 cm
Distinctive, large passerine: several unique features, including *brown plumage with white wing-patches and tail tip (prominent in flight); yellow legs, bill and skin around eye*. **Voice** Vocal and noisy. Fluty medium- to low-pitched *tuuu, chyuu* or *tweh* and similar, often repeated, sometimes with squawking quality, but also an amazing range of low- to high-pitched squealing, clicking, croaking, gurgling and whistling notes, often strung together in disjointed sequence as song. Mimics other birds, occasionally causing confusion. **SHH** Introduced (S Asia) throughout all four Comoro islands, up to *c*.1,100 m, including islets off Mayotte. Also Madagascar, Mauritius, Rodrigues, Reunion, Agalega. Found in virtually all habitats including towns and villages, primary forest, agricultural land, grassland, mangroves and open coasts. Strident voice and bold, commensal habits make this a familiar species.

PLATE 79: COMOROS: LANDBIRDS X

House Sparrow *Passer domesticus*   16 cm
♂ *distinctive but* ♀ *resembles Comoro and Madagascar Fodies*; see those species for distinctions. **Voice** *A harsh chissk, short weesk or abrupt, lower-pitched chup, chup-up, or short, coarse trill, chukukukukuk*. Song a series of *chup* or *chip* notes; chorus of *chissk* notes given by many birds together, e.g. at roosts. **SHH** Introduced (Eurasia); typically commensal and range may fluctuate: on Grande Comore and Mohéli, mainly in coastal towns and villages and a few inland sites; on Mayotte, restricted to Petite Terre, NE coastal Grande Terre and Mbouzi islet. Absent from Anjouan. Also Madagascar, Seychelles, Mauritius, Reunion, Rodrigues. Highly social and gregarious. Flight rapid and direct, on whirring wings.

Comoro Fody *Foudia eminentissima*   14 cm
Differs from Madagascar Fody by *larger and longer bill, more striking pale tips to wing-coverts usually producing clearer wingbars, and larger size*. Voices also differ. ♂ moults into ♀-like non-breeding plumage, *c*.Apr–Aug. Breeding-plumaged ♂ may be distinguishable by distribution and shade of red areas, but other features should confirm, because of variability in Madagascar Fodies. Differences dependent on island, as follows, each population of Comoro Fody compared to Madagascar Fody: Grande Comore (*consobrina*) ♂ deeper crimson on head, usually lacks red on rump, and has black mask extending to point behind eye; Mohéli birds (*eminentissima*) more heavily streaked on upperparts, with especially broad wingbars, more olive belly and very heavy bill; Anjouan birds (*anjuanensis*) with more olive belly and very heavy bill; Mayotte (*algondae*) ♂ with more orange head, and belly more olive with yellowish centre. Non-red birds recall ♀ House Sparrow, but *more olive, with longer bill*. **Voice** On Grande Comore, a high-pitched, buzzing *tsieeew* or *skeeez*, harsh *cha-cha-cha-...* and a soft trill. Territorial song a string of *jif* or *chwi* notes strung together in rapid succession. Elsewhere, voice poorly known: on Mohéli and Mayotte a flat, medium-pitched *chiup* or *chep*, sometimes extended to a trill; on Anjouan call *tiry-ry*, and song a far-carrying *ch-r-r-r-r-r-r-r, pya, pya, pya, pya, pya*; also on Mayotte a rolling *tsiiuu*, sometimes combined with song phrase, *twiderit*, all repeated. **SHH** Endemic to Comoros. Found throughout three W islands, especially in humid uplands but also other wooded habitats; on Mayotte, locally common in non-forested, dry lowlands in five separate areas of Grande Terre, especially on S & E coasts, and Mbouzi islet. Commonest on Grande Comore. Arboreal, feeding on invertebrates and fruit in canopy and on branches. Not closely related to Aldabra or Forest Fodies as formerly suggested; some or all four races could be treated as separate species.

Madagascar Fody *Foudia madagascariensis*   13 cm
Care needed to distinguish from Comoro Fody, which see. A *fody with entirely red underparts must be a Madagascar Fody*, but ♂ *may show patchy red* very like ♂ Comoro Fody. ♂ moults into ♀-like non-breeding plumage (Jun–Oct). Gregarious habits and preference for non-forested country, where it feeds mainly on seeds, often on ground, also suggest this species. House Sparrow larger, with more grey-brown plumage and no trace of greenish or buffish-yellow. Sibilant *tsip* call not given by any confusion species. **Voice** A very *high-pitched, slightly chirping tsip or tsikip*; also a lower *chup* or *chutut*. Breeding ♂ has trilled call, or buzzing *zzz-zzz-zzz*. Song lacks regular form: wheezes, chirrups, squeaks and *tsip* calls, with up to *c*.1 s gap between phrases: *tsitsitsitstit tiptiptip tititititit chipchipchip*. **SHH** Endemic to Malagasy region; introduced to Comoros, as to Seychelles, Mauritius, Reunion, Rodrigues, Glorieuses, Juan de Nova, Agalega; native to Madagascar. On Comoros, common almost everywhere except in closed forest: on open ground with grass and herbaceous growth, cultivation, towns and gardens; on Mt Karthala, uses clearings and heath zone. Highly gregarious seed-eater, flocks sometimes in hundreds; also takes insects.

Bronze Mannikin *Spermestes cucullatus*   9 cm
Only tiny finch-like bird on Comoros: other thick-billed small passerines (fodies and House Sparrow) much larger, streaked on upperparts, and lack pied head and underparts pattern of Bronze Mannikin. **Voice** Shrill, buzzy *dzeep, peejee*, or rolled *drrreep*; also a sparrow-like chirping or dry chatter, sometimes disyllabic. In flight, a repeated *tsrii*, and a sharp *tsek* in alarm. Song a rather deep, slow and measured chuckling or sequence of *c*.8 *chi* and *chu* notes. **SHH** Africa and Comoros, where found from coasts up to 1,100 m on all four islands, although possibly more coastal on Mayotte (Grande Terre alone). Inhabits open country, cultivation, plantations, thickets, roadsides, clearings, herbaceous marshes and gardens; never in evergreen forest. Usually in single-species flocks of 8–20, sometimes up to 50, hopping briskly on ground or making short, whirring flights.

PLATE 80: MAURITIUS AND RODRIGUES: LANDBIRDS I

Feral Pigeon *Columba livia*   33 cm
The typical 'town pigeon'; tight flocks of pigeons in urban areas or flying high likely to be this species. Plumage very variable. Pink Pigeon and Madagascar Turtle Dove have wings shorter and/or broader; Spotted and Laughing Doves smaller and less gregarious. **Voice** A long, moaning *o-o-orr*, repeated monotonously, or a hurried, rambling *coo-roo-ooo-ooo*. **SHH** Native to Eurasia and N Africa, but introduced worldwide including Mauritius and Rodrigues. Locally abundant, especially in markets, parks and any sites providing food scraps and nesting ledges; also cliff habitats in wilder areas. Flight powerful and fast, often in tight flocks, on long, pointed wings.

Pink Pigeon *Nesoenas mayeri*   40 cm
Largest pigeon on Mauritius; plumage distinctive, with *pale pink or whitish head, neck and breast, and broad russet tail*; also red bill and legs. Madagascar Turtle Dove is smaller and darker, with different tail pattern, shorter tail, and faster, more direct flight on narrower, more pointed wings. **Voice** A low monotone *coo-coo-cooooo-oooo* or repeated *coo-woo-wah*, often from a prominent perch; rather similar to Madagascar Turtle Dove but tends to have downward, not upward, inflection at end of longest *coo*. A deep grunting *wah-wah* in flight or on landing. **SHH** Endemic to Mauritius; wild population in 1991 *c.*10 birds but now occurs throughout much of Black River Gorges NP in native evergreen forest, adjacent plantations and secondary forest or open areas. Introduced to Ile aux Aigrettes. Feeds mainly in trees but often visits ground; undertakes long flights above canopy. **EN**

Madagascar Turtle Dove *Nesoenas picturata*   28 cm
Only sturdy, short-tailed, dark dove on Mauritius. Grey head, dark reddish-brown upperparts, and tail with only *slightly paler (not white) corners* distinguishes it from all other species. Pink Pigeon is larger, with broad rufous tail. Spotted and Laughing Doves are smaller and slimmer, with white tail-corners and three-toned upperwing in flight. Spotted has spotted neck-patch and much more variegated, duller brown upperparts; Laughing has grey wing-patch, and pinkish head and neck. **Voice** Low-pitched, double *coo*: *coo cooooo-uh*, the second *coo* louder, longer and rising at end; first often not noticed. **SHH** Endemic to Malagasy region. Nominate race introduced from Madagascar to Mauritius. Common in wooded habitats and nearby open areas; feeds mainly on ground.

Spotted Dove *Spilopelia chinensis*   29 cm
Buff-fringed brown upperparts and prominent white-spotted black hind-collar diagnostic. *In flight, shows white tail-corners and three-toned wings* (dark flight-feathers, largely brown coverts with broad grey area across greater coverts), and 'wriggling' wingbeats. Juv lacks spotted hind-collar. Laughing Dove most similar, but has plainer, much more rufous upperparts. Madagascar Turtle Dove larger, stockier, with reddish-brown upperparts, grey head and pale (but not white) tail-corners in flight. Zebra Dove much smaller and prominently barred. **Voice** A distinctive *2–4-note cooing*, with emphasis, and often slight trill, on second or (usually) third syllable, *coo-coo-CRROOO* or *coo-CRROOO-CROOO*. **SHH** SE Asian race *tigrina* introduced to Mauritius; common in most habitats with woody vegetation, including villages, edges of cultivated areas, forest and plantations.

Laughing Dove *Spilopelia senegalensis*   27 cm
Smaller than Spotted and Madagascar Turtle Doves, larger than Zebra Dove. Pinkish head and *rufous upperparts with blue-grey wing-patch* diagnostic; *in flight shows blackish primaries and contrasting white tail-corners*. Neck-patch inconspicuous, composed of black mottling. **Voice** Distinctive, with laughing quality: a soft, rather musical, bubbling *cu-cu, cu-oo* or *curucu-cu-cucoo*, usually 4–6 notes. **SHH** Introduced (Africa and Asia) to Mauritius; recently established and probably still expanding from gardens and woody cover in SW and may ultimately occupy any human settlements, woodland and savanna, but in native range not a forest bird.

Zebra Dove *Geopelia striata*   21 cm
Very distinctive, *tiny, greyish-brown dove, with extensive black barring, otherwise pink or grey plumage* and blue skin round eye. In flight, *long, wedge-shaped tail with narrow white tips; tapering shape* creates a longer white fringe compared to white corners of Spotted and Laughing Doves. Bounding flight on take-off distinctive. **Voice** A *medium- to high-pitched, bubbling, quick cooruruwuh* or *wurruwawa*, first note cooing and last two sharper, final note sometimes repeated; in display, a harsher *caaw-caaw-caaw*. **SHH** Introduced (SE Asia) to Mauritius and Rodrigues, also Seychelles, Reunion, Glorieuses, Juan de Nova, Agalega. Common in any habitat with scattered woody vegetation, including urban areas. [Alt: Barred Ground Dove]

PLATE 81: MAURITIUS AND RODRIGUES: LANDBIRDS II

Rose-ringed Parakeet *Psittacula krameri*   41cm (♂), 37 cm (♀)
Much like Echo Parakeet but more *lightly built and a paler shade of green*, with longer tail, longer, thinner wings and faster, more direct flight. Other parrot species often escape from captivity but are not currently established. **Voice** *Shrieking calls* distinctive. Most frequent a raucous, far-carrying *kee-ak kee-ak*; many other calls, varying in pitch and tempo but most with similar strident tone. Especially vocal at roost and in flight. Perched birds also utter various whistling and subdued chattering or twittering notes. **SHH** Introduced (Asia) to Mauritius; also Seychelles and probably Reunion. Mostly secondary forest and savanna of dry coastal lowlands but can occur anywhere including humid upland forest preferred by Echo Parakeet. Gregarious, flocks up to 30 and many more at roosts. [Alt: Ring-necked Parakeet]

Echo Parakeet *Psittacula eques*   39 cm
Easily confused with Rose-ringed Parakeet, but is *stockier and darker green*; imm and all *ad ♀ Echo separable by black bill* (very youngest have orange bills). Echo has slower flight on *shorter and much broader wings*; tail also broader and shorter; consequently Echo is more manoeuvrable in forest canopy and can turn in much tighter circles. Calls very distinctive. **Voice** A *low-pitched, nasal chaa-chaa, chaa-chaa*, c.2 notes/s, deeper and hoarser than Rose-ringed, although becomes higher-pitched when alarmed or excited; usually given in flight and accompanied by rapid, shallow wingbeats with wings below horizontal. Various quiet whistles and chatters from perched birds, also a drawn-out creaking *yowl* and short, sharp *ark!* **SHH** Endemic to Mauritius, formerly on brink of extinction: wild population in 1985 *c.*20 birds but now widespread in Black River Gorges National Park. Restricted to native forest, mainly in uplands but occasionally down to 150 m. Normally alone or in pairs or small flocks; does not form large communal roosts. **EN** [Alt: Mauritius Parakeet]

Mauritius Cuckoo-shrike *Coracina typica*   22 cm
No other comparable-sized passerine on Mauritius has similar *pale grey, black and white* (♂) or *rufous* (♀ and subad) plumage. Juv strikingly different (pinkish-brown, scaled and streaked) but still distinct from other Mauritian species. **Voice** Often found by ♂ *song, a series (c.12 notes) of short whistles*, usually increasing in volume and frequency then levelling off; reminiscent of Whimbrel. Call a loud rasp usually followed by a loud *kek*. **SHH** Endemic to Mauritius, restricted to native forest (not scrub, but occasionally in nearby exotics) in SW, from uplands down to 250 m, perhaps lower. Strictly arboreal. **VU**

Red-whiskered Bulbul *Pycnonotus jocosus*   20 cm
Much *smaller than Mauritius Bulbul*. Plumage unique, with *crest, contrasting face pattern with white cheek-patch*, black patches on breast-sides, white underparts, red undertail-coverts and white tip to tail. 'Whiskers' are glossy, hair-like red feathers behind eyes. Juv lacks red. **Voice** Slow, fluty or whistling song with rising and falling notes: *churee-chuwee-churiwiwee* or *plu chi chireree* or *pli chu tiriwee*; repeated 10–20 times, every 1–5 s, on varying pitch. Contact notes of similar quality, *pettigrew*; quick to signal alarm, with drawn-out *lerrr* or shorter *chip*. **SHH** Introduced (Asia); commonest bird on Mauritius (absent Rodrigues), where found throughout in woody vegetation, even if low and patchy, and on sparsely vegetated islets. Also Reunion, Juan de Nova, formerly coralline Seychelles and Comoros.

Mauritius Bulbul *Hypsipetes olivaceus*   28 cm
Large size, long, square-ended tail, and restless, aggressive behaviour with frequent and often strident vocalisations gives distinctive character. *Entirely dark greenish-grey plumage with pinkish-orange bill* of ad unique on Mauritius. Juv duller grey-brown with brown bill, but still distinctive locally. Mauritius Cuckoo-shrike is grey or rufous, and Common Myna dark brown with white flashes in wing and tail; both have shorter, more rounded tails. Red-whiskered Bulbul much smaller and distinctively plumaged. **Voice** Very noisy. ♂ song or ♀ response based on strident, repeated (with variation) *chiriyoririk...*, often several calling together; also shorter, *chirk-chilk-chip-whip-chilk....* 'Cat' call, typical of genus, a loud and conspicuous, hoarse mewing *weeeer* or *cheeeeer*, lasting *c.*1 s. **SHH** Endemic to Mauritius, in native and neighbouring exotic forest but not softwood plantations; mainly in uplands down to 200 m, rarely lower; takes much fruit as well as animal prey, occasionally wandering to visit favoured foodplants outside usual range.

PLATE 82: MAURITIUS AND RODRIGUES: LANDBIRDS III

Mascarene Swiftlet *Aerodramus francicus*   11 cm
Only swift on Mauritius, but see Mascarene Martin. **Voice** *Usually silent except at colonies* where gives a quiet twitter or very weak 'scream', or remarkable echolocating sound, used to orientate in darkness of colonies: rapid, double click, sometimes a rattle. Calls and clicks occasionally in the open, perhaps especially in mist. **SHH** Endemic to Mauritius and Reunion. Singly or in small groups. Nests colonially in caves, in Mauritius usually lava tubes, often in entirely exotic habitats including sugarcane fields; congregates around cave entrances but hawks over all kinds of terrain throughout mainland. **NT**

Mascarene Martin *Phedina borbonica*   15 cm
Only hirundine on Mauritius; larger than Mascarene Swiftlet, with *shorter, broader wings, earth-brown rather than blackish-brown upperparts* and lacks paler rump-patch. **Voice** *Bubbling or buzzing crreeeww or gzeee-gzeee or phree-zz*, also a shorter *chep*. Song, in flight or perched, a complex warble. Tone may differ between Madagascar and Mauritius/Reunion birds, but requires more investigation. **SHH** Endemic to Malagasy region. Nominate race resident on Mauritius and Reunion; paler, more clearly streaked *madagascariensis* on Madagascar. Fairly common and widespread on Mauritius. Typical hirundine, pursuing insects in flight singly or in small groups over almost all habitats. Often perches on wires. Flight fast and powerful. [Alt: Mascarene Swallow]

Mascarene Paradise Flycatcher *Terpsiphone bourbonnensis*   16 cm
Small size and rufous upperparts and tail at all ages eliminate all other species. In Mauritian race *desolata*, ♂ has upperparts slightly more rufous, underparts paler and more gloss on hindneck than in nominate race (Reunion). **Voice** *Distinctive whistling or grating*, often first means of detection in dense vegetation. Long, noisy, medium-harsh buzz or *chrr*, often extended to harsh scolding. Song a high fluid warble *ker-chirkirerchire-ker-chirkirerikire* or *ker-chikirikiri*. May combine buzz and song notes. **SHH** Endemic to Mauritius and Reunion. Rare and very local on Mauritius, in native and exotic forest with deep shade and open understorey. Small numbers in forests of SW (Black River Gorges National Park) and adjacent C plateau, and important population in exotic plantations of Plaine des Roches on E coast. Catches insects in flight in mid-air or more often from vegetation surfaces, launching from a perch. Races rather similar, but DNA study indicates long isolation from each other.

Rodrigues Warbler *Acrocephalus rodericanus*   14 cm
Only warbler or thin-billed insectivore on Rodrigues. **Voice** Song a *short, melodious sequence of whistled and fluted notes*, 10–15 elements and segments repeated after 3–5 s. Calls sharp and sparrow-like, *chipipi, chitik* or a low, quiet, squeaky *crio-crio-crio...*, and a harsh *chhchchchch* and *chhhtrrrr*. **SHH** Endemic to Rodrigues; formerly on brink of extinction: wild population in 1960s probably <10 birds, but now in most hills and valleys in C uplands with population estimated in thousands, due to recovery of exotic and native forest cover. Inhabits dry thickets, woodland and forest, now mainly composed of exotic plant species, especially Rose-apple *Syzygium jambos*. **EN**

Mauritius Olive White-eye *Zosterops chloronothos*   10 cm
Only Mauritian bird with conspicuous white eye-ring; olive wings, calls and lack of white rump diagnostic. **Voice** Characteristic *call an explosive pit pit pit*, often repeated in a trill or 'apoplectic' splutter; also a ringing *ch-rrreee* or abrupt *tche*. Song rarely heard: short, hurried, warbled phrases interspersed with *pit* call; less tuneful than Mauritius Grey White-eye. **SHH** Endemic to Mauritius. Rare, restricted to native upland forest, scrub and adjacent nectariferous exotics, especially Rose-apple *Syzygium jambos* and Bottlebrush *Callistemon citrinus*, of SW uplands and adjacent C plateau. Commonest on S-facing slopes from Alexandra Falls to Combo; scarce in closed-canopy forest and absent from relatively intact Macchabé–Brise Fer forest. Introduced to Ile aux Aigrettes. A restless nectar- and insect-feeder; most common view is in fast flight or perching briefly. **CR**

Mauritius Grey White-eye *Zosterops mauritianus*   11 cm
Only small, grey passerine on Mauritius: differs from much rarer Mauritius Olive White-eye by grey wings and back, lack of white eye-ring, short, straight bill, *white rump* (displayed by habit of keeping *tail cocked and wings drooped*) and calls. **Voice** Call a downward-inflected, *chirping or slightly warbled p-tree*, sometimes strung together, running up and down scale over 10–30 s. Also gives *chip* calls or piercing, aggressive squeaks. Sweet warbling song often contains mimicry and is preceded by calls. **SHH** Endemic to Mauritius, where *by far the commonest native bird*, found in any habitat with trees or shrubs, including many non-native habitats where other native birds absent. Noisy and sociable, found in groups, usually 3–6 but up to 100.

PLATE 83: MAURITIUS AND RODRIGUES: LANDBIRDS IV

House Crow *Corvus splendens*   40 cm
Only large black landbird on Mauritius; a small crow, glossy black with dark grey on neck and breast, and squawking calls. **Voice** *High-pitched, nasal kehhh or kerrrr or caaah*, singly or repeated; also *cah-cah-cah-cah*.... **SHH** Introduced (S Asia but widely established elsewhere) to Mauritius, where common in Port Louis area and especially rubbish dump at Roche Bois, but may occur anywhere. Has occurred on Rodrigues without becoming established, and further ship-assisted arrivals possible anywhere in region, indicated by occasional records on Seychelles, Reunion and Madagascar. Commensal, favouring rubbish dumps for feeding and town parks with tall trees for breeding and roosting; also ports and neighbouring scrub and grassland. Flight slow and direct, on shallow wingbeats, but sometimes aerobatic.

Common Myna *Acridotheres tristis*   25 cm
Distinctive, large passerine: several unique features easily distinguish it, including *brown plumage with white wing-patches and tail tip (prominent in flight); yellow legs, bill and skin around eye*. **Voice** Vocal and noisy. Fluty medium- to low-pitched *tuuu*, *chyuu* or *tweh* and similar, often repeated, sometimes with squawking quality, but also an amazing range of low- to high-pitched squealing, clicking, croaking, gurgling and whistling notes, often strung together in disjointed sequence as song; any of these components also used as calls; at roosts, many birds call together in cacophonous medley. Mimics other birds, occasionally causing confusion. **SHH** Introduced (S Asia) to Mauritius and Rodrigues; also Madagascar, Seychelles, Comoros, Reunion, Agalega. Found in virtually all habitats including towns and villages, primary forest, agricultural land, grassland, savanna, mangroves and open coasts. Strident voice and bold, commensal habits make this a familiar species.

Common Waxbill *Estrilda astrild*   12 cm
Very small size and long tail, combined with brown upperparts, paler underparts, *red eye-patch and red bill* unique on Mauritius and Rodrigues. Juv has black bill, black line below red lores and indistinct barring; juv Scaly-breasted Munia has short tail and lacks red feathers or bill, and usually seen with distinctive ads. **Voice** Medium- to high-pitched *tewk* and hesitant *tetetete* or *tewtewtew*, in alarm a nasal *jaa* or longer *weeez*. In groups, weak twittering, sparrow-like chirping and shrill *tseep*. Song ends in a long buzzy upslurred note, *ji-ji-kweeezz*. **SHH** Introduced (Africa) to Mauritius and Rodrigues; also Seychelles, Reunion, Juan de Nova, possibly Madagascar. Found in wooded savanna, tracks and clearings in forest and plantations, sugarcane and other cultivation, gardens and roadsides. Gregarious except when breeding, in tight flocks of 10–40, rarely 100, which fly off and move *en masse*.

Scaly-breasted Munia *Lonchura punctulata*   12 cm
Tiny, seed-eating passerine, usually appearing *brown, with thick bill and short tail*; unique scaled underparts of ad not apparent in brief or distant view. Common Waxbill is similar size but has long tail, high-pitched calls and red bill. **Voice** *A short, rather husky whistle, often heard in flight*; sometimes preceded by a shorter note, *kit-teeeeee*. Song a virtually inaudible melody of high, flute-like whistles and low-pitched slurred notes. **SHH** Introduced (Asia) to Mauritius; also Reunion. Found in dry, open savanna, cultivated areas, orchards, forest clearings and roadsides; avoids forested uplands (native or plantation) although sometimes flies over or visits clearings. Often overlooked but common on Mauritius; inconspicuous on ground where usually feeds, and often seen in flight, calling. Gregarious, often forming small flocks, which may include Common Waxbills, Madagascar Fodies and other seed-eaters. [Alt: Spice Finch]

Yellow-fronted Canary *Crithagra mozambica*   13 cm
Distinctively marked, with *bright yellow underparts and rump, white-tipped tail, yellow face with black eyestripe and malar stripe*. Could be confused with rare 'flavistic' Madagascar Fody, which has equally bright yellow underparts and rump, but also black bill and oval eye-patch, all-yellow head and nape, and very different voice. **Voice** Calls *tssp* or *swee-eet* given in flight; song a mellow *tew-tedit* or variants, sometimes extended to a calm warble *tewdidiwtewitewewteditit*.... Can accurately mimic other small passerine species. **SHH** Introduced (Africa) and common in dry lowlands of Mauritius and Rodrigues; also Reunion, Agalega. Found in cultivated areas (including sugarcane), lightly and heavily wooded savanna, plantations of *Araucaria*, *Casuarina* and other exotic trees, parks and gardens. Forages mainly on ground but uses, and often sings from, elevated perches. Flight bounding and erratic. Seen singly, in pairs, or flocks where seeds abundant.

PLATE 84: MAURITIUS AND RODRIGUES: LANDBIRDS V

House Sparrow *Passer domesticus* 16 cm
♂ *distinctive but* ♀ *resembles Village Weaver and Mauritius and Madagascar Fodies*; see those species for differences. **Voice** A *harsh chissk, short weesk or abrupt, lower-pitched chup, chup-up, or short, coarse trill, chukukukukuk*. Song a series of *chup* or *chip* notes; chorus of *chissk* notes given by many birds together, e.g. at roosts. **SHH** Introduced (Eurasia) to Mauritius and Rodrigues; also Madagascar, Seychelles, Comoros, Reunion. Typically commensal, common in towns and villages, and follows man to remote camps or shelters in forest but also roadside cuttings away from habitation, and open woodland and cultivated areas.

Village Weaver *Ploceus cucullatus* 17 cm
Breeding ♂ *distinctive*; ♀, imm and non-breeding ♂ confusable with House Sparrow or Madagascar Fody, but *larger, longer-billed and yellower on underparts*; also *red eyes* in most ads. Smaller Yellow-fronted Canary has brighter plumage, complex face pattern and yellow rump. **Voice** Many calls, usually harsh alarms directed at predators, and flock contact call, a short *tsuk*. Song an extended, descending chatter, *cheee cheee shrrr zzzzzrrr cheee ch ch ch*. **SHH** Introduced (Africa); also Reunion. Common on Mauritius (not Rodrigues) in villages, cultivation, lowland savanna, *Casuarina* belts and other habitats with scattered trees. Gregarious and sociable. Nests colonially: conspicuous, woven, ovoid structures with entrance, hanging from tip of branch.

Rodrigues Fody *Foudia flavicans* 12 cm
Breeding ♂ from common Yellow-fronted Canary and rare 'flavistic' Madagascar Fody by *orange-red face*; also from canary by very different head pattern with *black eye-patch*. ♂ moults into ♀-like non-breeding plumage; timing variable but usually in austral winter. ♀, imm and non-breeding ♂ best distinguished *from Madagascar Fody by slender bill, long tail and near-constant habit (at all ages) of quivering partly open wings as if begging; plumage also greyer*. ♀ House Sparrow has much thicker bill. **Voice** An *agitated chip repeated very frequently*; also *chew* given by ♂. Song a long, pleasant *assortment of clear whistles and trills*, with clicks and scolding notes. **SHH** Endemic to Rodrigues. Population in c.1970 <10 pairs, but now in most hills and valleys in C uplands with population in thousands, due to recovery of exotic and native forest cover; uses *Syzygium jambos* thickets but not usually *Eucalyptus* monoculture. Conspicuous and vocal, acrobatic arboreal passerine, often nectar-feeding; locally abundant, singles to family parties; larger flocks mostly of brown-plumaged birds. **VU**

Mauritius Fody *Foudia rubra* 13 cm
Care needed to distinguish from moulting or part-red ♂ Madagascar Fody, which may show identical extent of red to ♂ Mauritius Fody, but latter has *shorter tail and finer bill, more olive (less brown) plumage*, and striking pale tips to wing-coverts, especially median coverts, producing *clear wingbars* except in heavily worn plumage. In breeding ♂, *black eye-patch extends back to a point* in Mauritius Fody but is rounded in Madagascar Fody; moults into ♀-like non-breeding plumage mainly Apr–Aug); ♀, imm and non-breeding ♂ can also resemble ♀ House Sparrow or Village Weaver; former is browner with thicker bill, latter yellower and much larger, and both lack clear wingbars. Calls usually distinctive. **Voice** A *distinctive, clear plick*, sometimes prolonged into a splutter. **Song** *jif-jif jif-jif-jif*, also sings with a quiet, brief, buzzing or wheezing, 'mutter'. Other calls a slow *chew*, or a harsh buzz often repeated and followed by a piercing squeak. **SHH** Endemic to Mauritius. Rare and very local in wettest forests, scrub and neighbouring plantations of SW, Piton Savanne to Montagne Cocotte (c.100 pairs); introduced to Ile aux Aigrettes. Arboreal, seen singly, in pairs or family groups; often seeks insects on tree trunks or takes nectar. **EN**

Madagascar Fody *Foudia madagascariensis* 13 cm
Care needed to distinguish from other fodies; see Mauritius and Rodrigues Fodies for differences. *A fody with entirely red underparts must be a Madagascar Fody*, but ♂ *can show patchy red very like* ♂ *Mauritius Fody*, while rare 'flavistic' birds recall Rodrigues species. ♂ moults into ♀-like non-breeding plumage (Jun–Oct). House Sparrow larger, with more grey-brown plumage and no trace of greenish or buffish-yellow; see also Village Weaver and Yellow-fronted Canary. **Voice** A very *high-pitched, slightly chirping tsip or tsikip*; also a lower *chup* or *chutut*. Breeding ♂ has trill call, or buzzing *zzz-zzz-zzz*. Song lacks regular form: wheezes, chirrups, squeaks and *tsip* calls, with up to c.1 s gap between phrases: *tsitsitsitsitstit tiptiptip tittittitit chipchipchip*. **SHH** Endemic to Malagasy region; introduced to Mauritius and Rodrigues, and elsewhere; native to Madagascar. Common almost everywhere with open ground, grasses and herbaceous growth including forest, grassland, savanna, cultivation, towns and gardens. Highly gregarious, flocks sometimes in hundreds.

PLATE 85: REUNION: LANDBIRDS I

Feral Pigeon *Columba livia*   33 cm
Typical 'town pigeon'; flocks of pigeons in urban areas or flying high up probably this species. Plumage very variable. For distinctions from Madagascar Turtle Dove and Zebra Dove, see those species. **Voice** A long, moaning *o-o-orr*, repeated monotonously, or a hurried, rambling *coo-roo-ooo-ooo*. **SHH** Native to Eurasia and N Africa, but introduced worldwide including Reunion; also Madagascar, Seychelles, Comoros, Mauritius, Rodrigues. Locally abundant especially around populated coastal belt, in markets, parks and any sites providing food scraps and nesting ledges; may also feed in agricultural land, other open areas and has reverted to ancestral cliff habitats. Flight powerful and fast, often in tight flocks, on long, pointed wings; at distance confusable with waders such as Grey Plover.

Madagascar Turtle Dove *Nesoenas picturata*   28 cm
Only sturdy, short-tailed, dark dove on Reunion. Grey head, dark reddish-brown upperparts, and *tail with only slightly paler (not white) corners* distinguishes it from all other species. Feral Pigeon, although variable, never shows similar colour pattern; Zebra Dove much smaller. **Voice** Low-pitched, double *coo: coo cooooo-uh*, the second *coo* louder, longer and rising at the end; first often not noticed. **SHH** Endemic to Malagasy region. Nominate race introduced from Madagascar to Reunion, and to Seychelles (Amirantes and granitics), Mauritius and Agalega; endemic races on Comoros and Aldabra group. Uncommon (but probably increasing) up to 1,800 m in wooded habitats and nearby open areas. Feeds mainly on ground, from where often flushed in forest with noisy wingbeats, but flight otherwise silent, very fast and usually low.

Zebra Dove *Geopelia striata*   21 cm
Very distinctive, *tiny, greyish-brown dove, with extensive black barring, otherwise pink or grey plumage* and blue skin round eye. In flight, *long, wedge-shaped tail with narrow white tips*; tapering shape creates a long white fringe on either side. Bounding flight on take-off distinctive. **Voice** A *medium- to high-pitched, bubbling, quick cooruruwuh or wurruwawa*, first note cooing and last two sharper, final note sometimes repeated; in display, a harsher *caaw-caaw-caaw*. **SHH** Introduced (SE Asia) to Reunion, and to Seychelles, Mauritius, Rodrigues, Glorieuses, Juan de Nova, Agalega. Common to 1,500 m in semi-open country or scattered woody vegetation, including urban areas; in forest, often in clearings. [Alt: Barred Ground Dove]

Rose-ringed Parakeet *Psittacula krameri*   41cm (♂), 37 cm (♀)
Only parrot on Reunion; escapes of other species may occur but pale green plumage, long tail, neck collar in ♂, red bill and shrieking calls distinctive. **Voice** Most frequent call is a raucous, far-carrying *kee-ak kee-ak*; many other calls, varying in pitch and tempo, but most with similar strident tone. Most vocal at roost and in flight. Perched birds also utter various whistling and subdued chattering or twittering notes. **SHH** Introduced (Asia and Africa) to Reunion, and to Seychelles and Mauritius. Rare, with scattered records near N & W coasts; self-sustaining population possibly not established. In secondary forest and savanna of dry coastal lowlands. Gregarious when feeding and, especially, roosting. [Alt: Ring-necked Parakeet]

Common Myna *Acridotheres tristis*   25 cm
Distinctive, large passerine: several unique features easily distinguish it, including *brown plumage with white wing-patches and tail tip (prominent in flight); yellow legs, bill and skin round eye*. **Voice** Vocal and noisy. Fluty medium- to low-pitched *tuuu, chyuu* or *tweh* and similar, often repeated, sometimes with squawking quality, but also an amazing range of low- to high-pitched squealing, clicking, croaking, gurgling and whistling notes, often strung together in disjointed sequence as song; any of these components also used as calls; at roosts, many birds call together in cacophonous medley. Mimics other birds, occasionally causing confusion. **SHH** Introduced (S Asia) to Reunion; also Madagascar, Seychelles, Comoros, Mauritius, Rodrigues, Agalega. Found in virtually all habitats including towns and villages, agricultural land, grassland, savanna, mangroves and open coasts; absent from most native forest and above 1,500 m. Strident voice and bold, commensal habits make this a familiar species on Reunion. Larger, blacker Hill Myna *Gracula religiosa* recently reported apparently feral in SE Reunion but not fully established.

PLATE 86: REUNION: LANDBIRDS II

Mascarene Swiftlet *Aerodramus francicus*   11 cm
Only swift on Reunion, but see Mascarene Martin. **Voice** *Usually silent except at colonies* where gives a quiet twitter or very weak 'scream', or remarkable echolocating sound, used to orientate in darkness of colonies: rapid, double click, sometimes a rattle. Calls and clicks occasionally in the open, perhaps especially in mist. **SHH** Endemic to region; also Mauritius. Found throughout Reunion, feeding up to highest summits, with colonies widely scattered. Single or in small groups; larger numbers near nesting colonies, which are in caves, and may be in entirely exotic habitats; congregates around cave entrances but hawks over all kinds of terrain throughout mainland. **NT**

Mascarene Martin *Phedina borbonica*   15 cm
The only hirundine on Reunion; larger than Mascarene Swiftlet, with *shorter, broader wings, earth-brown rather than blackish-brown upperparts* and lacks paler rump-patch. **Voice** Bubbling or buzzing *crreeeww* or *gzeee-gzeee* or *phree-zz*; also a shorter *chep*. Song, in flight or perched, a complex warble. Tone may differ between Madagascar and Mauritius/Reunion birds, but needs more investigation. **SHH** Endemic to Malagasy region. Nominate race resident on Mauritius and Reunion; paler, more clearly streaked *madagascariensis* on Madagascar. Scarce but occurs almost throughout, up to 1,500 m (rarely 2,400 m) but particularly in E and around Cilaos; rare in dry W. Typical hirundine, pursuing insects in flight, singly or in small groups; often perches on wires. Flight fast and powerful, swoops interspersed with fluttering, and distinctive calls. Rarely perches on ground. [Alt: Mascarene Swallow]

Reunion Cuckoo-shrike *Coracina newtoni*   22 cm
No other comparable-sized passerine on Reunion has similar *pale grey, black and white* (♂) or *brown/barred* (♀ *and subad*) *plumage*. Juv strikingly different, resembling juv Mauritius Cuckoo-shrike (Plate 81). **Voice** Often found by ♂ song, commonly described as *soft tweet-weet-weet-weet-weet*, or *pipipi-wheetit* with strong emphasis on *whe*, and variants. Aggressive call a harsh flat *crik* or *gek*; harsher 'ripping' *shrrr* in alarm. **SHH** Endemic to Reunion; very rare and restricted to native forest of N on Roche-Ecrite massif, Plaine des Chicots, Plaine d'Affouches, parts of Cirque de Dos d'Ane and massif de la Grande Montagne at 1,000–1,800 m. Strictly arboreal, singly or in pairs. **CR**

Red-whiskered Bulbul *Pycnonotus jocosus*   20 cm
Obviously smaller than Reunion Bulbul. Plumage unique, with *crest, contrasting face pattern with white cheek-patch*, black patches on breast-sides, white underparts, red undertail-coverts and white tip to tail. 'Whiskers' are glossy, hair-like red feathers behind eyes. Juv lacks red. **Voice** Slow, fluty or whistling song with rising and falling notes: *churee-chuwee-churiwiwee* or *plu chi chireree* or *pli chu tiriwee*; repeated 10–20 times, every 1–5 s, on varying pitch. Contact notes of similar quality, *pettigrew*; quick to signal alarm, with drawn-out *lerrr* or shorter *chip*. **SHH** Introduced (Asia) to Reunion, and to Mauritius, Juan de Nova, formerly coralline Seychelles and Comoros. Has expanded very rapidly and now common and widespread, throughout except high mountain areas.

Reunion Bulbul *Hypsipetes borbonicus*   24 cm
Large size, long, square-ended tail, and restless, aggressive and noisy behaviour gives distinctive character. Entirely *dark greenish-grey plumage with pinkish-orange bill and white eye* of ad unique on Reunion. Juv duller grey-brown with brown bill, but still distinctive locally. Reunion Cuckoo-shrike is grey or rufous-brown, and Common Myna dark brown with white flashes in wings and tail; both have shorter, more rounded tails. Red-whiskered Bulbul is distinct. **Voice** Loud chuckling or chattering; ♂ song very varied and sometimes melodious, typically consists of various chuckling phrases, and sometimes mimetic. *'Cat' call*, typical of genus, a loud and conspicuous, hoarse mewing *weeeer*. Alarm a loud *kek*. **SHH** Endemic to Reunion, in native and exotic forest but not tree-heath or softwood plantations, up to 1,800 m. Strictly arboreal, singly, in pairs or small groups.

Red-billed Leiothrix *Leiothrix lutea*   14 cm
Bright red bill, yellow-and-orange throat and breast and wing-panel, and strongly forked tail very distinctive. Sexes similar, ♀ duller. Juv less colourful but orange wing-panel and generally olive-grey plumage distinctive. Voice confusable with Red-whiskered Bulbul, which occurs in same areas. **Voice** Monosyllabic contact call constantly uttered: guttural, slightly nasal, short, abrupt *zhirk* or *shreep* notes, with various shorter, harder calls. Warbling ♂ song long and complex, a loud, cheerful warbling, similar to Red-whiskered Bulbul but more prolonged and musical. **SHH** Introduced (Asia) to Reunion, mainly in humid E, N & W flanks, in forest and thickets especially at 600–700 m, perhaps to 1,500 m, probably expanding. Usually in small flocks; secretive and difficult to observe.

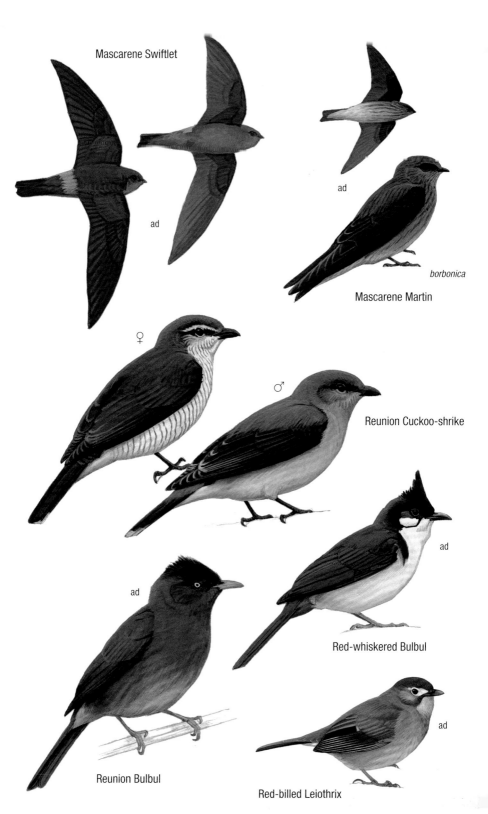

PLATE 87: REUNION: LANDBIRDS III

Reunion Stonechat *Saxicola tectes*   12 cm
Variable but always distinctive on Reunion, where it is *the only chat or chat-like bird*. ♂ *dark brown to blackish above, paler below;* ♀ *more uniformly brown above*. **Voice** A *vigorous tek or tek tek,* often directed at an intruder, sometimes with initial, plaintive *hweet*. Alarm a longer, more rapid *trre trre*. Song, given from perch or in flight, very complex, rapid, high-pitched *tak... tak... tsstwirchrkchirtskchriik*. **SHH** Endemic to Reunion, where widespread in and around all types of native forest and heath vegetation, usually above 800 m but to sea level in SE and locally in NW; reaches highest summits, using areas devoid of vegetation, taking food scraps left by hikers, and locally also in orchards, pastures, plantations, secondary forests, riversides, and vicinity of habitations and other haunts of people such as tracks, picnic sites and even gardens in upland towns; much wider range of habitats than in most stonechats and relatives. Conspicuous, tame and inquisitive. Characteristic *tek* call often heard, accompanied by head-bobbing, wing-flicking and tail-flirting.

Mascarene Paradise Flycatcher *Terpsiphone bourbonnensis*   16 cm
Small size and rufous upperparts and tail at all ages eliminate all other species. In ♂ nominate race (Reunion), upperparts slightly browner (less rufous), grey underparts duller and gloss on head less extensive on hindneck compared to Mauritian race *desolata*. **Voice** Distinctive *whistling or grating* often first means of detection in dense vegetation. Long, noisy, medium-harsh buzz or *chrr*, often extended to harsh scolding. Song variable, high-pitched, fluid, with simple repetition of warbling, buzzing or whistled notes, e.g. *ker-chirkirerchire-ker-chirkirerikire* or *ker-chikirikiri*. **SHH** Endemic to Reunion and Mauritius. Widespread on Reunion (unlike on Mauritius), occurring in most wooded areas, locally at sea level (where often along rivers) up to treeline at 1,800–2,000 m, especially in mixed native forest. Catches insects in flight in mid-air or more often from vegetation surfaces, launching from a perch. Races rather similar, but DNA study indicates long isolation from each other and may be treated as two species.

Reunion Olive White-eye *Zosterops olivaceus*   11 cm
Only Reunion bird with conspicuous white eye-ring, *olive wings*, calls and lack of white rump diagnostic. **Voice** Frequently uttered, an *emphatic tchip-tchip-tchip*, single *tsip* or series of *tu-tu-tu* notes; sometimes a trill *wrrri*. Song a loud, fast warble, with much bill-snapping, louder than Reunion Grey White-eye, with *tu* and *tchip* notes interspersed. **SSH** Endemic to Reunion, where common in native forest containing favoured nectar-bearing plants (also likes exotic *Fuchsia* species), mainly above 500 m up to 2,600 m; at lowest levels, often in wooded ravines with exotic *Syzygium jambos*. Closely related to Mauritius Olive White-eye which is also a nectar and insect feeder, with long, curved bill. Aggressive and not social: mainly in pairs rather than flocks, with congregations only where attracted by rich food source, which is often vigorously defended.

Reunion Grey White-eye *Zosterops borbonicus*   11 cm
Variable; four morphs with largely exclusive distributions, brown to grey, but always differs from Reunion Olive White-eye by *grey or brown coloration, lack of white eye-ring, short, straight bill, white rump* (displayed by habit of keeping *tail cocked and wings drooped*) and calls. **Voice** A *plaintive eeee or double eeeee-eeee*. When agitated, gives emphatic chips in rapid series or followed by *eeeeeee*. Song a long and elaborate loud series of warbling with contact calls, sometimes containing mimicry, running slightly up and down scale. **SHH** Endemic to Reunion, where the *commonest native bird,* found in almost any habitat with trees or shrubs, including many non-native habitats where other native birds scarce or absent. A noisy, sociable, arboreal species found in groups, usually 3–6 but up to 100. Several races described but genetic groups, morphs and named races do not match each other, and variation is currently considered to represent a form of polymorphism.

PLATE 88: REUNION: LANDBIRDS IV

House Sparrow *Passer domesticus* — 16 cm
♂ *distinctive but* ♀ *resembles Village Weaver, Madagascar Fody, Red-billed Quelea and Pin-tailed Whydah*; see those species for differences. **Voice** *A harsh chissk, short weesk or abrupt, lower-pitched chup, chup-up, or short, coarse trill, chukukukukuk.* Song a series of *chup* or *chip* notes; chorus of *chissk* notes given by many birds together, e.g. at roosts. **SHH** Introduced (Eurasia) to Reunion, and to Madagascar, Seychelles, Comoros, Mauritius and Rodrigues. Typically commensal, common in towns and villages, but may occur in forest dwellings, roadside cuttings away from habitation, and open woodland and cultivated areas; less common at higher altitudes.

Village Weaver *Ploceus cucullatus* — 17 cm
Breeding ♂ *distinctive*; ♀, imm and non-breeding ♂ confusable with House Sparrow, Madagascar Fody, Red-billed Quelea and Pin-tailed Whydah, but *larger and longer-billed* than any of these, and most ads have *red eyes*. Further distinguished by yellower underparts; quelea has red, pink or yellow bill, and whydah more heavily striped head. Cape and Yellow-fronted Canaries smaller with brighter plumage and much smaller bills; Yellow-fronted has complex face pattern and yellow rump; Cape has grey nape and unstreaked back (♂) or streaked underparts (♀ and juv). **Voice** Many calls, frequently harsh alarms directed at predators, and flock contact call, a short *tsuk*. Song an extended, descending chatter, *cheee cheee shrrr zzzzzrrr cheee ch ch ch*. **SHH** Introduced (Africa) to Reunion, and to Mauritius. Common in villages, cultivation, savanna, *Casuarina* belts and other habitats with scattered trees; mainly in coastal areas and lowlands, locally higher in disturbed non-forest habitat as on Plaine des Cafres at 1,300 m. Gregarious and sociable. Nests colonially: conspicuous, woven, ovoid structures with entrance, hanging from tip of branch, often of Coconut Palm or *Casuarina*.

Red-billed Quelea *Quelea quelea* — 12 cm
♀, imm and non-breeding ♂ confusable with House Sparrow, Village Weaver, Madagascar Fody and Pin-tailed Whydah, but have *large, red, pink or yellow bill, red or orange eye-ring, and yellowish edges to remiges*; Pin-tailed Whydah has small, duller pink bill and heavily striped head, while weaver, fody and sparrow have duller bills. **Voice** When perched, a short *chirt chirt*; call on take-off or in response to another flying overhead *tseep tseep*. Also a high thin sizzle, grating *jaaaaaaaaa* and chattering rapid *jjjjjjjjjjj*. Song a preliminary chatter, warbled *tweedle toodle tweedle* and long, high-pitched whistle. **SHH** Introduced (Africa) to Reunion; a recent addition to the region's avifauna, currently restricted to dry lowlands of W coast, St-Paul to St-Pierre, where still uncommon and local in grassland, woody savanna and patches of *Casuarina*; several colonies established and expansion can be expected. A seed-eater, occurring in vast flocks (millions) in Africa, but this is not known on Reunion where grains are not abundant. Mixes with other seed-eaters.

Madagascar Fody *Foudia madagascariensis* — 13 cm
Only fody on Reunion; red plumage unique to ♂ (rare flavistic birds show yellow where normally red, so still distinctive), apart from much smaller, differently marked (and rare) Red Avadavat. ♂ moults into ♀-like non-breeding plumage (Jun–Oct), when confusable with other streaked, brownish seedeaters. ♀ House Sparrow is larger, with more grey-brown plumage and no trace of greenish or buffish-yellow. ♀ or non-breeding ♂ Village Weaver yellower and much larger. See also Red-billed Quelea and Pin-tailed Whydah. Sibilant *tsip* call not given by any confusion species. **Voice** A very *high-pitched, slightly chirping tsip or tsikip*; also a lower *chup* or *chutut*. Breeding ♂ has trill call, or buzzing *zzz-zzz-zzz*. Song lacks regular form: wheezes, chirrups, squeaks and *tsip* calls, with up to *c*.1 s gap between phrases: *tsitsitsitsitstit tiptiptip titititititit chipchipchip*. **SHH** Endemic to Malagasy region; introduced to Reunion, and to Seychelles, Comoros, Mauritius, Rodrigues, Glorieuses, Juan de Nova, Agalega; native to Madagascar. Almost ubiquitous below *c*.2,000 m in open areas, grassy and herbaceous growth including forest (especially if patchy or degraded), grassland, savanna, cultivation, towns and gardens. Often gregarious, flocks sometimes in hundreds.

PLATE 89: REUNION: LANDBIRDS V

Common Waxbill *Estrilda astrild*   12 cm
Very small size and long tail, combined with brown upperparts, paler underparts, *red eye-patch and red bill* unique on Reunion. Juv Scaly-breasted Munia has short tail, lacks red, and is usually seen with distinctive ads. **Voice** Medium- to high-pitched *tewk* and hesitant *tetetete* or *tewtewtew*, in alarm a nasal *jaa* or longer *weeez*. In groups, weak twittering, sparrow-like chirping and shrill *tseep*. Song ends with a long buzzy upslurred note, *ji-ji-kweeezz*. **SHH** Introduced (Africa) to Reunion; also Seychelles, Mauritius, Rodrigues, Juan de Nova, possibly Madagascar. Found throughout up to c.2,000 m in wooded savanna, clearings in forest and plantations, sugarcane and other cultivation, gardens and roadsides. Gregarious except when breeding, in tight flocks of 10–40, rarely 100.

Red Avadavat *Amandava amandava*   10 cm
Breeding ♂ *spectacular*. ♀ and non-breeding ♂ dull but made distinctive by *wings, rump, uppertail-coverts and tail showing red or black with white spots* as in breeding ♂; *bill also red*. Juv lacks red, but has *pale double wingbars*. **Voice** Distinctive, high-pitched, shrill monosyllables, often rapidly repeated, in sequences varying in volume, length of notes and repeat rate. Song sweet and varied but fairly short, high-pitched, mainly a descending, soft liquid twitter. **SHH** Introduced (Asia) to Reunion, where long-established but rare, restricted to dry S & W. Found in open or densely wooded grassland and agricultural land, mainly sugarcane; in pairs or small groups.

Scaly-breasted Munia *Lonchura punctulata*   10 cm
Tiny seed-eating passerine, usually appearing *brown, with thick bill and short tail*; unique scaled underparts of ad not apparent in brief or distant view. Common Waxbill similar size but has longer tail; also high-pitched calls and red bill. See also Red Avadavat. **Voice** *A short, rather husky whistle*, often heard in flight; sometimes preceded by a shorter note, thus *kit-teeeeee*. Song a quiet melody of high, flute-like whistles and low-pitched slurs. **SHH** Introduced (Asia) to Reunion; also Mauritius. Found in bushland, cultivation or grassy habitat, especially near coast, locally inland. Commonest in dry W coast savannas and typically avoids forested uplands (native or plantation). Inconspicuous on ground where it usually feeds; often seen in flight, calling. Gregarious. [Alt: Spice Finch]

Pin-tailed Whydah *Vidua macroura*   12 cm + 20 cm tail-streamers (breeding ♂)
Breeding ♂ *unique*. In other plumages, differs from other streaky brown seed-eaters by *boldly striped face pattern; bill pink, or blackish in breeding ♀ and juv*, white inner half of outer tail-feathers shows in flight. **Voice** Both sexes give a harsh chatter; other calls include a low *peeee* and in flight a double *chip-chip*. Song an uneven, jerky series of measured, single, sibilant notes, interspersed by trills and jingling or whistles. **SHH** Introduced (Africa) to Reunion. Recently established along N & W coasts, where scarce but spreading inland; a few in S. Found in parks, gardens, sports grounds, cultivation and other short grassy areas. Singly or in flocks. A brood parasite, in Africa laying in nest of estrildids including Common Waxbill, also introduced to Reunion.

Cape Canary *Serinus canicollis*   14 cm
Obviously *smaller than Village Weaver*. Confusable mainly with Yellow-fronted Canary; ♂ Cape differs by *grey ear-coverts, neck and upper mantle contrasting with yellow areas on head*; ♀ duller. Juv heavily streaked on underparts and shows little yellow. Yellow-fronted Canary always has bolder facial pattern and brighter yellow rump. **Voice** High-pitched, cheerful *chriiiii... chriii...*, slightly lower *peeer... peeer...* or trilled *pit-it-it-it*. Song a short sequence of mellow phrases, *chiriliu... chirilwi... treeootseee...*, or more complex and rapid high-pitched, tinkling phrases with much variation. **SHH** Introduced (Africa) to Reunion, where, unlike other exotics, *confined to mountains, plains and cirques* mainly at 600–1,500 m, a few up to 2,700 m. Scarce in open habitats including cultivation, tree-heath and forest interrupted by clearings and paths; occasionally sugarcane.

Yellow-fronted Canary *Crithagra mozambica*   13 cm
Both sexes have *bright yellow underparts and rump, white-tipped tail, and striking face pattern*; see Cape Canary. ♀ Village Weaver is larger, much duller and lacks striking head, rump and tail patterns. Rare 'flavistic' Madagascar Fody has black bill and oval eye-patch on otherwise all-yellow head and nape, and very different voice. **Voice** Calls *tssp* or *swee-eet* given in flight; song a mellow *tew-tedit* or variants, sometimes a calm warble *tewdidiwtewitewewteditit...*. May accurately mimic other small passerines. **SHH** Introduced (Africa) to Reunion; also Mauritius, Rodrigues, Agalega. Common in coastal lowlands up to c.1,500 m, especially in lightly wooded savanna and *Casuarina*. Forages mainly on ground but sings from elevated perches. Flight bounding and erratic.

PLATE 90: OUTER ISLANDS I: AGALEGA AND MOZAMBIQUE CHANNEL

Madagascar Turtle Dove *Nesoenas picturata* 28 cm
Easily distinguished from Zebra Dove, also present on Agalega: *larger, stocky, short-tailed, dark dove* with distinctive dark reddish-brown upperparts, grey head and *tail has pale (not white) corners*. **Voice** Low-pitched, double cooing: *coo cooooo-uh*, the second *coo* louder, longer and rising at the end; first often not noticed. **SHH** Endemic to Malagasy region: native to Madagascar, Seychelles, Comoros; nominate race (Madagascar) introduced to Mauritius, Reunion, Seychelles. Agalega population, close to extinction, may be introduced Madagascar race *picturata* but could be distinct (and native): suggested to be smaller, with greenish gloss on upperparts. Feeds mainly on ground from where often flushed in forest with noisy wingbeats, but flight otherwise silent, very fast and usually low.

Zebra Dove *Geopelia striata* 21 cm
Only pigeon or dove on Juan de Nova and Glorieuses. On Agalega, requires separation from Madagascar Turtle Dove: a very distinctive, *tiny, greyish-brown dove, with extensive black barring, otherwise pink or grey plumage* and blue skin round eye. In flight, *long, wedge-shaped tail with narrow white tips*; tapering shape creates a long white fringe on either side of tail. Bounding flight on take-off distinctive. **Voice** Medium- to high-pitched, *bubbling, quick cooruruwuh or wurruwawa*, first note cooing and last two sharper, final note sometimes repeated; in display, a harsher *caaw-caaw-caaw*. **SHH** Introduced (SE Asia) and common on Agalega, Juan de Nova and Glorieuses; also Seychelles, Mauritius, Rodrigues, Reunion. [Alt: Barred Ground Dove]

Madagascar Fody *Foudia madagascariensis* 13 cm
Only fody or streaky 'brown' seed-eating passerine on the Outer Islands. ♂ moults into ♀-like non-breeding plumage (probably Jun–Oct). Yellow-fronted Canary also occurs on Agalega; see that species. **Voice** A very *high-pitched, slightly chirping tsip or tsikip*; also a lower *chup* or *chutut*. Breeding ♂ has trill call, or buzzing *zzz-zzz-zzz*. Song lacks regular form: wheezes, chirrups, squeaks and *tsip* calls, with up to *c.*1 s gap between phrases: *tsitsitsitsitstit tiptiptip titititititit chipchipchip*. **SHH** Endemic to Malagasy region; introduced and common on Grande Glorieuse, Juan de Nova and Agalega, as well as Seychelles, Comoros, Mauritius, Reunion, Rodrigues; native to Madagascar. Gregarious seed-eater, flocks sometimes in hundreds; also takes insects. Recent reports of Village Weaver (see Mauritius, Reunion) on Agalega require confirmation.

Yellow-fronted Canary *Crithagra mozambica* 13 cm
Only finch on Agalega, although somewhat similar size and shape to Madagascar Fody; *bright yellow underparts and rump, white-tipped tail, yellow face with black eyestripe and malar stripe* distinguishes it. More similar to rare 'flavistic' male Madagascar Fody, which has equally bright yellow underparts and rump, but black bill and oval eye-patch, all-yellow head and nape, and very different voice. **Voice** Calls *tssp* or *swee-eet* given in flight; song a mellow *tew-tedit* or variants, sometimes extended to a calm warble *tewdidiwtewitewewteditit...*. May accurately mimic other small passerines. **SHH** Introduced (Africa) to Agalega, where common in *Casuarina* groves; also to Mauritius, Rodrigues and Reunion. Forages mainly on ground but uses, and often sings from, elevated perches. Flight bounding and erratic. Seen singly, or in pairs or flocks that form where seeds abundant.

Common Myna *Acridotheres tristis* 25 cm
Distinctive, the *only large passerine on Agalega*. Brown with white wing-patches and tail tip (prominent in flight) unique; yellow legs, bill and skin around eye. **Voice** *Vocal and noisy*. Fluty medium- to low-pitched *tuuu, chyuu* or *tweh* and similar, often repeated, sometimes with squawking quality, but also an amazing range of low- to high-pitched squealing, clicking, croaking, gurgling and whistling notes, often strung together in disjointed sequence as song; any of these components also used as calls; at roosts, many birds call together in cacophonous medley. Mimics other birds, occasionally causing confusion. **SHH** Introduced (S Asia) and common on Agalega; also Madagascar, Comoros, Seychelles, Mauritius, Rodrigues, Reunion. Strident voice and bold, commensal habits make this a familiar species.

PLATE 91: OUTER ISLANDS II: MOZAMBIQUE CHANNEL

Barn Owl *Tyto alba* — 35 cm
Only owl on Europa, often detected by its distinctive screech call, familiar in much of world. **Voice** Loud, toneless, grating screech *ksscchh*. Hisses and "snores" when at rest and makes loud bill-snaps when disturbed at or near nest. **SHH** Cosmopolitan; presumed native on Europa, as on Madagascar and Comoros; introduced to Seychelles. Probably uses all terrestrial habitats on Europa, where eats mainly Black Rats *Rattus rattus*, but some specialise on seabird chicks.

Olive Bee-eater *Merops superciliosus* — 30 cm
Only bee-eater recorded on the Outer Islands, although others could occur as vagrants; very distinctive at all ages. **Voice** *Sequence of rapid, hollow and rolling medium-pitched notes, krioor krioor krioor* or *kriew kreiw kriew*, often repeated and in chorus from flocks; also single *prioop*. **SHH** Africa and Malagasy region, mainly Madagascar, Comoros; regular on Juan de Nova, where breeding proven in 2013, and probably vagrant to Glorieuses. [Alt: Madagascar Bee-eater]

Broad-billed Roller *Eurystomus glaucurus* — 29 cm
Only roller recorded on the Outer Islands. Very distinctive at all ages, dark plumage highlights broad, bright yellow bill and, in flight, largely turquoise tail. **Voice** Rarely if ever heard outside breeding range on Madagascar. **SHH** Africa and Malagasy region; endemic, nominate race breeds in Madagascar and migrates to Africa; some stop over on Mozambique Channel islands (Glorieuses, Juan de Nova, Europa, as well as Comoros) in Mar–May and Oct–Dec. Could also overshoot to Outer Islands, E of Madagascar, as also recorded on Mauritius and Reunion. Noisy and conspicuous, spending long periods perching prominently on treetop or other exposed perch.

Madagascar Bulbul *Hypsipetes madagascariensis* — 24 cm
No obvious confusion species on Glorieuses; absent on other Outer Islands. **Voice** Very vocal. Calls assumed as in Madagascar and Comoros: varied but include a squeaky *kee kee kee kee*; 'cat-call', a high-pitched, whining *keeeeer* or *chair*, a clucking *tut tut tut tut tut* or *chup-chup-chup-chup* interspersed with dropping *tisilyu*. Gatherings produce a *continual hubbub of chattering*. Song a lengthy, babbling medley of extended and repeated calls. **SHH** Endemic to Malagasy region, including Glorieuses (Grande Glorieuse only), where probably rare, once suspected to be extinct but seen recently; also Madagascar, Comoros, coralline Seychelles. A conspicuous, noisy, aggressive and curious inhabitant of any habitat with trees or shrubs. Glorieuses population has been treated as an endemic race but there are no consistent differences from *rostratus* of Aldabra.

Pied Crow *Corvus albus* — 45 cm
Only crow on the Outer Islands. Often soars, when confusable with raptor, but distinguished by black-and-white plumage and typical corvid silhouette, with relatively short wings and stout bill. **Voice** Flat, loud, low-pitched *caah*, *caw* or *craaah*; also a low *cuh... cuh... cuh* from flocks and a range of other bubbling or gurgling noises. **SHH** Africa and Malagasy region, including islands of Mozambique Channel: Glorieuses (both islands), Juan de Nova and Europa. Also Madagascar and coralline Seychelles. Common except possibly on Juan de Nova where few data available; often around human settlements, beach crest and coastal grassland. A scavenger and predator, taking seabirds and eggs, but presumed native and not considered a threat to native fauna.

PLATE 92: OUTER ISLANDS III: MOZAMBIQUE CHANNEL

Barn Swallow *Hirundo rustica* — 18 cm
Only swallow regularly recorded on the Outer Islands, although vagrants of other species could occur, particularly Mascarene Martin, from which separated by *blue-black upperparts, white underparts and underwing-coverts, and long tail-streamers*. **Voice** Contact call *witt-witt*, given at any time, sometimes run together as series of *tswitt* notes. Song a more prolonged bubbling twitter, not recorded in Malagasy region. **SHH** Breeds Eurasia and North America, spending boreal winter in sub-Saharan Africa, Asia and South America. Rare migrant across much of Malagasy region; migrant visitor (especially Oct–May) to Mozambique Channel islands, with few records on Juan de Nova and Glorieuses, but apparently more regular on Europa. A typical hirundine, foraging for flying insects in very agile flight.

Red-whiskered Bulbul *Pycnonotus jocosus* — 20 cm
No confusion species on Juan de Nova; absent elsewhere on Outer Islands, so no overlap with Madagascar Bulbul on Glorieuses. **Voice** Slow, fluty or whistling song of rising and falling notes: *churee-chuwee-churiwiwee* or *plu chi chireree* or *pli chu tiriwee*; repeated 10–20 times, every 1–5 s, on varying pitch. Contact notes of similar quality, *pettigrew*; quick to signal alarm, with drawn-out *lerrr* or shorter *chip*. **SHH** Introduced (Asia) to Mauritius and Reunion, formerly coralline Seychelles and Comoros; very common on Juan de Nova.

Madagascar Cisticola *Cisticola cherina* — 11 cm
Only brown, streaky, insectivorous bird resident on the Outer Islands (Glorieuses only); also marked by tiny size and short, graduated, black-and-white-tipped tail. Calls and prominent perching habits (often flicking tail) also distinctive. **Voice** Song a very high-pitched, metallic *tint... tint... tint* or *pik... pik... pik... pik...*, repeated at 0.5–1 s intervals; also a high-pitched, metallic, repeated *triit* or *chiit* or *chreep*, slightly trilled and modulated. Both given in song-flight, and in duet when may be delivered more rapidly. Alarm note *tic*. **SHH** Endemic to Malagasy region; abundant on Glorieuses in grassy or bushy areas (reported on Grande Glorieuse, not documented on Ile du Lys); also Madagascar, coralline Seychelles (Cosmoledo and Astove). Conspicuous, often perching on tops of grass stems or bushes, alarm-calling on sight of observer, but sometimes hiding effectively in grass tussocks.

Souimanga Sunbird *Nectarinia souimanga* — 11 cm
Only sunbird, or long-billed passerine, on the Outer Islands (Glorieuses only). **Voice** Not described on Glorieuses, but probably as elsewhere. Characteristic call, especially in alarm, a high-pitched, drawn-out, peevish (or mewing) *chweee*; also a high-pitched *chink* or staccato *chip*. Song a high-pitched twittering *plit... chit... chirrit ... chirrititiwitittirrit*; last phrase very complex and sometimes extended to 1–1.5 s. **SHH** Endemic to Malagasy region. Nominate race (as on Madagascar) common in all habitats with woody vegetation on Glorieuses (Grande Glorieuse and Ile du Lys); also Madagascar, coralline Seychelles (Aldabra group). Typical small sunbird: active and restless in seeking insects and nectar.

Madagascar White-eye *Zosterops maderaspatanus* — 11 cm
Only white-eye on the Outer Islands, where restricted to Glorieuses amd Europa. **Voice** Presumed same as in Madagascar: flocks emit medley of nasal *tew, ter* or *ter-tew* notes, or a distinctive, querulous, rising *wrrrri* or *trrrrri*. Song a medium-pitched warble, running slightly up and down scale, e.g. *pita weer weer whit weer weer witwit weer*, recalling a canary. **SHH** Endemic to Malagasy region. Nominate race (as in Madagascar) is commonest bird on Glorieuses (Grande Glorieuse and Ile du Lys). Less common on Europa, where endemic race *voeltzkowi* is the only landbird taxon endemic to the Outer Islands; closely recalls nominate race, but longer-tailed. Also Madagascar, Comoros, coralline Seychelles (Aldabra group). A typical white-eye: arboreal, gregarious and found in all wooded habitats.

Common Waxbill *Estrilda astrild* — 12 cm
No confusion species on Outer Islands (Juan de Nova only); much smaller than Madagascar Fody, which is brown in most plumages but also lacks red bill. **Voice** Medium- to high-pitched *tewk* and hesitant *tetetete* or *tewtewtew*, in alarm a nasal *jaa* or longer *weeez*. In groups, weak twittering, sparrow-like chirping and shrill *tseep*. Song ends with a long buzzy upslurred note, *ji-ji-kweeezz*. **SHH** Introduced (Africa) to Malagasy region including Juan de Nova; also Seychelles, Mauritius, Rodrigues, Reunion, possibly Madagascar.

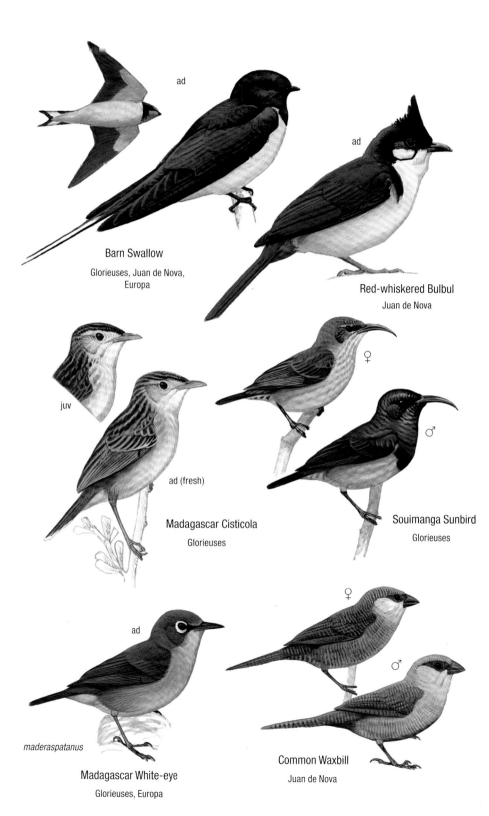

PLATE 93: VAGRANTS

Wandering Albatross *Diomedea exulans*   115 cm
Largest seabird. Huge bill, usually pale pink with yellowish tip. Ad from other albatrosses in region (see Plate 1) by *white back, white wings with black flight feathers and a variable amount of black on outer wing-coverts*, decreasing with age. Underwing white with narrow black trailing edge, broad black wingtip and black forewing to carpal joint. Often shows black edges to white tail. ♀ has less white in upperwing and retains some dark marks on body, tail and head. Imm from giant petrels by chocolate-brown body and white face-mask; underwing as ad. Underparts progressively whiten with age, becoming speckled. **Voice** Normally silent at sea. **SHH** Breeds on islands in South Atlantic and Indian Oceans. Vagrant to Madagascar, Mauritius and Reunion, reported at sea in austral winter N to $c.25°S$, to S & W of Madagascar on transects from Reunion S to subantarctic islands. Highly pelagic, often following ships, feeding on carrion more than other congeners but rarely joins feeding frenzies. Flight effortless, on stiff, outstretched wings, legs projecting well beyond tail. Amsterdam Albatross *D. amsterdamensis*, Tristan Albatross *D. dabbenena* and Antipodean Albatross *D. antipodensis* (in the Wandering Albatross group), and Northern and Southern Royal Albatrosses *D. sanfordi* and *D. epomophora* could also occur as vagrants; see specialist literature for identification. **VU**

Black-browed Albatross *Thalassarche melanophris*   88 cm
Combination of orange bill, white head with black brow and white underwing outlined with broad dark borders, especially along leading edge is diagnostic save for unrecorded (and unlikely) Campbell Albatross *T. impavida*, which has smaller eye-patch and yellow (not dark) eye. Imm has grey flecking on white underwing-panel, dark grey neck collar; bill greyish with black tip. Indian Yellow-nosed Albatross (Plate 1) more slender, has narrow dark border to white underwing and black bill. Shy Albatross (Plate 1) has much narrower black borders to underwing and more contrast between grey back and dark upperwing. **Voice** Normally silent at sea. **SHH** Breeds on islands in Southern Ocean. Vagrant Madagascar and Mascarenes, Jun–Sep, much rarer than Indian Yellow-nosed and Shy. Follows ships, sometimes occurs in coastal waters. **NT**

Sooty Albatross *Phoebetria fusca*   85 cm
Medium-sized, uniform sooty-brown albatross, from giant petrels by *slender body, long narrow wings, long wedge-shaped tail and dark bill with thin yellow line on sulcus*. Imm similar to ad but duller. Light-mantled Albatross has paler back, with distinct contrast between hood and mantle, and is less slender, with broader wings and at close range a blue line on bill. **Voice** Normally silent at sea. **SHH** Breeds on islands in S Atlantic and Indian Oceans, dispersing N in austral winter to 30°S, well S of Madagascar. Vagrant Madagascar and Mascarenes, May–Jul. Usually solitary and rarely follows ships, but associates with cetaceans. Highly manoeuvrable flight, rarely flapping wings. **EN**

Light-mantled Albatross *Phoebetria palpebrata*   85 cm
From Sooty Albatross by *dark wings, head and tail dark contrasting with paler grey back and underparts, and blue line on sulcus* (but difficult to see and absent in imm). See Sooty Albatross for other differences. **Voice** Normally silent at sea. **SHH** Breeds circumpolar S of 35°S and regular N to 30°S in austral winter, but generally to S of Sooty Albatross. Vagrant to Rodrigues and possibly Mauritius, May–Jul. May follow ships. Like Sooty, extremely manoeuvrable and agile in flight, rarely flapping wings. **NT**

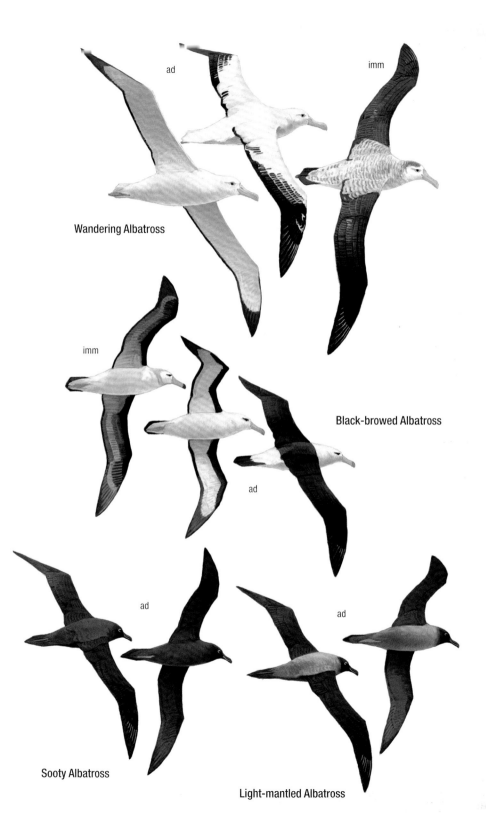

PLATE 94: VAGRANTS

Southern Fulmar *Fulmarus glacialoides*   48 cm
Distinctive pale, stocky petrel, from other seabirds of region by *high forehead, rounded wings and long square tail*, white head and underparts, and white wings (above and below) with dark trailing edge and outer wing. Bill pink with black tip and blue nasal tubes. From gulls by white patch on upperwing and *shearing, stiff-winged flight*. **Voice** Generally silent at sea. **SHH** Circumpolar in seas well S of subantarctic islands of S Indian Ocean; non-breeders N to c.30°S. Probable vagrant; specimen labelled as from Reunion, but origin not proven. Shallow wingbeats, fluttering flight, followed by glides on stiffly held wings. Will investigate ships but rarely follows them. Surface feeder but will perform shallow dives.

Antarctic Prion *Pachyptila desolata*   28 cm
From Slender-billed Prion (with great difficulty) by broader base to bill, *less pale face with well-marked dark eye-patch and ear-coverts, clearer grey collar* on white underside, *better-defined dark 'M' on upperwing* and slightly broader black tail tip. **Voice** Generally silent at sea. **SHH** Widespread in S Indian and Atlantic Oceans, and regular off S Africa in austral winter. Vagrant to Madagascar in austral winter, unidentified prion records from Mascarenes and S Seychelles probably this or next species. Fairly erratic flight, rapid wingbeats followed by glides, wings bowed. Prion taxonomy unresolved and identification extremely difficult: six species recognised (sometimes more), two accepted in Malagasy region based on specimen records. Broad-billed *P. vittata*, Salvin's *P. salvini* and Fairy *P. turtur* claimed without proof; of these, Salvin's is most likely.

Slender-billed Prion *Pachyptila belcheri*   28 cm
Smallest, palest prion; from others by narrow blue bill, appearing straight-sided from above. Black tip to tail often restricted to central tail-feathers, outer feathers white and *very broad white supercilium, narrow black eye-patch, and indistinct pale grey collar* on white underparts. Dark 'M' across upperwing comparatively ill-defined on central wing. **Voice** Generally silent at sea. **SHH** Breeds islands in S Indian and Atlantic Oceans, dispersing N, including coasts of SE Africa and recorded N to 26°S, Jul–Sep, on transects from Reunion S to subantarctic islands. Vagrant to Madagascar and Mauritius. Unidentified prion records from Mascarenes and S Seychelles probably this or previous species. Rapid flight, low over surface with fluttering wingbeats and short glides; rocks from side to side in *most erratic flight of any prion*, frequently changing direction. Rarely follows ships. Pelagic but sometimes feeds in shallow inshore waters.

Grey Petrel *Procellaria cinerea*   50 cm
Robust petrel with long narrow wings, fairly long wedge-shaped tail and slender, pale, dark-tipped bill. *Distinctive combination of grey upperparts, white underparts and dark underwing*. From Cory's Shearwater (Plate 4) by uniform grey, not brown, upperparts and dark underwing. **Voice** Generally silent at sea. **SHH** Breeds islands in S Atlantic and Indian Oceans and Australasia; probably vagrant in Malagasy region, where reported Reunion. Direct effortless flight, gliding with occasional shallow wingbeats. Solitary, associates with cetaceans, follows fishing boats. **NT**

White-chinned Petrel *Procellaria aequinoctialis*   55 cm
Large, sturdy, long-winged petrel with heavy, pale-tipped bill. From giant petrels (Plate 1) by smaller size and *uniform brownish-black coloration, except for small white chin patch*, variable in size. Slightly paler outer underwing. Feet barely project beyond short, slightly wedge-shaped tail. Spectacled Petrel has striking white spectacles and dark tip to bill. **Voice** Generally silent at sea. **SHH** Breeds islands in S Atlantic and Indian Oceans, and Australasia, dispersing mostly S of 30°S. Vagrant to Mauritius, May. Powerful flight with deep wingbeats and occasional low glides on slightly bowed wings. Gregarious, often follows fishing boats. **VU**

Spectacled Petrel *Procellaria conspicillata*   55 cm
Large, sturdy, long-winged petrel. From White-chinned Petrel by *broad white bands forming crescent from base of bill to side of rear crown, and dark bill tip*. **Voice** Generally silent at sea. **SHH** Breeds Tristan da Cunha. Vagrant Reunion, 19th century (when range possibly wider). Powerful flight with deep wingbeats and occasional low glides on slightly bowed wings. Gregarious, often follows fishing boats. **VU**

PLATE 95: VAGRANTS

Streaked Shearwater *Calonectris leucomelas* — 48 cm
Large, broad-winged shearwater with long neck; from other seabirds by *brownish upperparts and pale underparts*. From Cory's Shearwater (Plate 4) by *mainly white head with dark brown streaks increasing in intensity towards hindcrown and neck*, outlining white eye-ring, upperparts and upperwing mainly grey-brown with *broad whitish fringes to feathers giving a scaly appearance*, and dark mark on underside of primary coverts. Long, slender, pale horn bill tipped darker. Imm Kelp Gull flies higher, constantly flapping with bowed, flexible wings. **Voice** Generally silent at sea. **SHH** Breeds islands in NW Pacific, Mar–Nov, dispersing S to Australian waters, and entering Indian Ocean, mostly in N. Vagrant recorded off SW Madagascar, Dec. Gregarious, often in mixed flocks, attracted to fishing vessels. Lazy, direct low-level flight on slightly bowed wings.

Short-tailed Shearwater *Puffinus tenuirostris* — 42 cm
Medium-sized shearwater with long, narrow, slightly rounded wings and short, slender stubby bill. From potential (could be more frequent?) vagrant Sooty Shearwater *P. griseus* (with great difficulty) by steeper forehead with shorter bill, *more contrast between hood and body, duller whitish or pale brown underwing* with rectangular (not wedge-shaped) pale panel on underwing-coverts; flight of Short-tailed more buoyant and wingbeats less fluid, more relaxed. From Wedge-tailed and Flesh-footed Shearwaters by pale underwing, smaller size, *shorter bill and tail, narrower wings and more hurried flight with stiffer wingbeats*. **Voice** Generally silent at sea. **SHH** Breeds islands around Australia, dispersing May–Sep to Pacific. Vagrant to Rodrigues, Jun; one, Madagascar, either this species or Sooty Shearwater. Rapid wingbeats and glides, smooth shearing flight mostly low over waves. Gregarious, often in association with other seabirds, often follows fishing boats.

Little Shearwater *Puffinus assimilis* — 27 cm
Small, compact, black-and-white shearwater with short rounded wings and short tail. Slightly smaller and paler than Tropical Shearwater (Plate 4), with shorter bill, wings and tail, *much whiter face than other small shearwaters, blacker upperparts, whiter undertail* and paler underwing with narrower trailing edge. Legs bluish with pink webs. **Voice** Generally silent at sea. **SHH** Breeds islands of all oceans S of Tropic of Capricorn. Vagrant Mauritius. Generally solitary, avoiding ships. Low, hurried flap-and-glide flight, with very rapid wingbeats interspersed by glides, wings held arched with wingtips parallel to surface.

Kerguelen Petrel *Aphrodroma brevirostris* — 35 cm
Medium-sized, bull-necked petrel. From other medium-sized, all-dark petrels by *body slightly paler than head, slate-grey underwing with silvery sheen to flight feathers and whitish leading edge to inner underwing*. Rare dark-morph Soft-plumaged Petrel (Plate 2) very similar, but more slender, has longer bill, grey underwing and different flight. **Voice** Generally silent at sea. **SHH** Breeds on subantarctic islands of Indian and Atlantic Oceans, dispersing N to $c.33°$S. Subject to occasional 'wrecks' and vagrant N to Somalia, May–Sep; one vagrant, Reunion. Highly pelagic, solitary, rarely follows ships. In strong winds, *flight unique, flying in effortless arcs high above surface and often gliding into wind, maintaining position with falcon-like jerks of wing*.

Black-winged Petrel *Pterodroma nigripennis* — 30 cm
Small, long-tailed, slender petrel, with long, fairly broad wings and boldly contrasting underwing. From larger Barau's Petrel (Plate 2) by blackish eye-patch contrasting with pale grey nape and crown, and bold blackish band on underwing from carpal angle to central inner wing. Soft-plumaged Petrel is similar but lacks distinctive underwing pattern. **Voice** Silent in region. **SHH** Breeds tropical and subtropical islands of SW Pacific, dispersing to NW Pacific, Jul–Nov, reaching 40°N. Vagrant Mauritius. Gregarious, but rarely follows ships. Rapid flight, sometimes soaring high and hanging stationary facing the wind. In strong winds, rocks erratically side to side.

PLATE 96: VAGRANTS

Rockhopper Penguin *Eudyptes chrysocome*   60 cm
Only penguin recorded in the region. Small penguin with long, stiff black-and-yellow plumes. Short, stubby reddish bill; reddish eyes and legs. Imm lacks crest, bill dull brown, throat grey. **Voice** Harsh staccato barks and brays. **SHH** Breeds islands in Southern Ocean, pelagic in austral winter. Vagrant S Madagascar. Proceeds on land with stiff-legged hops but can walk normally. Often treated as two species, Northern *E. moseleyi* and Southern *E. chrysocome*. Malagasy region records not assigned; Northern is larger with longer, thicker yellow crest, head and upperparts tinged bluish-slate. **VU** (Southern) or **EN** (Northern).

Red-billed Tropicbird *Phaethon aethereus*   41 cm + 50 cm tail-streamers
Large tropicbird; ad from other species (Plate 6) by *white tail-streamers and large red or orange-red bill; only tropicbird with both black outer primaries and primary coverts and black eye-patches joining around collar*. Mantle, back, scapulars and rump have fine black barring appearing grey at distance. Imm lacks tail-streamers; from imm Red-tailed by extensive black on primaries; with difficulty from imm White-tailed by somewhat denser barring above, broader black eye-stripes meeting on nape. **Voice** Generally silent at sea. May give loud, shrill screams. **SHH** Breeds on islands in tropical and subtropical Atlantic Pacific and N Indian Oceans. Vagrant Seychelles, Jun– Oct; unconfirmed reports from Madagascar. Powerful, pigeon-like flight action with slow wingbeats on broad wings. Plunge-dives, but flying-fish sometimes taken in flight.

Cape Gannet *Morus capensis*   87 cm
Typical black-and-white gannet with a *black tail. Juv blackish-brown with narrow white streaks and spots*, much blacker than juv Australasian Gannet, white spots on underparts soon becoming mainly white. Ad from ad Masked Booby by yellow not white head, and grey not yellow bill; lacks mask. Ad white-morph Red-footed Booby has white tail. **Voice** Unlikely to be heard in region. **SHH** Breeds on coasts of South Africa and Namibia, Sep–Apr, dispersing NE along coast. Vagrant Madagascar and Reunion (juvs), Mar–Apr. Coastal, rarely far from land. Powerful stiff-winged flight with rapid wingbeats and glides. Often follows fishing boats. **VU**

Australasian Gannet *Morus serrator*   87 cm
Similar to Cape Gannet; ad only separable when perched or swimming at close range; has *white tail with black centre, shorter gular stripe and less black on face*; imm has *larger white spots on dark feathers*. **Voice** Not likely to be heard in region. **SHH** Breeds austral summer Australia and New Zealand, dispersing N to Tropic of Capricorn. Vagrant Mauritius, Feb. Coastal, rarely far from land. Powerful stiff-winged flight with rapid wingbeats and glides. Often follows fishing boats. Catches prey by plunge-diving.

PLATE 97: VAGRANTS

Pink-backed Pelican *Pelecanus rufescens* — 146 cm
Only pelican recorded in the region. Huge, mainly greyish-white with all-yellow bill, pouch and orbital skin, brown eyes, and grey legs and feet. In flight, shows brownish-grey secondaries and dark wingtips above and below. Juv is much browner than ad, bill pale yellow, pouch grey and feet greyish-pink. **Voice** Non-breeders generally silent. **SHH** Breeds tropical and subtropical Africa. Former resident of Madagascar and Seychelles (St Joseph's Atoll), now only vagrant to Madagascar. Favours sheltered coasts, large lakes and rivers. A solitary hunter, catching fish with a snapping action.

Great Cormorant *Phalacrocorax carbo* — 100 cm
Large cormorant with thick neck and angular head. Ad from Reed Cormorant (Plate 7) by much larger size, shorter tail and longer neck, *yellow patch of bare skin around eye and bill base*; race *lucidus* recorded in region has white chin extending to throat and upper breast. Dull grey bill with black culmen. Imm has dull white extending from throat to undertail-coverts. **Voice** Non-breeders generally silent. **SHH** Widespread from Arctic to tropics. Vagrant Seychelles, Jan. Mainly coastal. Goose-like flight with shallow wingbeats. When perched, often opens wings in characteristic 'wing-drying' posture.

Eurasian Bittern *Botaurus stellaris* — 75 cm
Large bittern, mainly tawny-buff, mottled and barred; from imm Black-crowned Night Heron (Plate 9) by black crown to nape contrasting with neck and throat; black moustache. In flight has broad rounded outer primaries, dark-barred rufous flight-feathers contrasting with pale buff-speckled wing-coverts and rump. Imm less boldly marked, browner crown less extensive towards nape, browner moustachial streak, upperwing-coverts paler, less buff and more strongly vermiculated, and has narrow pointed outer primaries. **Voice** Non-breeders generally silent. **SHH** Breeds Eurasia at 40–60°N, dispersing S to equator. Vagrant Seychelles S to Alphonse, Oct–Nov. Favours freshwater marshes. Solitary and skulking, only taking flight when disturbed.

White Stork *Ciconia ciconia* — 122 cm
Large stork, mainly white with contrasting black greater coverts and flight-feathers, like Yellow-billed Stork (Plate 11), but *legs red* not pale pink, *bill red* not yellow, lacks red on face. Imm has duller plumage and bare parts. **Voice** Non-breeders generally silent. **SHH** Breeds Europe to Turkey, spending boreal winter in sub-Saharan Africa to South Africa. Vagrant Seychelles, Dec–Apr. Favours open grassland. Flies with slow wingbeats, neck outstretched and slightly lowered; soars readily.

African Sacred Ibis *Threskiornis aethiopicus* — 80 cm
Large black-and-white ibis. From very similar Madagascar Sacred Ibis (Plate 12) by black tips to primaries and secondaries forming a *narrow black trailing edge to wing in flight*, thicker bill (though not obvious on a lone bird) and *dark (not white) eye*. **Voice** Non-breeders generally silent. Flight call a harsh croak. **SHH** Mainly resident in sub-Saharan Africa to South Africa. Vagrant to Aldabra. Less restricted to estuarine and coastal sites than Madagascar Sacred Ibis. Flies with rapid wingbeats, neck outstretched.

PLATE 98: VAGRANTS

Black-headed Heron *Ardea melanocephala* 92 cm
Large heron; from Grey and Madagascar Herons (Plate 8) by *black crown and back of neck, white chin* (variable in extent), white spots on foreneck (sometimes lacking), short deep bill and slate-grey legs. Distinctive in flight with *two-tone underwing, rear half black*. Juv mainly grey with brownish-grey neck. **Voice** A raucous croak or loud nasal *kuark*. **SHH** Breeds sub-Saharan Africa. Vagrant Madagascar. Often found away from water in grassland, cultivation and forest clearings, but also at coasts and freshwater sites.

Goliath Heron *Ardea goliath* 150 cm
From Madagascar, Grey and Purple Herons (Plate 8) by *enormous size with very long tibia*, massive blackish bill, slate-grey upperparts and rich rufous underparts. Juv has head/neck paler and duller, with rusty buff edging above and paler underparts. **Voice** A deep, raucous *kowoork-kowoork-worrk-worrk*. **SHH** Resident mainly sub-Saharan Africa. Vagrant Madagascar. Fresh water, hunts in deeper water than other herons. Slow flight, legs dangling.

Intermediate Egret *Ardea intermedia* 69 cm
From Great Egret (Plate 10) by relatively *short yellow bill often tipped blackish*, yellow-brown tibia, blackish-brown tarsus and feet, and *gape line restricted to in front of eye*; smooth S-shape to neck, less kinked. Cattle Egret much smaller with shorter bill, Little Egret also smaller with yellow toes and thin black bill. Ad breeding has long white plumes on back and breast, brighter bill, bright green lores, red eye and pinkish tibia. **Voice** Sometimes gives a low *quawrk*. **SHH** Sub-Saharan Africa and S & SE Asia. Vagrant Seychelles, Mar–Oct. Frequents freshwater margins. Buoyant flight with slow wingbeats.

Little Egret *Egretta garzetta garzetta* 64 cm
Nominate race of Little Egret reaches Malagasy region as a vagrant; resident birds, race *dimorpha*, sometimes treated as a separate species, 'Dimorphic Egret'; see that form (Plate 10) for differences of both Little Egret races from other non-vagrant herons. Both are medium-sized, all-white, slender, long-necked herons with long black bills; separation from *dimorpha* very difficult; nominate nearly always white-plumaged with yellow feet and *blue-grey lores*; when breeding, feet and lores become yellowish, orange or bright pink. In *dimorpha*, dark birds sometimes outnumber white (especially on coast), and some intermediates bluish-grey speckled white; *feet yellow or olive, this colour sometimes extending up tarsi*, perhaps diagnostically, and lores generally bright yellow turning deep pink when breeding, as do feet. Outside courtship period, yellow lores of *dimorpha* usually distinct from blue-grey of nominate, but birds with bluish facial skin recorded in Madagascar suspected to be *dimorpha*. Bill of *dimorpha* reputedly longer and thicker-based than nominate, but no consistent difference confirmed. **Voice** A harsh croak when disturbed. **SHH** Wide breeding distribution in Eurasia and Africa. Near-annual vagrant to granitic Seychelles and Amirantes in all months. Frequents fresh water and sheltered coasts. Leisurely flight, sometimes gliding.

Indian Pond Heron *Ardeola grayii* 46 cm
Typical pond heron; ad breeding from Squacco Heron (Plate 8) by head and neck unstreaked pale yellowish-buff with long white nape plumes and *deep reddish-brown (not buff) mantle and scapulars*. Often shows diagnostic *dark line on lores with yellow stripe above*. Upper breast heavily streaked grey-brown, pale ground colour to crown. In flight *uniform grey-brown back* contrasts with pure white wings and tail. Imm has breast more spotted than streaked; dark shafts to primaries, two outermost smudged grey and next three with grey tips. Imm/winter Madagascar Pond Heron (Plate 8) is darker, more boldly streaked below, with streaked mantle. Imm Squacco slightly paler but plumage very similar in tone and perhaps not safely separated in field. **Voice** A deep croak when disturbed. **SHH** Breeds S Asia. Vagrant Seychelles, Sep–Feb. Mainly freshwater margins. Flight fast and agile.

Little Bittern *Ixobrychus minutus minutus* 35 cm
Larger, paler and much less rufous at all ages than Madagascar race *podiceps* (see Plate 9); wings more pointed. Similar to Yellow Bittern, but *blackish (not brown) back and whitish to grey-buff (not yellow-buff) wing patches*. Imm mainly dark blackish-brown with pale mid-wing (in Yellow, pale in inner wing extends across the back); upperparts dark buff-brown with paler streaking (pale with darker markings in Yellow). **Voice** Flight call a sharp rasping *eerk...eerk*. **SHH** Nominate race breeds Europe to S Asia, some migrating to sub-Saharan Africa in austral summer. Vagrant Seychelles Nov (assumed nominate), and may occur in Madagascar.

Cinnamon Bittern *Ixobrychus cinnamomeus* 41 cm
Small bittern; ad from other small herons (Plate 9) by *uniform dark rufous wings and back*. ♂ usually has white malar stripe and darker stripe on foreneck, underparts warm rufous-buff with black spots on sides of breast, greenish-orange bill and yellowish-green legs. ♀ is slightly darker and browner above, and duller below; imm similar to ♀ but more heavily streaked brown on breast and belly, with prominent buff fringes to scapulars/wing-coverts. Yellow Bittern and Little Bittern (Plate 9) have pale wing-coverts contrasting sharply with black flight-feathers. **Voice** In flight calls *krek-krek-krek*. **SHH** Breeds S Asia. Vagrant granitic Seychelles, Oct–Jan. Freshwater margins, sometimes on coasts. Solitary and skulking. In flight wingbeats are rapid and jerky.

PLATE 99: VAGRANTS

Black-necked Grebe *Podiceps nigricollis*   33 cm
Medium-sized, chunky grebe. Non-breeding ad from Little Grebe (Plate 13) by *steep forehead peaking behind eye and small sharp uptilted bill, black patch on ear-coverts* contrasting with white chin and sides to nape. In flight, white secondaries and innermost primaries on an otherwise all-dark wing. Ad breeding blacker with yellow plumes behind eye, chestnut flanks and red eyes. Juv has diffuse, dark chestnut streaking on sides of neck and ear-coverts. **Voice** Generally silent. Alarm call *whit-whit-whit*. **SHH** Breeds Europe to EC Asia, Africa. Vagrant Seychelles, Dec (where Little Grebe not recorded). Frequents shallow inland waters and sheltered bays. Flight weak and fluttering, but strong over distance. Dives to feed.

Eurasian Wigeon *Anas penelope*   51 cm
Short-necked duck; from other possible vagrant ducks by *large rounded head, short bill and white belly-patch*. ♀ greyish or dull rufous-brown, finely mottled, with paler head, blackish around eye. Breeding ♂ has chestnut head with creamy-buff forehead and crown, pinkish-grey breast, white line along wing, white patch on rear flanks and black undertail-coverts. Eclipse ♂ is more rufous, with breast and sides chestnut. In flight, *white patch on upper forewing and green speculum*, pale grey underwing with darker leading edge and flight-feathers. Pale blue-grey bill with black tip. Imm similar to ♀, white belly often mottled and speculum duller. **Voice** ♂ a loud whistle, *wheeeooo*, ♀ a low *karrr*. **SHH** Breeds N Palearctic wintering S to E Africa. Vagrant Seychelles, Feb–Mar. Freshwater sites, sometimes coasts. Straight, rapid flight.

Northern Pintail *Anas acuta*   66 cm (♂), 51 cm (♀)
Slender, elegant duck, ♀ from other vagrant ducks by mottled brown body, finer and paler on breast with *uniform pale brown head*. Eclipse ♂ has greyer upperwing and longer grey scapulars. ♂ has distinctive brown head and hindneck, and white breast extending up sides of neck. *In flight has distinctive silhouette* with long thin neck, and bronze-green speculum bordered buff above and white below. Imm has plainer, darker upperparts, more spotted underparts and cinnamon head and neck. See ♀ Mallard for other differences. **Voice** ♂ a long *greee*, ♀ a deep quack. **SHH** Breeds across most of North America and Eurasia, wintering S to E Africa. Vagrant Seychelles, Nov–Feb, often 2–4 birds. Frequents mainly shallow fresh water, sometimes sheltered coasts. Swift flight.

Northern Shoveler *Anas clypeata*   51 cm
Medium-sized duck, from other vagrant ducks by *distinctive long spatulate bill*. Non-breeders mottled brown and buff, head paler, grey-green speculum, grey-brown bill with orange at sides and base, ♀ browner and ♂ has rufous tones to flanks. In flight appears front-heavy, ♂ having *bluish forewing and green speculum with white border*, and whitish underwing. **Voice** Mainly silent. **SHH** Breeds Holarctic, wintering S to E Africa. Vagrant Seychelles, Oct–Jan. Favours freshwater areas, occasionally on coasts. Rapid flight.

Ferruginous Duck *Aythya nyroca*   41 cm
Small, uniform diving duck, from others except Madagascar Pochard by *peaked crown, generally rufous plumage and white undertail*; ♂ has white eye and is brighter rufous; ♀ duller and browner with brown eye. In flight dark plumage contrasts with white wingbar, belly and undertail. ♀ Tufted Duck is darker, browner, head with short tuft, has shorter blue-grey bill with variable white blaze at base; may show white undertail, but never as extensive. From very similar Madagascar Pochard (Plate 14), with difficulty, by smaller size, peaked head, much slenderer and shorter bill. **Voice** ♂ calls *gek* or whistles, ♀ gives a quiet *krrr*. **SHH** Breeds W Europe to Mongolia, wintering to S, vagrant to E Africa. Vagrant Seychelles, Nov–Dec and Apr–May. Prefers shallow fresh water. Feeds by both dabbling and diving. **NT**

Tufted Duck *Aythya fuligula*   46 cm
Medium-sized, compact diving duck with short neck, large rounded head and *short broad bluish-grey bill with black tip*. ♂ distinctive; see Ferruginous Duck for differences from that and other species. **Voice** ♂ may give a low whistle; ♀ may growl in flight. **SHH** Breeds across N Palearctic, wintering S to equator; scarce E Africa. Vagrant Seychelles, Dec. Frequents freshwater sites, occasionally sheltered coasts. Rapid, straight flight. Dives or sometimes feeds at surface.

PLATE 100: VAGRANTS

Osprey *Pandion haliaetus* — 58 cm
Distinctive large, long-winged, raptor with *uniform grey-brown upperparts, white underparts and dark breast-band. Head white except dark eye-band. Distinctive flight silhouette*, with sharp bend at carpal joint, short, square, banded tail and black carpal-patches on underwing. Imm paler brown than ad, with speckled back and wing-coverts. Pale-morph Booted Eagle lacks all-brown upperwing and back, has white uppertail-coverts, white underwing-coverts, blackish flight-feathers, longer tail and broader wings. Pale-morph European Honey-buzzard has obvious dark trailing edge to underwing, wings shorter and broader. In Madagascar from imm Madagascar Fish Eagle (Plate 17) by smaller size, pale underwing and dark eyestripe. **Voice** Non-breeders silent. **SHH** Breeds North America, N Eurasia, wintering to S including sub-Saharan Africa to South Africa. Vagrant to Madagascar, Seychelles and Comoros in boreal winter. *Hunts over water, plunging to take fish.* Powerful wingbeats, gliding with carpal joint pushed forward.

European Honey-buzzard *Pernis apivorus* — 58 cm
Small-headed hawk with variable plumage, but always shows *black carpal-patches.* ♂ is ashy-grey above with broad black trailing edge, barred below, bars well spaced with black tips to primaries. ♀ is darker brown, has barring less distinct and lacks contrasting trailing edge. Dark morph is dark brown below. Pale morph mainly white below, coarsely streaked on breast. Imm usually darker on tail with 4–5 bands, wingtips more extensively dark, with heavier barring and yellow cere. Booted Eagle has paler undertail and pale patch across upperwing-coverts and scapulars. Black Kite (Plate 17) more uniformly dark, lacking obvious barring on underwing, with pale diagonal bands across upperwing-coverts and forked tail. Harriers have longer tails and slenderer bodies and wings. From Madagascar Buzzard (Plate 16) by lack of dark bar on underwing marginal coverts, dark rump, longer tail and narrower head; Madagascar Cuckoo-hawk (Plate 16) has conspicuous dark barring on primaries and short, feathered tarsi. **Voice** Non-breeders silent. **SHH** Breeds N Europe to Siberia, spending boreal winter mainly in sub-Saharan Africa. Vagrant Seychelles (S to Amirantes), possibly also Madagascar. Soars, wings held straight out, tail fanned.

Booted Eagle *Hieraaetus pennatus* — 45 cm
Small eagle, from medium-sized raptors by *square-cut tail with pointed corners, all morphs with broad buff band on upperwing-coverts*, buff scapulars and a pale crescent on uppertail-coverts. Pale morph has mainly white underbody and underwing-coverts contrasting with black flight-feathers and grey tail darker at tip and in centre. *Small white patches at junction of neck and wing show up as 'headlights'.* Intermediate morph below has a rufous wash to body and wing-coverts. Black Kite (Plate 17) has more 'aimless' flight with shallow wingbeats, slimmer longer, more angular wings, dark uppertail-coverts, longer notched tail and no 'headlights'. Harriers are longer-winged. **Voice** Non-breeders silent. **SHH** Breeds SW Europe to Mongolia, spending boreal winter S to sub-Saharan Africa and Indian subcontinent. Vagrant Seychelles, Nov. Favours forest, mixed and open woodland. Rapid flight with deep, powerful wingbeats, frequently gliding and soaring.

Long-crested Eagle *Lophaetus occipitalis* — 53 cm
Distinctive blackish eagle. From other raptors *by long, loose crest; in flight shows white underwing-coverts and conspicuous white wing-patches above and below near carpal joint.* **Voice** Highly vocal. A loud high-pitched scream, *keeeee-eh*, or long series of sharp calls, *kik-kik-kik-kik-keeeeh*, repeated for several minutes. **SHH** Breeds sub-Saharan Africa. Vagrant to Madagascar. Woodland margins, adjacent marsh and grass areas, plantations and cultivation; spends long periods perched.

PLATE 101: VAGRANTS

Western Marsh Harrier *Circus aeruginosus* — 48 cm
Long-winged, long-tailed, slender raptor. ♂ from other harriers by *rufous underparts and chest streaked dark brown, contrasting with grey wings and tail, and a grey crown*. Black primaries contrast with grey secondaries and dark brown back, rump and upperwing-coverts. ♀ and imm distinctive, dark brown, lacking white rump, with red-brown tail, creamy-buff crown, throat and (♀) forewing, and often a pale patch on the breast. Imm browner overall with darker underwing-coverts, no breast-band, rarely an obvious creamy forewing and more golden-rufous fringes to crown. Black Kite (Plate 17) has slightly forked not rounded tail, more 'fingered' wingtips and less angled wings. **Voice** Normally silent; occasional chattering *kek-kek* or wailing *pee-yu*. **SHH** Breeds Europe to C Asia, spending boreal winter S to sub-Saharan Africa. Vagrant to granitic Seychelles, Jan–Feb. Favours wetlands but other habitats often used on migration. Flies low, wings held in a shallow 'V', pausing before dropping suddenly onto prey; also soars high.

Pallid Harrier *Circus macrourus* — 45 cm
Slim, narrow-winged, long-tailed harrier. ♂ *pearl grey above with a black wedge in wingtips*, but no black bar on closed wing, white below. ♀/imm very similar to resident harriers (Plate 16) but slimmer, showing only three projecting primaries; *dark cheek-crescent contrasting with pale collar and darker neck*. Imm is unstreaked pale rufous-buff below. Potential vagrant ♀/imm Montagu's Harrier *C. pygargus* is darker on belly, lacks pale collar and dark on neck. **Voice** Non-breeders silent. **SHH** Breeds Ukraine to China, spending boreal winter S to sub-Saharan Africa and India. Vagrant Seychelles. Favours open grassland. Hunts low, pouncing on prey. **NT**

Black-winged Kite *Elanus caeruleus* — 33 cm
Small, distinctive raptor with broad pointed wings, short square-cut tail, *white plumage and black wingtips*. Imm has rufous-tinged breast, dark grey upperparts feathers tipped white, and brownish crown. Male Pallid Harrier is much larger and longer-tailed. **Voice** Generally silent. **SHH** Breeds sub-Saharan Africa to SE Asia. Vagrant to Madagascar, Sep–Oct. Usually in open grassland with scattered trees. Graceful flight, soars and glides with raised wings, frequently hovering. [Alt: Black-shouldered Kite]

Black Kite *Milvus migrans migrans* — 55 cm
Distinctive nominate race of Black Kite, reaches Malagasy region as a vagrant; resident birds, race *parasitus*, sometimes treated as a separate species, 'Yellow-billed Kite', which see (Plate 17) for differences between both Black Kite races and other non-vagrant raptors. *M. m. migrans* is larger than *parasitus*, with paler head streaked markedly darker, darker brown body above and below, and more conspicuous pale under-carpal patch; dark bill not diagnostic, as juv *parasitus* has this too. Western Marsh Harrier holds more pointed wings in a shallow 'V' and tail is not forked. Dark-morph Booted Eagle (Plate 100) has larger head, unbarred rounded tail, pale uppertail-coverts and scapulars, and lacks pale 'window' on underwing. **Voice** Typically a whining *piiieeerrrrrr*, starting high-pitched, becoming higher still, ending in lower, slightly trilled whistle. **SHH** Breeds Palearctic from Europe to N Asia and Japan, spending boreal winter to South Africa and SE Asia. Vagrant Seychelles, Dec–Mar. Frequents a wide variety of habitats. Soars and glides with wings held flat or slightly arched, never raised in a 'V'.

PLATE 102: VAGRANTS

Lesser Kestrel *Falco naumanni* 28 cm
Slim falcon; from resident kestrels and Common Kestrel by long wings, *tail projecting only slightly beyond wingtips, and white claws*. ♂ *distinctive; unspotted red-brown back, narrow blue-grey wing-panel*, buff breast with small black spots, blue-grey hood and no moustache. *Paler underwing with darker wingtip than Common Kestrel.* ♀/imm from Common Kestrel by pale cheeks, no dark eyestripe, less streaked and spotted above and below, and underwing whitish. Imm has rump and undertail-coverts often tinged grey, more spotting on underwing-coverts, pale barring on flight-feathers. **Voice** A diagnostic trisyllabic shrill *kye-ki-ki* or *tye-tye*. **SHH** Breeds S Europe to Mongolia, spending boreal winter mainly in sub-Saharan Africa and S Arabia. Vagrant Seychelles and Comoros, Oct–May. Favours open areas, taking insects by stooping after a brief hover.

Common Kestrel *Falco tinnunculus* 30 cm
See Lesser Kestrel for distinctions from that species and local species. *At rest tail projects well beyond wingtips.* ♂ *has small black spots on rufous back*, grey finely streaked head with dark moustache and heavy black spotting on pale buff underparts. *Both sexes have black claws*. Imm more heavily streaked below, paler red-brown above than ♀; imm ♂ may show grey uppertail and rump barred blackish. **Voice** A shrill *kee-kee-kee-kee-kee*. **SHH** Breeds Africa, across Eurasia, sedentary in S & W but nominate race migratory in N & E. Vagrant Seychelles, Jan. In open woodland, grassland, cultivation, wetlands and urban areas. Frequently hovers, tail fanned. Rapid continuous flight, long wings held straight out with little gliding.

Saker Falcon *Falco cherrug* 50 cm
Large, long-tailed falcon with broad blunt-tipped wings. Larger and longer-tailed than Peregrine, which has more pointed wings, and is barred not streaked on underparts. Ad from imm Peregrine by tawny-brown upperparts and inner wing with rufous fringes contrasting with darker grey flight-feathers and barred tail; also streaking on underparts heaviest on flanks and thighs, and often has dark 'trousers'. Underwing pale with contrasting dark forewing. At rest long tail extends beyond wingtips, folded uppertail appearing uniform brown. Imm has darker brown upperparts and head, more heavily marked crown emphasising distinct supercilium, and is more heavily streaked below. **Voice** Generally silent. **SHH** Nominate race (most likely in region) breeds E Europe to Mongolia, spending boreal winter S to E Africa, where locally common Feb–Apr. Vagrant Seychelles, Dec. Favours forest next to open areas. Rapid flight with slow, shallow wingbeats, hunting low or from perch. Sometimes hovers. **EN**

'Tundra' Peregrine Falcon *Falco peregrinus calidus* 44 cm (♂), 40 cm (♀)
Paler migratory race of Peregrine Falcon, differing from resident *radama* on Madagascar and Comoros by *larger size, paler and more finely barred breast, larger white cheek-patches* and narrower moustache. Imm paler brown above, whiter below with brown streaking and with obvious pale supercilium. Large, thickset falcon with relatively short tapered tail and pointed wingtips. From Sooty and Eleonora's Falcons by same distinctions as resident birds (Plate 20); see also Saker Falcon. **Voice** Generally silent, may give a gruff, slightly muffled barking. **SHH** Species of worldwide distribution, this race breeding Lapland to NE Siberia, a long-distance migrant to SE Asia, New Guinea, S Africa. Vagrant Seychelles (probably this race), Mauritius (confirmed *calidus*), possibly Reunion, Nov–Dec; may be regular (overlooked) in region. Frequents coasts and open habitats. Rapid flight with rapid, stiff wingbeats, gaining height rapidly.

PLATE 103: VAGRANTS

Red-footed Falcon *Falco vespertinus* — 30 cm
Medium-sized falcon with red-orange cere and legs. ♂ from Amur Falcon by *dark underwing and uniform grey cheeks* without distinct moustache. ♀ *has orangey breast, underwing-coverts, crown and nape*, blue-grey upperparts, black flight-feathers, narrowly barred tail and black moustache, lacking distinct dark streaks or spots of Amur. Imm has pale underparts streaked reddish-brown, heavily barred underwing, distinct buff wash to underwing-coverts, *sandy-brown crown lacking grey tones, short untapered moustache*, mantle not as dark and a narrower dark trailing edge than imm Amur. Dark-morph Eleonora's and Sooty Falcons (Plate 20) are larger, longer-winged and -tailed, lack chestnut belly and thighs, and have yellow not red legs. Imm Eurasian Hobby has unbarred upperparts, heavily streaked on breast, no trace of rufous wash on head or breast, entirely black central tail-feathers and more uniformly barred darker underwing. **Voice** Generally silent. May give *kew kew kew* in flight. **SHH** Breeds E Europe to China, spending boreal winter mainly in S Africa. Vagrant Seychelles, Nov–Dec; much scarcer than Amur Falcon. Favours open grassland with trees and cultivated areas. Fast, agile flight, sometimes gliding and often hovers. **NT**

Amur Falcon *Falco amurensis* — 28 cm
Medium-sized falcon with red-orange cere and legs; see Red-footed Falcon for differences from that species. ♀ has white ground colour to underparts – Eurasian Hobby and Eleonora's Falcon (Plate 20) both pale but not pure white – with heavy black streaking turning to blotches on lower flanks. White cheeks bordered at rear by second collar-mark, similar to Hobby but moustache more tapered, less broad. **Voice** Flight call a shrill *kew-kew-kew*. **SHH** Breeds NE Asia, spending boreal winter in E Africa to South Africa. Annual Seychelles, from Bird I south to Alphonse, mainly Nov–Jan; vagrant Comoros. Frequents open areas with adjacent trees including grassy airstrips. Rapid flight, sometimes hovering.

Eurasian Hobby *Falco subbuteo* — 30 cm
Elegant, long-winged falcon with medium-length square-cut tail, wings projecting beyond tail at rest. Eleonora's Falcon (Plate 20) is much larger, more heavily built, wingtips fall level with tail at rest, and head smaller in proportion; less contrasting more rounded pale cheeks, upperparts less blue-grey more dark brown, moustache more pointed, crown often paler than sides of head, supercilium less obvious or non-existent, underparts streaking narrower and less dense and no 'trousers'; in flight dark underwing-coverts contrast with paler flight-feathers; first-winter has less white more buff on cheeks and throat, upperparts browner (no blue-grey tones), with more numerous, broader pale V-shaped (not rounded) fringes. Imm Sooty Falcon (Plate 20) larger, paler with slate-grey upperparts, buffy wash on thinly streaked breast and no supercilium. Red-footed and Amur Falcons can also show rufous 'trousers', but ♂ distinctive and ♀/imm have barred upperparts and tail. **Voice** Generally silent. **SHH** Nominate race breeds across Eurasia, spending boreal winter to C & S Africa and S Asia. Vagrant Seychelles, Oct–Jan. Favours woodland and cultivated areas. Dashing flight with stiff wingbeats interspersed by long glides, feeding almost entirely on wing.

PLATE 104: VAGRANTS

Buff-banded Rail *Gallirallus philippensis* — 29 cm
Robust rail with distinctive head pattern and fairly long tail. *Rufous facial stripe and hindneck, white chin, grey throat and buff breast-band*, rest of underparts barred black and white. Imm duller and less boldly marked. **Voice** A harsh squeaky *crek* when flushed. Also a low growl and a harsh *creek* repeated 4–5 times, probably territorial. **SHH** Breeds Australasia. Vagrant Mauritius. Walks with tail erect, which it flicks persistently. Occurs in all types of freshwater wetlands as well as mangroves and coasts.

Corn Crake *Crex crex* — 28 cm
Large, tawny-buff rail, streaked above with rows of black spots, *grey-white flanks with chestnut barring* and plain face with blue-grey supercilium. In flight shows striking bright chestnut-red wings. Imm more buff above and below, with little or no barring on flanks. **Voice** Non-breeders silent. **SHH** Breeds Europe to China, spending boreal winter in E Africa to South Africa. Vagrant Seychelles, Oct–Jan. Favours dense grassland, cultivation and meadows. Tame but skulking. Weak flight usually over short distances, legs dangling, but much stronger over distance.

White-breasted Waterhen *Amaurornis phoenicurus* — 32 cm
Distinctive ad has *black crown, hindneck, back, wings, tail and flanks, white face, throat, breast and belly, and bright cinnamon-brown undertail-coverts*. Bill and legs greenish. Imm has whitish face and breast flecked black or brown, dark olive-brown upperparts, duller undertail-coverts and yellow-green eye. **Voice** Non-breeders silent. **SHH** Nominate breeds Asia, also Sri Lanka and Maldives. N populations are partial migrants including W to NW Arabian Peninsula. Vagrant Seychelles, Dec. Freshwater margins, damp scrub and forest margins. Fluttering flight.

Little Crake *Porzana parva* — 19 cm
Small crake with long primary projection and long tail. ♂ from Baillon's Crake (resident in Madagascar and potential vagrant elsewhere; Plate 24) by *sparse white streaking on mantle and scapulars, less extensive barring on flanks, red base to green bill and long primary projection*, extending conspicuously beyond tertials. ♀/imm from imm Baillon's with difficulty by whitish not barred greyish breast, hint of red bill base and long primaries. Imm similar to ♀ with whiter face, supercilium and underparts, more extensive barring on sides, whitish chest, and only a hint of red at bill base. Spotted Crake is larger and copiously spotted. **Voice** Alarm a sharp *tyuik*; contact calls, subdued single notes or trills. **SHH** Breeds S Europe to NW China, spending boreal winter to S, and vagrant to E Africa. Vagrant Seychelles, Dec. Frequents still or slow-moving fresh water with dense vegetation. Strong flight.

Spotted Crake *Porzana porzana* — 23 cm
Small, brown crake, from others by *extensive white flecks on upperparts, white spots on neck and breast, pale buff undertail-coverts*, and short yellow bill with red at base in ad. Striped Crake smaller, dark olive-brown above, has obvious white streaking on back and dark rufous-buff undertail. **Voice** Non-breeders generally silent. **SHH** Breeds Europe to China, spending boreal winter to S including sub-Saharan Africa to South Africa. Vagrant Seychelles, Nov–Feb. Frequents freshwater wetlands with dense vegetation and shallow muddy water. Flies reluctantly, keeping to dense vegetation. Sometimes swims.

Striped Crake *Aenigmatolimnas marginalis* — 20 cm
Small crake, distinctive with *rufous lower flanks and undertail, whitish throat and belly*, dark brown wing-coverts striped white and greenish legs with large toes. ♂ has dark cinnamon head, neck and breast, and white throat. ♀ has dark grey forehead and hindneck, pale grey face, breast grey scalloped white and flanks grey edged white. Imm duller and less patterned, with rufous tinge to face and breast, and blue-grey legs. **Voice** Generally silent. **SHH** Intra-African migrant, uncommon in E Africa. Vagrant Seychelles, Dec. Favours grassy areas and shallow pools. Secretive and crepuscular. Rapid flight.

Lesser Moorhen *Gallinula angulata* — 23 cm
Small moorhen; ad from Common Moorhen (widespread in Malagasy region, Plate 26) by *bright yellow bill* and *pointed red frontal shield*, lack of red garter around top of legs and *less conspicuous white flanks*. ♀ has pale grey face, throat and underparts. Imm from Common Moorhen by dusky-yellow bill and from imm Allen's Gallinule (resident Madagascar, vagrant elsewhere, Plate 26) by plainer brown upperparts lacking dark centres to feathers and bluish tinge of Allen's. **Voice** Similar to Common Moorhen but quieter. **SHH** Sub-Saharan Africa; intra-African migrant, uncommon in E Africa. Vagrant Seychelles and possibly Comoros. Favours wetlands and temporarily flooded grassland. Shyer than Common Moorhen; swims and flies less readily.

PLATE 105: VAGRANTS

Eurasian Oystercatcher *Haematopus ostralegus*   43 cm
Very distinctive; *large wader with black head, back and tail contrasting with white underparts*. Red bill, relatively short pink legs, red eye and eye-ring. In flight, shows broad white wingbar and white rump contrasting with black wings and broad black tail-band. Ad non-breeding has white frontal neck-band. Race *longipes* (most likely in region) is browner above than nominate, back noticeably paler than head and chest, and has longer bill with nasal groove extending > half length. Imm has narrower tail-band, dull brownish-black upperparts, pale fringes on back and wing-coverts, pinkish-grey legs, reddish-brown eye with indistinct dull orange eye-ring and extensive black tip to bill, which is orange-red at base. **Voice** A loud shrill *kleep* and an insistent, accelerating *peep-peep-pip-pip-ip-ip-ip*. **SHH** Breeds across N Eurasia, with nominate in W and *longipes* in Russia to W Siberia, spending boreal winter in E Africa, Arabia and India. Vagrant Seychelles and Comoros throughout N winter. Frequents estuaries and sheltered bays. Flight strong and direct with shallow wingbeats.

Pied Avocet *Recurvirostra avosetta*   43 cm
Very distinctive *black-and-white wader, with strongly upturned black bill* and long blue-grey legs. Imm tinged brownish, upperparts mottled brownish and legs duller. Crab-plover (Plate 27) has shorter heavy bill. **Voice** A clear *kluit*, louder and sharper in alarm. **SHH** Breeds Europe to Mongolia spending boreal winter to S including E & S Africa. Vagrant Madagascar, Apr, and Seychelles, Nov–Dec. Frequents saltwater sites including estuaries, lagoons and mudflats. Feeds by picking or scything bill through water. Stiff-winged flight, legs trailing.

Eurasian Stone-curlew *Burhinus oedicnemus*   43 cm
Very distinctive; a thickset curlew-like wader with *pale, mottled, buff-brown plumage*. Large yellow eye. *In flight shows bold wing pattern, black flight-feathers contrasting with striking white patches*. Heavy bill yellow at base, tipped black; pale yellow legs. Imm has rufous-buff fringes to tertials and scapulars, less prominent wingbar and more prominent white tips to greater coverts. **Voice** Non-breeders normally silent. **SHH** Breeds Eurasia and N Africa; European and C Asian populations (nominate, *indicus* and *harterti*) most strongly migratory, nominate reaching E Africa. Vagrant Seychelles, Oct–Feb. Frequents open grassland. Flies low over the ground with shallow wingbeats. Walks with stealth or runs with head and neck jutted forward.

Collared Pratincole *Glareola pratincola*   26 cm
Pratincoles are distinctive short-legged waders with long wings and a long, deeply forked tail. Vagrant pratincoles lack conspicuous white line under eye of Madagascar Pratincole (Plate 27). In flight Collared shows *obvious contrast between pale brown upperwing-coverts and darker flight-feathers, white trailing edge and chestnut underwing-coverts*. Inner primaries have blackish outer webs and paler inner webs, the shaft forming a clear divide, and outer secondaries are paler than inner secondaries. Bill black with *slit-shaped nostril, often has extensive red base to lower mandible*. Juv has buff tips and dark chevrons on back and wing-coverts, and buff not white secondary tips. Other adult vagrant pratincoles are darker above with less contrast in upperwing, shorter-tailed (wings longer than tail), lack obvious white trailing edge and have less red at base of bill. Oriental has darker upperparts, no obvious white trailing edge, little red at base of bill, and inner primaries and all secondaries are uniformly dark; Black-winged has blackish underwing-coverts. However, immature very hard to separate from Oriental. **Voice** Shrill tern-like calls. **SHH** Nominate race breeds S Europe to Kazakhstan and Pakistan, spending boreal winter in sub-Saharan Africa to 5°N. Vagrant Seychelles, mainly Oct–Dec. Two African races are not long-distance migrants. Frequents mudflats, open areas and airstrips. Tern-like flight, catching insects in flight and perching in short grass for long periods.

Oriental Pratincole *Glareola maldivarum*   24 cm
See Collared Pratincole for differences. **Voice** A sharp, tern-like *kyit* or *tyik*. **SHH** Breeds Asia and NE populations are long-distance migrants S to Australia. Vagrant Madagascar, Seychelles and Mauritius mainly Oct–Dec, but also Mar–Apr. Frequents grassland and airstrips. Tern-like flight.

Black-winged Pratincole *Glareola nordmanni*   25 cm
See Collared Pratincole for differences. *Uniquely, shows black underwing-coverts in all plumages*. **Voice** In flight calls *kikeek* and *yup*, lower-pitched than Collared. **SHH** Breeds E Europe to Kazakhstan, spending boreal winter mainly in S Africa. Vagrant Seychelles, throughout N winter. Favours mudflats and grasslands. Tern-like flight. **NT**

PLATE 106: VAGRANTS

Blacksmith Lapwing *Vanellus armatus* — 30 cm
Very distinctive, large, boldly patterned lapwing. Imm duller, grey-brown replacing black, streaked on crown, with white chin and throat. Spur-winged and Sociable Lapwings have pale brown upperparts. **Voice** Vocal, metallic alarm *tink-tink*, recalling a blacksmith's hammer on anvil. **SHH** Breeds S & E Africa, mainly sedentary. Vagrant Europa, May. Frequents dry areas near freshwater, open country and mudflats.

Spur-winged Lapwing *Vanellus spinosus* — 27 cm
Distinctive wader with *large white cheek-patch and sides of neck, contrasting sharply with black crown, throat and breast.* In flight from Sociable Lapwing by white bar across wing-coverts contrasting with black primaries and secondaries, and broad black tail-band, lacking white secondaries of Sociable. Imm has pale scaly fringes on upperparts and wing, black areas tinged grey-brown and a smaller cap. **Voice** Highly vocal. Call a sharp repeated *kwik*. **SHH** Resident Africa, SE Europe, Middle East; latter populations migrate to Africa. Vagrant Seychelles, Nov–Apr. Favours wide variety of habitats, usually on dry ground near water including grassland, cultivation and mudflats.

Sociable Lapwing *Vanellus gregarius* — 29 cm
Distinctive lapwing with grey-brown crown, broad white supercilium and black eyestripe meeting in V-shape on nape. In flight shows black primaries and white secondaries. Ad breeding has black-and-chestnut belly-patch, black crown and unstreaked breast. Imm has broad pale buffish tips to feathers of upperparts and wings, breast more extensively streaked. See other lapwings for differences. **Voice** May give a short whistle. **SHH** Breeds Russia to Kazakhstan, spending boreal winter to S, reaching NE Africa. Vagrant Seychelles, Nov–Mar. Favours dry plains and open short-grassy areas. Flies low if disturbed, but at greater height over distance. **CR**

Little Ringed Plover *Charadrius dubius* — 17 cm
Small plover, non-breeding ad and imm like Common Ringed (Plate 28) but slimmer, more grey-brown above, with indistinct buff supercilium, *obvious yellow eye-ring and dark ear-coverts pointed (not rounded) at rear*; longish *dull pink (not orange) legs* in ad and finer black bill lacking paler base. *In flight shows no obvious wingbar*. From Madagascar Plover (Plate 29) by shorter legs, less scaly upperparts and thicker, shorter bill; thicker black eyestripe does not join collar. Breeding ♂ has black face-mask and collar, and brighter eye-ring, clearer white forehead with white line and black frontal bar across forecrown. Breeding ♀ has breast-band tinged brown and narrower eye-ring. Imm has back and wing-coverts scaled buff and greyish-yellow legs. **Voice** A distinctive *peeu* call, falling in pitch and a shorter *peep* or *pip*. **SHH** Race *curonicus* breeds across Eurasia spending boreal winter to S including to sub-Saharan Africa. Vagrant Seychelles, mainly Oct–Nov. Favours mainly open lowlands near fresh or brackish water, rarely coastal. Rapid flight.

Caspian Plover *Charadrius asiaticus* — 21 cm
Ad non-breeding from sand plovers (Plate 28) by *broad white supercilium flared behind eye*, dark grey-brown upperparts with pale fringes, long yellowish-brown to dull olive-grey legs and slimmer bill. At rest wingtips extend well beyond tail. *In flight shows short white wingbar and mainly white underwing* with grey greater coverts, toes extending beyond tail. Breeding ♂ has broad rufous breast-band bordered black at lower margin, forehead and throat white with narrow brown stripe through eye. Breeding ♀ has grey-brown breast-band lacking black border. Imm has pale fringes on upperparts. Oriental Plover is longer-necked, has more upright stance, longer paler legs, less sharply demarcated supercilium, indistinct rufous fringes to upperparts (paler and broader in Caspian), no wingbar and all-dark underwing; breeding ads very pale-headed. **Voice** Flight call a soft piping *tik, tik* and a sharper *kuhit*. **SHH** Breeds Caspian Sea to China, spending boreal winter in E & S Africa. Vagrant Seychelles, mainly Oct–Dec. Favours open grassy areas, including airstrips. Rapid flight with long run on landing. Hunts in 'run-stop' manner

Oriental Plover *Charadrius veredus* — 25 cm
See Caspian Plover for distinctions from that species and sand plovers (Plate 28). **Voice** Flight call a shrill *tyip, tyip, tyip*. Also a short, piping *klink*. **SHH** Breeds S Siberia to NE China, spending boreal winter S to Australia. Vagrant Seychelles, mainly Oct–Nov. Favours open grasslands including grassy airstrips. Runs swiftly and has rapid, erratic flight, climbing high.

PLATE 107: VAGRANTS

Jack Snipe *Lymnocryptes minimus* — 19 cm
Small snipe; from others by *shortish but deep-based bill* only slightly longer than head, *breast and flanks heavily streaked not barred*, and bold yellowish stripes on back contrasting with very dark mantle and scapulars. *Head pattern distinctive: pale double supercilium* (supercilium and lateral stripe), a dark crescent below the eye joins dark eyestripe behind eye and no crown-stripe. In flight, note flight-feathers paler than primary coverts/greater coverts and *all-dark wedge-shaped tail*. **Voice** Mainly silent, but may give a barely audible wheeze when flushed. **SHH** Breeds Scandinavia to Siberia, spending boreal winter to S within Eurasia, some reaching tropical Africa mainly N of equator, but vagrant to E Africa. Vagrant Seychelles, Oct–Mar. Favours freshwater margins with vegetation. Crouches, rising reluctantly with fast mainly straight flight and occasional zigzag. Unique bouncing feeding action.

Pin-tailed Snipe *Gallinago stenura* — 25 cm
Pale, short-billed, short-tailed snipe. From Common Snipe by less obvious pale stripes on back, *supercilium flared widely in front of eye, tapering behind* (broader than eyestripe at bill base). In flight, feet project beyond tail, wings rounded, shows pale wing-panel, indistinct pale trailing edge and uniformly barred underwing. Tail shows little white and has rufous terminal band with pin-like outer feathers. See Great Snipe for differences. Madagascar Snipe (Plate 27) is much larger (size of Great Snipe) with shorter wings. Separable only in hand from potential vagrant Swinhoe's Snipe *G. megala*, which has 18–26 (usually 20) tail-feathers, whereas Pintail has 24–28, outer 7–9 pairs narrow and pin-shaped. **Voice** A monosyllabic *quack* when flushed. **SHH** Breeds across Russia, spending boreal winter in India, SE Asia and Maldives. Vagrant Seychelles, Oct–Jan. Often favours drier sites than other snipes, but also muddy shores and marshland.

Great Snipe *Gallinago media* — 28 cm
Large chunky snipe with medium-sized bill. From other snipes by *obvious white tips to wing-coverts* (showing as white bars in centre of wing), *heavily barred underparts extending to belly and more white in outer tail-feathers, obvious when spread on landing*. **Voice** A low *ech* when flushed. **SHH** Vagrant Seychelles, Oct–Nov. Breeds Scandinavia to C Russia, spending boreal winter in sub-Saharan Africa, but scarce in E Africa. Favours damp grassland and marshes, but frequently occurs in drier habitat than other snipe. Much slower, heavier and more direct flight than most other snipes, taking flight just in front of disturber (unlike Common Snipe) and landing quickly. **NT**

Common Snipe *Gallinago gallinago* — 26 cm
Long-billed, long-tailed snipe, bill at least twice head length. At rest, tail extends well beyond primary tips, *pale crown-stripe, creamy-buff supercilium narrower than eyestripe at base of bill*, prominent creamy-buff stripes on mantle and scapulars, heavy barring on underparts and flanks, and white belly. *In flight shows broad white trailing edge to secondaries*, inconspicuous wingbar, broad pale bands on underwing, and toes project slightly beyond tail. Tail tipped rufous with very little white at sides. Pintail Snipe lacks broad white trailing edge, has less extensive white on belly, pale supercilium broader than eyestripe at base of shorter bill and has less erratic escape flight showing all-dark underwing, toes projecting conspicuously. Jack Snipe is smaller with shorter bill, no crown-stripe, flanks streaked not barred, and has less erratic escape flight. Great Snipe is larger and plumper with heavy barring below, white outer tail-feathers and striking white bars on greater coverts. **Voice** In flight a distinctive, rasping *ketch*. **SHH** Nominate race breeds across N Eurasia, spending boreal winter to S including sub-Saharan Africa and common in E Africa. Vagrant Seychelles, Oct–Mar, and possibly Reunion. Frequents freshwater margins. Rises rapidly with erratic flight. About one-third of Seychelles snipe reports have not been confirmed to species level; more confirmed records have been accepted as Common Snipe than all other snipe species combined.

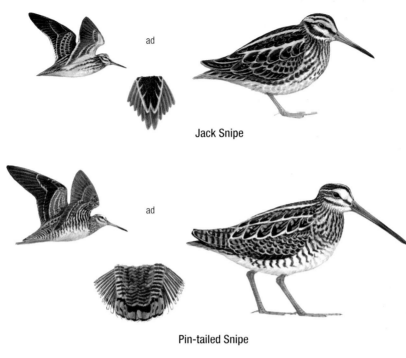

ad

Jack Snipe

ad

Pin-tailed Snipe

ad

Great Snipe

ad

Common Snipe

PLATE 108: VAGRANTS

Black-tailed Godwit *Limosa limosa* 40 cm
Distinctive large wader; from all others by *broad white wingbars and rump, black flight-feathers and tail*, unbarred white underwing, toes trailing conspicuously. *Long, straight bill, pink at base, black at tip*. Breeding ♂ has pink-orange head, neck and upper breast, contrasting with white belly, barred black on flanks and belly. Breeding ♀ paler below with white flecks. Bar-tailed Godwit (Plate 30) has slightly uptilted bill, shorter legs and lacks strong wing and tail pattern. **Voice** Flight call a quiet *kuk kuk*. **SHH** Breeds Iceland and temperate Europe to C Russia, spending boreal winter S to sub-Saharan Africa. Vagrant Madagascar, Seychelles and possibly Comoros, mainly Oct–Dec. **NT**

Little Curlew *Numenius minutus* 30 cm
Similar to Whimbrel (Plate 30), but *half the size, with proportionately shorter, more slender bill, less noticeably decurved. Dark crown with buff mid-stripe, buff eyestripe and pale lores*. In flight shows no white on rump, wings mainly dark with pale central panel, flight-feathers evenly black and unspotted with pale tips and grey-brown underwing-coverts. **Voice** A distinctive three-note *pe-pe-pe* rising slightly in pitch; alarm call a strident *tew-tew-tew*. **SHH** Breeds Siberia, spending boreal winter in Australasia. Vagrant Seychelles, Oct. Favours dry grassland and freshwater sites, rarely coastal. Flies with shallow, languid wingbeats.

Green Sandpiper *Tringa ochropus* 23 cm
Ad from Wood Sandpiper (Plate 30) by *darker, less spotted upperparts, darker breast sharply demarcated from white underparts, pale supercilium in front of eye*, and shorter greyish-green legs. *In flight shows blackish (not pale) underwing*, white rump, three broad black uppertail-bars concentrated towards tail tip with toes projecting only very slightly (Wood Sandpiper legs project more obviously). **Voice** A clear *pueet-wit-wit*. **SHH** Breeds N Eurasia, spending boreal winter to S including throughout sub-Saharan Africa. Vagrant Madagascar, Seychelles, mainly Oct–Jan, also Mar. Favours sheltered freshwater sites with fringing vegetation. When disturbed, climbs steeply, zigzagging then drops to ground.

Grey-tailed Tattler *Tringa brevipes* 25 cm
From other medium-sized waders by *uniform ash-grey upperparts* including upperwing and rump, *conspicuous white supercilium* contrasting with darker lores, long primary projection, wingtips reaching to or slightly beyond tail tip (unlike Common Sandpiper), straight bill, dark with pale yellowish base and fairly *short dull yellow legs*. Ad breeding is darker above, has more distinct speckling on neck and breast, grey barring on flanks and brighter yellow legs. Imm has white spotting on back and wing-coverts, and fine greyish barring on breast/flanks. **Voice** A two-note, rising *tu-wip* call and a ringing *tlee-tlee*. **SHH** Breeds Siberia to Kamchatka Peninsula, spending boreal winter S to Australasia. Vagrant Seychelles, Mauritius, Rodrigues, mainly Nov–Apr. Frequents wide range of habitats including inland freshwater sites, mudflats, mangroves and coasts. Rapid flight with flicking action. Bobs like Common Sandpiper.

Spotted Redshank *Tringa erythropus* 32 cm
Non-breeding ad from Common Greenshank (Plate 30) by grey upperparts with white spotting most conspicuous on wings, unmarked white below, red legs, *prominent white supercilium outlined by dark line across lores. In flight shows white oval on back, darker tail, dark wings with paler trailing edge*, feet projecting beyond tail. Long slender bill slightly drooping at tip with red base to lower mandible, *upper mandible all dark*. Ad breeding mostly sooty-black. Imm slightly browner on upperparts, underparts barred darker. See also Common Redshank. **Voice** A sharp, rapid *chu-wit*. Alarm call a short *chip*. **SHH** Breeds N Europe to NE Siberia, spending boreal winter to S including across sub-Saharan Africa, small numbers regular in E Africa. Vagrant Seychelles, Dec. Favours open freshwater and brackish sites, and sheltered muddy coasts. Often swims.

Common Redshank *Tringa totanus* 28 cm
Distinctive; uniform grey-brown above, from other species by red legs and *red base to both mandibles of straight rather stout bill*. Faint supercilium mainly between bill and eye, dark loral stripe and pale eye-ring. *In flight shows white wedge on lower back, broad white trailing edge to wings contrasting with black primaries, finely barred tail*, toes projecting only slightly. Ad breeding has brighter mottling on back and heavier spotting on face, breast and flanks. Imm is generally browner with orangey legs. **Voice** A loud yelping *tew-hu-hu* flight call. **SHH** Breeds Europe to Siberia, spending boreal winter to S including tropical Africa but uncommon in E Africa. Vagrant Seychelles, Sep–Feb. Mainly coastal but also in freshwater sites. Strong flight with rapid flicking wingbeats.

PLATE 109: VAGRANTS

Great Knot *Calidris tenuirostris* — 27 cm
Medium-sized, deep-chested wader; larger and straighter-billed than Curlew Sandpiper (Plate 31). From smaller Red Knot by *longer legs, longer, more tapered bill, longer neck and smaller head, streaked crown, nape and mantle, and variable black spots and chevrons on breast and belly.* Wingtips project beyond tail, supercilium and wingbar are less obvious, and rump is purer white. Ad breeding has bright rufous scapulars, dark mottled back and white underparts with large black spots, densest on upper breast. **Voice** Mainly silent, flight call *nyut-nyut*. **SHH** Breeds NE Siberia, spending boreal winter mainly in SE Asia to Australia but also W to India and Arabia. Vagrant Seychelles, Mauritius, Mar–Apr and Dec. On open coasts; with other waders. **VU**

Red Knot *Calidris canutus* — 25 cm
Stocky grey wader with short, heavy, straight bill, short legs, and clear white wingbar. See Great Knot for differences. Ad breeding has bright rufous face and underparts. **Voice** Mainly silent; may give a low, soft *knot*. Alarm call *kikikik*. **SHH** Circumpolar breeder, spending boreal winter S to S Africa, Tierra del Fuego, Australasia. Vagrant Seychelles, Oct. Entirely coastal, often associating with other waders. More rapid flight action than Great Knot.

Sharp-tailed Sandpiper *Calidris acuminata* — 21 cm
From Pectoral Sandpiper with difficulty by *greyish breast with darker streaks, fading to whitish lower breast and belly, white supercilium prominent behind eye outlining rusty brown cap, and marked pale eye-ring*. In flight shows white edges to rump and obvious wingbar. Ad breeding has warm brown fringes on upperparts, markings on breast extending to flanks and undertail-coverts, not cleanly demarcated from white belly like Pectoral. Juv has chestnut fringes to mantle, scapulars and tertials, bright chestnut cap, orange-buff flush to breast, streaking confined to upper breast and dark chevrons on sides of lower breast, flanks and undertail. Curlew Sandpiper is longer-billed with less prominent supercilium, clean white rump and blackish legs. **Voice** Twittering flight call. **SHH** Breeds N Siberia, spending boreal winter S to Australia, New Zealand. Vagrant Madagascar and Seychelles, mainly Oct–Feb. Favours mudflats or freshwater. Less upright stance than Pectoral Sandpiper, but similar flight.

Pectoral Sandpiper *Calidris melanotos* — 23 cm
Medium-sized sandpiper. See Sharp-tailed Sandpiper for separation. From other sandpipers by *clear breast-band ending abruptly on lower breast*, pale legs, long neck. In flight shows a pale indistinct wingbar, rump black in centre with white edges. Imm warmer brown above with conspicuous mantle 'V'. **Voice** In flight a harsh *chrrrt*. **SHH** Breeds N Siberia and N North America, spending boreal winter S to Australasia and South America. Vagrant Madagascar, Seychelles and Mauritius, Aug–Dec. Frequents coasts and grassland. Erratic snipe-like escape flight.

Buff-breasted Sandpiper *Tryngites subruficollis* — 19 cm
Small, long-necked shorebird *with pigeon-like head and short straight bill*; from Ruff by smaller size, *shorter straighter bill, unstreaked buff face, prominent dark eye surrounded by a pale eye-ring*, dark brown centres to broadly fringed scapulars, *buff feathering on lower mandible extends further onto bill than on upper mandible*, yellowish legs. In flight shows no obvious wingbar but has dark spot on underside of primary coverts. **Voice** Sometimes a low *krrrt* in flight. **SHH** Breeds Arctic North America, spending boreal winter mainly in South America. Vagrant Seychelles and Madagascar, mainly Oct–Jan. Grasslands, freshwater sites, mudflats and beaches. Often very tame. Walks with a high-stepping gait. **NT**

Ruff *Philomachus pugnax* — 30 cm (♂), 23 cm (♀)
Medium-sized wader with *long neck, small head and pot belly*, grey-brown above, whitish below, ♂ much larger than ♀. *Bill mostly straight, downcurved towards tip. Legs yellowish-green to dull orange.* See Buff-breasted Sandpiper for distinctions. **Voice** Sometimes a quiet *kuk*. **SHH** Breeds Europe to E Russia, spending boreal winter mainly in Africa, common in E Africa. Vagrant Madagascar, Seychelles, Comoros, Mauritius and Reunion, mainly Sep–Feb. Frequents freshwater wetlands, brackish waters and coastal pools, but not usually in intertidal habitats. Lazy flight, frequently gliding. Upright stance, head raised, tail lowered.

PLATE 110: VAGRANTS

Temminck's Stint *Calidris temminckii* 15 cm
Small, *long-tailed*, rather sluggish wader; from similar species (see Plate 31) by uniform upperparts, short, *slightly decurved bill, short yellowish-green legs and white outer tail-feathers*. See also Long-toed Stint. Ad breeding has rufous tints to crown and mantle, and distinctive black-centred scapulars with broad grey tips. Juv has fine black crescents on upper scapulars and warm buff scalloping on lower wing-coverts. Little Stint (Plate 31) has longer black legs, is dumpier (tail does not project beyond primaries) and more upright. **Voice** In flight a trilling *tirrrit*. **SHH** Breeds across Eurasia, spending boreal winter to S including sub-Saharan Africa mainly N of equator. Vagrant Seychelles, Sep–Dec. Outside breeding season frequents inland freshwater sites, wetlands and sheltered coasts. Usually solitary; avoids open areas unlike Little Stint, feeding more slowly, warily and, unlike Little, frequently towers when flushed.

Long-toed Stint *Calidris subminuta* 15 cm
From other small waders by *long neck, long pale yellowish legs*, long toes, and dark grey upperparts in non-breeding plumage. Little Stint (Plate 31) has black legs and a shorter bill. From Temminck's Stint by pale mantle braces, longer legs, longer neck, very long toes, *more upright posture and darker, more contrasting back*. **Voice** In flight a soft *chirrup*. **SHH** Breeds Siberia to Kamchatka, spending boreal winter mainly in SE Asia to Australia. Vagrant Seychelles, Nov–Jan. Mainly freshwater and brackish sites, sometimes sheltered mudflats. When flushed, tends to 'tower' like Temminck's Stint.

Broad-billed Sandpiper *Limicola falcinellus* 17 cm
Small, short-legged sandpiper, from similar species by *broad-based bill drooping at tip, double-forked white supercilium and contrasting dark eyestripe*. Ad breeding and juv are dark above with heavy black centres to back feathers, prominent mantle 'V', boldly streaked neck and upper breast, and strong black crown-stripes. **Voice** In flight a buzzing trill *chreeet*. **SHH** Breeds Scandinavia to E Siberia, spending boreal winter S to S Africa and Australia. Vagrant Seychelles, mainly Oct. On sheltered muddy shores and estuaries, usually with other waders.

Red-necked Phalarope *Phalaropus lobatus* 18 cm
Slim elegant wader, from other species by *ash-grey upperparts, black patch through eye and cheeks, and black needle-like bill*. Ad breeding has dark grey face, upperparts and breast, with white chin, rufous collar and sides of neck, and buffy stripes on dark back. Breeding ♂ duller than breeding ♀. Juv dark brown above with buff edges to coverts and pinkish wash to breast and throat (soon lost). **Voice** In flight a sharp *kirrit*. **SHH** Breeds mainly Arctic, spending boreal winter to S including Arabian Sea, regular off E Africa. Winters mainly at sea in areas of upwelling but also uses inland sites on migration. Vagrant Madagascar, Seychelles Apr, Oct–Nov. Flight over long distance similar to a small *Calidris*, though over short distances more flitting and erratic. *Swims with buoyant, tail-up stance, paddling and spinning.*

PLATE 111: VAGRANTS

Sooty Gull *Ichthyaetus hemprichii* 47 cm
Distinctive dark gull with very long bill. Ad from other species by **brown hood and breast, and greenish-yellow bill with black subterminal band and red tip**. First-year is paler brown above, lacking hind-collar, pale grey bill with black tip, dull grey-green legs; in flight, scaly upperwing, broad black tail-band contrasting with broader U-shaped white lower rump/uppertail-coverts patch, two obvious wingbars and broad white trailing edge. White-eyed Gull *Larus leucophthalmus* is a potential vagrant from Red Sea and Gulf of Aden; from Sooty Gull by slightly drooping, slimmer red bill tipped black, white crescents above and below eye is and paler grey upperparts/breast. First-year darker and browner than Sooty Gull, upperwing not as scaly and bill all dark. **Voice** Vagrants generally silent. **SHH** Breeds Red Sea to Persian Gulf, Pakistan and N Kenya. Vagrant Seychelles and Madagascar, Aug–Feb. Usually inshore and intertidal zones, and port areas. Slow, rather laboured flight.

Lesser Black-backed Gull *Larus fuscus* 56 cm (nominate), 63 cm (*heuglini*)
Ad from larger, shorter-winged Madagascar race of Kelp Gull (Plate 33) by **streaked grey on crown and neck in non-breeding plumage, white tips to primaries and pale yellow eye** (sometimes darker). First-winter has whiter rump and base of tail than Kelp Gull. Race *heuglini* ('Heuglin's Gull') is heavily built, larger than nominate ('Baltic Gull') with heavier bill having more obvious gonydeal angle, broader wings and longer legs, sometimes pinkish; paler slate-grey upperparts. Other races may occur including *barabensis* ('Steppe Gull'). **Voice** A loud *kyew-kee-kee*. **SHH** Nominate race breeds Scandinavia to White Sea, spending boreal winter to S including coasts of Africa to Mozambique; race *heuglini* breeds N Siberia, spending boreal winter on coasts of Middle East, S to E Africa; race *barabensis* breeds C Asia, spending boreal winter mainly SW Asia. Vagrant Seychelles, Dec–May; most records identified to race have been *heuglini*, but nominate also recorded, others indeterminate between *heuglini* and *barabensis*. Frequents sheltered coasts, sometimes inland. Flight graceful and buoyant. Systematics very complex; 'Heuglin's Gull' is sometimes treated as a separate species, and 'Steppe Gull' is sometimes treated as race of Caspian Gull *L. cachinnans*.

Common Black-headed Gull *Choicocephalus ridibundus* 38 cm
Small gull; from Grey-headed Gull (Plate 33) by **prominent white leading edge to black-tipped outer primaries, and dark grey undersides to primaries**. Ad breeding has chocolate-brown hood and brighter bare parts. First-year has blackish-brown trailing edge, brown carpal bar, black subterminal tail-band, dull yellowish-flesh bill with extensive dark tip, and dull yellowish-flesh to orange legs. **Voice** Vagrants generally silent. **SHH** Breeds temperate zones across Eurasia, spending boreal winter to S including E Africa. Vagrant Seychelles, Dec–Apr. Usually coastal, rarely inland. Agile flight with steady deep wingbeats, wings appearing pointed. Some Seychelles records might refer to potential vagrant Brown-headed Gull *C. brunnicephalus*; ad from Grey-headed Gull (Plate 33) with difficulty by chocolate not grey head in breeding season, paler upperwing and darker underwing; first-year almost identical but has broader black tail-band.

Sabine's Gull *Xema sabini* 36 cm
Very distinctive small gull, from others **by slightly forked tail, tricoloured wing and yellow-tipped black bill**. Ad breeding has dark grey hood but by Oct has white head with grey hindcrown and nape. Imm has similar wing pattern with scaly grey-brown upperparts to inner forewing, dark tail tip and blackish bill. **Voice** Generally silent. **SHH** Breeds Arctic tundra, spending boreal winter S to South Africa. Vagrant Madagascar, Apr. Mainly pelagic but may frequent inshore waters. Light tern-like flight; swims or feeds on land or mudflats with plover-like action.

PLATE 112: VAGRANTS

Gull-billed Tern *Gelochelidon nilotica*   38 cm
Sturdy, whitish, gull-like tern; from others by *short, heavy black bill and long legs*. Ad non-breeding has distinctive white head, with small blackish eye-crescent and wedge on ear-coverts, lacking dark crown and nape of other species; in flight, dark trailing edge to upper and undersides of primaries, and *grey rump and tail*, white in Sandwich Tern. Ad breeding has black cap extending to hindneck. **Voice** An unmistakable deep, low *ger-vick*. **SHH** Nominate race breeds S Europe to N Africa through Middle East to China, spending boreal winter to S including tropical Africa, locally common in E Africa. Vagrant to Madagascar and Seychelles, mainly Oct–Jan. Frequents sheltered coasts and freshwater habitats; also inland on drier fields. *Gull-like, slow, steady flight on long broad wings, tail closed*. Generally does not plunge-dive, but drops to water surface.

Sandwich Tern *Thalasseus sandvicensis*   41 cm
Large, graceful, whitish tern; from others by *short crest, flat crown and slender long black bill with a yellow tip*, pale grey upperparts/upperwing and white rump/tail. Ad breeding has complete black cap and more prominent crest. First-winter head resembles ad non-breeding, but browner streaked grey-brown on crown. Common Tern is much less whitish and has different bill/leg colours. Gull-billed Tern more gull-like, with uniform pale grey upperparts and tail, shorter, broader and less pointed wings, and a heavy all-black bill with distinct gonydeal angle. **Voice** A shrill *kirrick*, inflected upward. Alarm call *gwit*. **SHH** Nominate race breeds Europe to Caspian Sea, spending boreal winter S to South Africa. Vagrant Madagascar, Aug–Sept, and Seychelles, Oct–Dec and Mar. Almost entirely coastal. Gregarious, often with other tern species. Flight strong and swift with deep wingbeats, neck protruding, head and bill pointing downward, suddenly plunge-diving or dipping to surface.

White-cheeked Tern *Sterna repressa*   33 cm
Medium-sized, long-billed tern. Very similar to Common Tern (Plate 34); ad non-breeding has *rump and tail grey* (white on Common) with slightly paler mantle and upperwing, *broad diffuse dark trailing edge to underwing contrasting with pale central band (greater and median coverts) and grey lesser coverts*. Retains more extensive cap than Common Tern. Ad breeding has darker grey underparts and upperparts than Common, contrasting with prominent white cheeks and black cap, red bill tipped black and red legs. First-year has grey rump and tail concolorous with upperparts, extensive black carpal bar, dark tip to tail and black bill sometimes with a pale orange base. Common Tern has whiter underwing, white rump and tail, and less graceful flight. Whiskered Tern has shorter bill and shorter, squarer tail. **Voice** A loud grating *kee-eerr*. **SHH** Breeds Red Sea S to Kenya and Persian Gulf to W India, dispersing along coasts. Vagrant throughout Seychelles, Nov–Dec, and possibly Comoros. Frequents tropical coasts, mainly over coral reefs. Graceful bouncing flight like a marsh tern, dipping to surface.

Antarctic Tern *Sterna vittata*   38 cm
Medium-sized tern, very similar to Common Tern (Plate 34), White-cheeked Tern and potential vagrant Arctic Tern *S. paradisaea*; *ad breeding has bright red bill, somewhat thicker than Arctic Tern, with no black tip* (unlike Common and White-cheeked), short legs, white cheeks and uniform grey underparts and wings, darker than Common and Arctic but paler than White-cheeked; wing pattern like Arctic with very narrow black trailing edge to primaries. Rump is white, unlike White-cheeked Tern. Black cap reduced in non-breeding season to just behind eye, underparts white and bill dull reddish-black. Imm has *very distinctive chequered buff, dark grey-and-white upperparts, especially on tertials and coverts*. Whiskered Tern has short, shallowly forked tail, grey rump and is much smaller. **Voice** A high-pitched *trr-trr-kriah*. **SHH** Breeds on Southern Ocean islands, spending boreal winter N to South Africa. Vagrant Madagascar, Oct. Forages mainly inshore, feeding by plunge-diving, or in rough seas by contact-dipping.

Black Tern *Chlidonias niger*   23 cm
Non-breeding or imm from other terns by small size, *black cap covering ear-coverts and nape; characteristic dark patches at sides of breast*; rest of underparts white. White-winged Tern (Plate 35) lacks dark breast patches and has white rump. Common Tern (Plate 34) also lacks breast patches and is larger, longer-winged, with more forked tail. Ad breeding distinctive with *dark uniform grey upperparts, black head/body contrasting with white underwing-coverts* (ad br White-winged Tern has black underwing-coverts). **Voice** A harsh *k'shlet* or nasal *klit*. Alarm call *ki-ki-ki* or *kyeh*. **SHH** Nominate race breeds Europe to C Asia, spending boreal winter in Africa, mainly in W but reaching South Africa and vagrant Kenya. Vagrant Madagascar, Dec. In flight has deep rapid wingbeats on long slim wings, giving a rakish appearance. Coastal and marine at non-breeding sites including over predatory fish and tuna, but can occur in any wetland habitat.

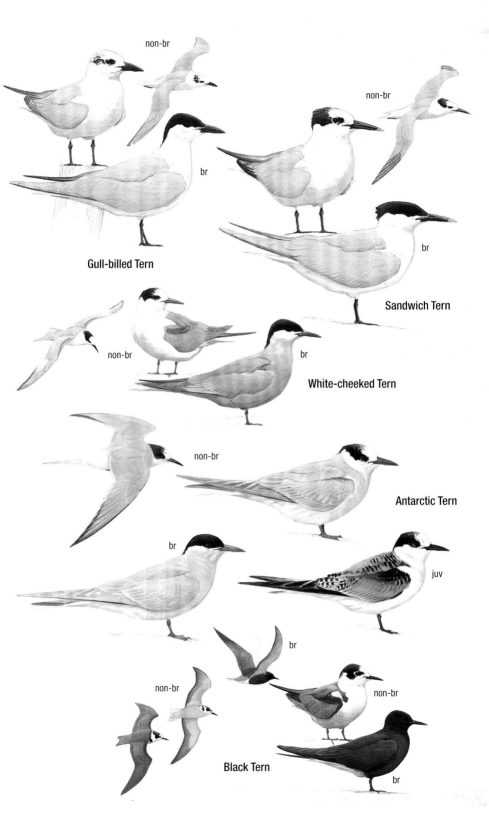

PLATE 113: VAGRANTS

European Turtle Dove *Streptopelia turtur*   28 cm
Medium-sized dove, from others by *oval neck-patch with 3–4 black bars against a whitish background*, red eye-ring, grey nape and crown, *chestnut or orange-buff edges to mantle, scapulars and upperwing-coverts, which have clear-cut black centres*; also blue-grey panel on greater coverts. Diamond-shaped tail is dark grey to black, edged white. Imm lacks neck-patch, but retains scaly upperparts. Oriental Turtle Dove *S. orientalis* could also occur; more robust, darker overall, especially on brownish-pink breast, brownish nape contrasting with grey crown and five or more neck-bars. **Voice** Non-breeders silent. **SHH** Nominate race breeds across Europe to W Siberia, spending boreal winter to S and vagrant E Africa. Vagrant Seychelles, Nov–Dec. Favours open woodland, feeding on ground. Flight rapid and agile, wings held well back with bursts of wingbeats. Often glides and spreads tail on landing.

Great Spotted Cuckoo *Clamator glandarius*   38 cm
Large cuckoo with broad blunt wings and long narrow wedge-shaped tail, very distinctive. Ad has dark grey upperparts, with numerous bold white spots, pale underparts with throat and sides of breast washed very pale yellow. First-winter blackish above with rusty-brown primaries. Jacobin Cuckoo is pied and much smaller. **Voice** A whistling *keeow-keeow-wow-wow*. Alarm call crow-like. **SHH** Breeds S Europe to Iran spending boreal winter to sub-Saharan Africa. Vagrant Seychelles, Oct–Jan. Frequents mainly open woodland. Shy and unobtrusive, flight strong and direct.

Jacobin Cuckoo *Clamator jacobinus*   33 cm
Very distinctive *medium-sized, crested, long-tailed pied cuckoo*. Ad is black above with a *prominent white patch at the base of primaries*, conspicuous in flight. Underparts white or greyish-white, sometimes washed buff. First-winter has dark brown upperparts. See Great Spotted Cuckoo. **Voice** A piping *pew-pew* leading into a loud laughing cackle. **SHH** Breeds sub-Saharan Africa S to E Africa and Indian subcontinent, spending boreal winter in Africa. Vagrant Seychelles, mainly Nov–Dec. Frequents mainly open woodland. Flight straight and rapid. Frequently flicks tail and when alarmed raises crest and lowers tail.

Lesser Cuckoo *Cuculus poliocephalus*   28 cm
Medium-sized, *relatively short-tailed cuckoo with short light bill, which is only slightly decurved*. Essentially identical to Madagascar Cuckoo (Plate 41), distinguished only by song, except hepatic morph ♀, lacking in Madagascar Cuckoo, which is barred brown above often with unmarked rufous crown/nape and rump/uppertail. Common Cuckoo is larger and longer-tailed with heavier bill, but smaller pigeon-like head, smoky-grey (not dark grey) back, thinner and wavier bars on underparts (10–13 in Lesser, 12–16 in Common) and a paler eye. **Voice** A 5–6-note whistle, but non-breeders generally silent. **SHH** Breeds Asia, spending boreal winter in India, Sri Lanka and E Africa. Vagrant Madagascar, Seychelles, mainly Nov–Jan; sometimes not distinguished from Madagascar Cuckoo. Frequents open woodland. Perches with a more upright stance than Common Cuckoo. Flight direct and rapid.

Common Cuckoo *Cuculus canorus*   33 cm
Largest of the 'typical' (*Cuculus*) cuckoos in the region, with *small head and large powerful decurved bill*. See Lesser Cuckoo for differences. **Voice** Non-breeders silent. **SHH** Four races recognised, breeding across most of Eurasia, spending boreal winter to S including sub-Saharan Africa. Near-annual vagrant Seychelles, mainly Oct–Dec. Most likely race in region is N Eurasian nominate; *subtelephonus* (paler above, narrower black bars below) of Turkestan to S Mongolia could occur. Frequents all types of woodland, reedbeds and open habitats. Flight direct and hawk-like. Very similar potential vagrant Oriental *C. optatus* (=*C. horsfieldi*) and Himalayan Cuckoos *C. saturatus* not ruled out in all records.

PLATE 114: VAGRANTS

Eurasian Scops Owl *Otus scops*   18 cm
Very small owl with a large head showing 'ear-tufts'. Variable but more grey-brown than resident scops owls, although identification challenging; most resident species have less evident 'ear-tufts', but posture and state of relaxation affect posture of ears greatly. Seychelles Scops Owl (Plate 63), which occurs in mountainous areas where Eurasian has never been recorded, has more rounded head and is darker, more rufous on facial disc and underparts. First-winter similar to ad, but paler with brown wash and shorter ear-tufts. **Voice** Non-breeders usually silent. **SHH** Breeds S Europe to C Asia and Middle East to NW Pakistan. Six races varying in ground colour and strength of markings; most winter in N tropics of Africa, some sedentary in S of breeding range. Most likely in region are nominate (breeds France to Turkey), *pulchellus* (E of nominate, slightly paler with narrower marks, larger more extensive pale nape-spots) or *turanicus* (Iran, Iraq, pale silver-grey, with narrow, sharply defined shaft-streaks), all recorded in E Africa. Vagrant Seychelles, Oct–Dec. Frequents open woodland. A reluctant flier.

Brown Fish Owl *Ketupa zeylonensis*   53 cm
Very distinctive, **large, flat-headed owl with shaggy, fairly short ear-tufts, short tail and large bill**. Legs and feet unfeathered. Ad fairly uniform buff, upper cheeks darker, streaked blackish-brown on ear-tufts, crown, back, wing-coverts and underparts, with a white patch on throat. Eyes bright yellow, bill pale horn and legs greyish. **Voice** A deep, moaning *boom-o-boom*, although vagrants likely to be silent. **SHH** Four races breed NE Mediterranean and Indian subcontinent to SE Asia. Vagrant Seychelles, origin unknown (possibly ship-assisted). Frequents woodlands near freshwater. Flight weak and slow, wings making quiet singing sound on downstroke. Hunts over water, dangling long legs, and often bathes.

European Nightjar *Caprimulgus europaeus*   27 cm
Large, greyish-brown, long-winged, long-tailed nightjar, from otherwise very similar Madagascar Nightjar (Plate 41) by larger size. In flight, ♂ shows white tips on outer tail-feathers and white patch across three outermost primaries. ♀ lacks white tips to outer tail-feathers while white spots on primaries may be absent or buffish. E birds paler. **Voice** Generally silent in non-breeding season. Alarm call *chuck* or *cheek-eek*. **SHH** Breeds Europe and NW Africa to L Baikal, spending boreal winter mainly in SE Africa. Vagrant Seychelles, Nov–Jan. European nominate race recorded Seychelles, and *unwini* (Iran to SW Asia) and *plumipes* (C Asia) are possible. Frequents open woodland. Flight agile with some gliding.

White-throated Needletail *Hirundapus caudacutus*   20 cm
Large swift with **bulky body and short, blunt tail**, all dark brown except **prominent white throat and forehead, and white rump-patch wrapped around tail**. Square tail appears slightly rounded when spread and pointed when closed. Alpine Swift (Plate 43) has a white belly. Little Swift (Plate 43) is much smaller with slimmer body and smaller head, less powerful flight. **Voice** Generally silent. **SHH** Breeds NE Asia, spending boreal winter S to Australasia. Vagrant Seychelles and Rodrigues, Oct–Nov and Apr. Highly aerial with powerful but sometimes also fairly slow flight, often gliding with wings held slightly downward.

Common Swift *Apus apus*   18 cm
Medium-sized swift. African Palm Swift (Plate 43) is slimmer and paler with longer tail and long narrow wings. African Black Swift (Plate 43) is very similar, but has paler secondaries and inner primaries, forming pale areas on upper- and underwings. First-winter slightly browner with scaly underparts and more prominent throat-patch extending to forehead. Race *pekinensis* is browner than nominate with more prominent throat-patch, pale forehead and pale-fringed wing-coverts. **Voice** Generally silent in non-breeding season. **SHH** Breeds W Europe to E Asia. Vagrant throughout Seychelles, mainly Sep–Dec, a few Apr–May. Nominate race (W) and *pekinensis* (E) both recorded; some others not definitely separated from similar species such as African Black Swift. Flight powerful with rapid wingbeats, often glides.

Pacific Swift *Apus pacificus*   18 cm
Medium-sized, elegant, **blackish swift**. **From others by well-defined white rump patch**, narrower than on Little Swift (Plate 43) wrapped around tail base, and whitish scaling on underparts. White throat-patch more prominent than on Common Swift. Long black, deeply forked tail often held closed. **Voice** Generally silent. **SHH** Breeds NE Asia, migrates S to Australia. Vagrant Seychelles (Bird I to St François), mainly Oct–Jan and May, possibly also Agalega; unrecorded elsewhere in African and Malagasy regions. Rapid, powerful flight.

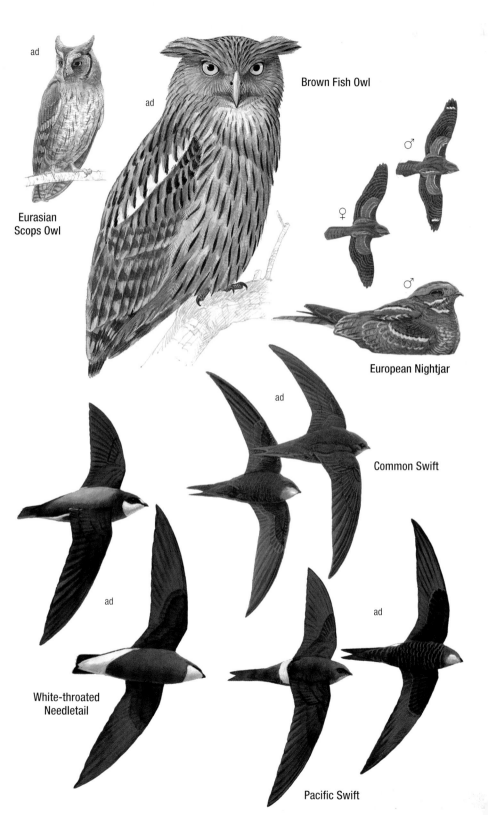

PLATE 115: VAGRANTS

Grey-headed Kingfisher Halcyon leucocephala   21 cm
Medium-sized kingfisher with *pale grey head, neck and breast, chestnut or pale tawny belly, black upper back, blue wings, rump and tail*. In flight, conspicuous blue bases to primaries especially on underwing. Heavy orange-red bill and red legs. Imm duller than ad, finely scaled on head, mantle and breast, with buff underparts, dull red legs and dull reddish bill tipped dark. **Voice** Calls include a loud, sharp *tchk-tchk-tchk*. **SHH** Sub-Saharan Africa. Equatorial populations sedentary, N & S populations more mobile; race *semicaerulea* (showing purer blues) breeds Arabian Peninsula and spends boreal winter S to Somalia. Vagrant Seychelles, Dec–Feb, race unknown. Frequents open woodland, riversides and cultivation. Perches on low branches and swoops to ground to take insects. Rapid flight.

Blue-cheeked Bee-eater Merops persicus   30 cm
Slender, graceful, bee-eater very similar to Olive Bee-eater (Plate 45); ad differs by *brighter green upperparts including crown, bright pale blue supercilium and cheeks* (sometimes white), yellow chin (not rufous) merging into rufous throat, and deeper rufous underwing. Eyes deep red in ♂, orangey-red in ♀. Imm very similar to Olive Bee-eater but crown and upperparts less tinged brown, throat with yellow tinge at top. **Voice** A repeated *dirrip*. Alarm call *dik-dik-dik*. **SHH** Nominate race (most likely in region) breeds Turkey to C Asia, spending boreal winter mainly in tropical E Africa. Vagrant Seychelles, mainly Nov–Dec but also Mar–May, invading granitics to Amirantes in large numbers in some years. Inhabits a wide variety of habitats, usually with trees, often perching on open branches or wires. Gregarious year-round.

European Bee-eater Merops apiaster   28 cm
Vividly coloured bee-eater, from others by combination of *all-yellow throat, pale yellow shoulder-patches and chestnut crown, back and inner wing-panel*. Non-breeding plumage duller, scapulars and rump greenish. Inner wing-panel bright rufous (unlike any other bee-eater). Sexes similar in pattern, ♂ brighter and ♀ has greener scapulars and wings, and less chestnut. First-winter less colourful, tail-streamers very short or absent, green tone to nape and back. **Voice** A liquid, melodious *prruip*, softer than Blue-cheeked Bee-eater. **SHH** Breeds SW Europe to SW Asia, spending boreal winter S of equator to South Africa; also breeds S Africa, spending austral winter in tropical Africa. Vagrant Madagascar, Seychelles and Europa, Oct–Dec. Cultivation, open woodland and riversides. Flight graceful, gliding on outstretched wings with shallow undulations.

European Roller Coracias garrulus   31 cm
Thickset roller, from Broad-billed Roller by *pale blue head and underparts, brown back and dark bill*. Autumn birds duller, many feathers brown-tipped and back more rufous. First-winter generally duller and whiter on face, duller brown on back. while throat and breast have a brownish hue, streaked white; bill pinkish at base. No similar species recorded. Race *semenowi* is slightly paler than nominate. **Voice** Generally silent. Flight call a short crow-like *krak*, often repeated. **SHH** Nominate race breeds mainly SW Europe to SW Siberia and race *semenowi* Iraq to NW China; most populations winter sub-Saharan Africa, with a large concentration in E Africa. Vagrant Seychelles, Comoros and Mauritius, mainly Oct–Dec. Frequents open woodland and wooded edges. Flight powerful and direct with shallow, rapid wingbeats and occasional glides. Swoops from perch to ground, ending with a glide. **NT**

Common Hoopoe Upupa epops   28 cm
Unlike any other bird other than very similar Madagascar Hoopoe (Plate 45), which is larger with thicker, straighter bill, and greater contrast between pinkish-white throat and darker face; latter presumed sedentary, so any hoopoes outside Madagascar more likely to be Common. Nominate race shows white-banded primaries, a white subterminal band to crest and pink head merges into greyish mantle. Race *senegalensis* has more white on secondaries and a pinkish mantle concolorous with head. Race *africana* (sometimes split as African Hoopoe) darker orange with entirely black primaries, a large white patch on secondaries and lacks white subterminal band to crest. Sexes similar, except *africana* ♀ which is smaller and duller with a whitish belly and dusky-streaked flanks. First-winter (all races) darker, more brownish, greyer on head and body, with a shorter crest. **Voice** Non-breeders generally silent. **SHH** Nominate race breeds Europe to C Asia, spending boreal winter mainly S of Sahara; *africana* throughout most of S half of Africa; *senegalensis* breeds Senegal to Somalia. Vagrant Seychelles (*africana*, *senegalensis* and possibly nominate), Aug, Oct, Mar and (unidentified hoopoe) Juan de Nova. Frequents open areas, feeding on ground, flying to trees if disturbed. Flight slow, erratic and undulating.

PLATE 116: VAGRANTS

'Grey-throated' Martin *Riparia paludicola chinensis*   12 cm
One of the Asian races of Brown-throated Martin (Plate 48), which is sometimes split. Differs from race *cowani* (resident Madagascar) *by greyer throat and breast demarcated from white belly, and paler upperparts*. Sand Martin has white throat separated by brown breast-band from cleaner white underparts, and Mascarene Martin is larger and heavier, with prominent dark brown streaking on underparts, and faster, more direct flight. **Voice** Calls reported in India a short, twangy *CHYip* and a metallic, spluttering trill. **SHH** Species breeds sub-Saharan Africa, Madagascar and C & S Asia (this race) to Philippines. This race partial migrant from C Asia, with vagrants reaching Arabia and Maldives; vagrant Seychelles, Dec, believed to be *chinensis*. Flight rather slow, stiff-winged and fluttering, sometimes twisting and turning.

Sand Martin *Riparia riparia*   12 cm
Uniform brown above, white below with well-defined brown breast-band across white throat and breast, grey-brown underwing-coverts and squarish tail. First-winter has pale edges to feathers of upperparts and buff wash to throat. Brown-throated Martin has grey throat and breast, and no breast-band. **Voice** A harsh rasp of 1–2 syllables, *tschr*, given perched or in flight. **SHH** Breeds across N Hemisphere, migrating S, including nominate race to sub-Saharan Africa. Vagrant Madagascar, Seychelles, Comoros, Glorieuses and Europa, Oct–Jun, mainly Oct–Dec. Open areas, often associated with water. Flight direct with rapid wingbeats and frequent glides.

Wire-tailed Swallow *Hirundo smithii*   18 cm
Distinctive long-tailed swallow, with glossy blue upperparts, forehead and crown, and white underparts except for blue-black patch on breast-sides. From other species by *chestnut cap, pure white underparts and long wire-like outer tail-feathers*, which can be difficult to see in the field, sometimes broken and shorter in ♀ and first-winter; the latter is duller blue above, washed buff below and cap dull brown. **Voice** A Barn Swallow-like twitter or a short *chit*. **SHH** Sub-Saharan Africa and S Asia. Vagrant to Aldabra, Feb. Frequents wetland, grassland and urban areas, often roosting in reedbeds. Fast, darting flight.

Common House Martin *Delichon urbicum*   14 cm
From other swallows *by conspicuous white rump, white underparts and dark glossy blue upperparts*, appearing black at a distance. Moderately forked blackish tail. First-winter browner and duller, lacking gloss, and white parts washed grey. **Voice** A shrill *prit* or a soft twitter. **SHH** Breeds Europe to N Asia, winters sub-Saharan Africa; has bred South Africa and Namibia. Vagrant Seychelles, Comoros and Europa, Oct–Dec and Mar–Jun. Hunts over open areas, often high up (*c*.50 m). Flight slower with more gliding than Barn Swallow.

Lesser Striped Swallow *Cecropis abyssinica*   17 cm
Distinctive, small, blue-and-white hirundine, from other species by *heavily striped underparts, rufous-chestnut crown and rump, and long, deeply forked tail*. Juv duller than ad, some black on head, tawny wash to breast and flanks, and tawny tips to tertials, wing-coverts and secondaries. Barn Swallow is unstriped. Mascarene Martin is also streaked below but is brown above with short, shallowly forked tail, and lacks pale rump. **Voice** Highly vocal in breeding range, including vigorous warbler-like song of 9–10 squeaky notes and a loud wheezing *tee-tee-tee*. **SHH** Sub-Saharan Africa; mostly resident but some post-breeding movements and to N and S of range during rains. Vagrant Madagascar and Europa, Dec–Jan. Frequents open woodland, grassland and forest margins. Agile, darting flight, sometimes fluttering and gliding, feeding low above ground or at mid-levels, often over water.

PLATE 117: VAGRANTS

Bimaculated Lark *Melanocorypha bimaculata* — 17cm
Large, heavy-billed, short-tailed lark; from other species by size, long primary projection, *boldly patterned head with obvious white supercilium, and black patches on sides of upper breast that often join in centre*. In flight, brownish-grey underwing and white tail tips not sides. Heavy finch-like bill is mainly horn, yellowish at base with dark culmen. Greater Short-toed Lark is smaller with white sides to tail and short primary projection. **Voice** A short *tripp-tripp* or *churup* and a short, liquid *plip*. **SHH** Breeds W to C Asia, spending boreal winter S as far as NE Africa. Vagrant Seychelles, Nov. In open treeless areas. Strong flight, tail often fanned. On ground, runs in rapid dashes.

Greater Short-toed Lark *Calandrella brachydactyla* — 14 cm
Only small, pale lark recorded in the region. From other species also by *variable light streaking on breast-sides forming a dark smudge on the neck-sides, broad whitish supercilium below brownish or rufous cap, and white outer tail-feathers*. Stout, pointed, yellowish-horn bill. Madagascar Lark (Plate 48) much more heavily streaked, especially below. Bimaculated Lark much larger with long primary projection, heavier bill and white-tipped tail. **Voice** A short, dry *tchirrup* or *drrit-drrit*. **SHH** Breeds S Eurasia; race in region may be *longipennis* of Ukraine to Manchuria, spending boreal winter mainly in S Asia but vagrant to Kenya. Vagrant Seychelles, Nov–Feb. Frequents open sandy or stony areas, dry mudflats and fields. Flight strong, low and undulating. Walks with a jerky gait making short erratic runs.

Richard's Pipit *Anthus richardi* — 18 cm
Large, long-legged, long-tailed pipit, from other pipits by *upright stance, limited streaking on upper breast forming dark wedge on side of lower throat*, long, stout thrush-like bill and very long hindclaw (longer than hind toe). **Voice** A grating *shreep*. **SHH** Breeds N Asia to Siberia, spending boreal winter S to Indian subcontinent and SE Asia. Vagrant Seychelles, Jan. Powerful undulating flight, often briefly hovering prior to landing.

Tree Pipit *Anthus trivialis* — 15 cm
Slender, medium-sized pipit. Smaller than Richard's, with much more streaking on breast; from Red-throated by *unstreaked rump, fine streaking on flanks and pale buff wash to flanks, breast and sides of white throat, and dark eyestripe beyond eye*; also by call. *Stout, dark brown bill with pinkish base to lower mandible. Legs flesh* with short hindclaw. Red-throated Pipit has no olive tones above, white stripes on darker mantle, more heavily streaked flanks on whiter background and longer hindclaw. **Voice** *A short sharp bzzzet*, sometimes drawn out, or a softer *zeep*. **SHH** Breeds W Europe to Siberia, spending boreal winter in sub-Saharan Africa and India. Annual Seychelles, mainly Nov–Dec; vagrant Europa. Frequents woodland edges, fields and grassy areas. Strong undulating flight.

Red-throated Pipit *Anthus cervinus* — 15 cm
Medium-sized pipit with *whitish underparts and broad, blackish flanks-streaking*. Darker above than Tree Pipit with *white mantle braces, streaked on rump*, often with black wedge at side of throat. Fine bill with yellowish base to lower mandible. *Legs yellowish or brownish-flesh* with long hindclaw. Breeding ♂ has distinctive *brick-red throat and breast* sometimes retained in non-breeding plumage. Breeding ♀ is less intensely coloured on breast, chin and supercilium, which are buffish sometimes with a pink tinge. First-winter also has strong black throat wedge and white ground colour to breast, flanks and belly. Tree Pipit has olive tones to upperparts, unstreaked rump, lightly streaked buffish flanks and shorter hindclaw. **Voice** Flight call is *a high-pitched tzeez, more drawn out than Tree Pipit*. **SHH** Breeds N Eurasia to Alaska, spending boreal winter mainly in sub-Saharan Africa S to Tanzania. Vagrant Seychelles and Europa, Oct–Mar. Frequents water margins, grassy areas, coasts and damp cultivation. Sometimes in loose groups. Undulating flight, often fairly high.

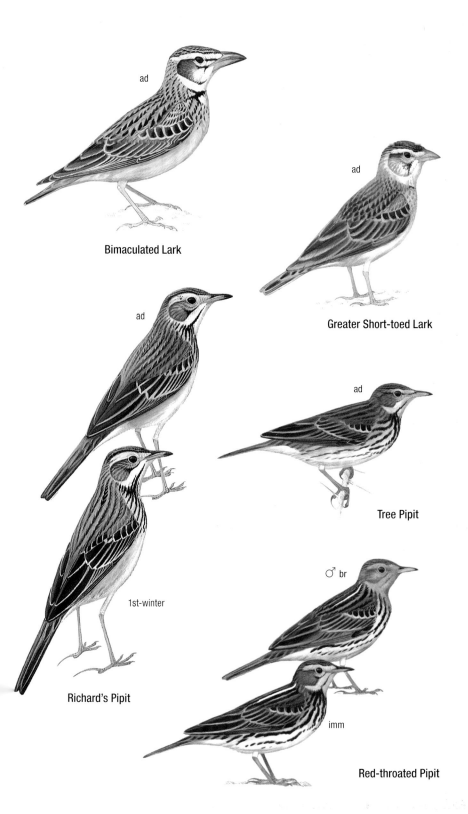

PLATE 118: VAGRANTS

White Wagtail *Motacilla alba* — 19 cm
Non-breeding ♂ has *white forehead, chin and throat, black breast-band, white underparts and grey crown, nape and upperparts*. Narrow white wingbars and tertial fringes. Breeding ♂ has neater all-black crown and throat. ♀ less clean-cut, face washed greenish-yellow, grey crown, lores and cheeks mottled dusky with dusky mantle, rump and flanks. First-winter ♂ similar to non-breeding ♀, blacker on crown. From first-winter Citrine Wagtail by longer tail, *less clear bold white wingbars, lack of clear pale band at rear of ear-coverts and obvious dark breast-band*, not (at most) smudges on sides of breast-band. **Voice** A distinctive *chi-sik*. **SHH** Breeds Eurasia to Alaska, spending boreal winter S to sub-Saharan Africa. Near-annual Seychelles, Oct–May; vagrant Reunion, Dec. Nominate race most likely, but Asian *dukhunensis* (with broader wingbars sometimes forming panel) reaches NE Africa and Socotra, and could occur. Frequents a wide variety of open habitats, often near buildings, grassy areas and water. Flight deeply undulating. Walks while constantly wagging tail and bobbing head.

Yellow Wagtail *Motacilla flava* — 19 cm
From other wagtails by *relatively short tail, olive-green or brownish upperparts* (greyer in E races), and *usually pale yellow below*. Pale supercilium does not extend around ear-coverts, unlike Citrine Wagtail, and wingbars less well marked. Grey Wagtail has blue-grey upperparts and much longer tail. Many races possible, usually only separable in breeding ♂; *flava* ('Blue-headed') has blue-grey crown separated from darker grey face by long white supercilium; *beema* ('Sykes's') has paler grey head and a broader supercilium, white chin, usually a white submoustachial stripe and white cheek-patch; *lutea* ('Yellow-headed') has entire head bright yellow; *feldegg* ('Black-headed') has top of head and ear-coverts black. **Voice** Call a shrill, drawn-out monosyllabic *tsweep*, occasionally disyllabic and some variation among races. **SHH** Breeds N Eurasia, spending boreal winter mainly in sub-Saharan Africa and S Asia. Near-annual Seychelles and vagrant Europa, Oct–May. Races *flava*, *lutea*, *beema* and *feldegg* all confirmed Seychelles and other races are possible. Favours a variety of damp habitats or in short grass. Bounding flight over distance, sometimes making short flights to catch insects. Less horizontal stance than other wagtails.

Citrine Wagtail *Motacilla citreola* — 17 cm
Most like Yellow Wagtail, but adult distinctive with grey upperparts. Breeding ♂ has *all-yellow head and underparts with black half-collar on lower nape and sides of neck*. ♀ has *broad yellow supercilium curving around greyish ear-coverts to yellow throat* and underparts, and greenish-grey crown. Usually shows pale centre to ear-coverts and may show buff forehead (absent in Yellow Wagtail). Non-breeding ♀ and imm very pale buff-yellow to whitish on underparts; from Grey Wagtail by *white undertail, black not pinkish legs, double white wingbar* and shorter tail. From Yellow Wagtail by broad white or yellow supercilium surrounding ear-coverts, broader white wingbars, dull ash-grey not olive-green upperparts, sides of breast and flanks washed grey, and pale lores. **Voice** A distinctive throaty, rather harsh *peep* or disyllabic *tit-tit*. **SHH** Breeds E Europe to China with nominate in W and N of range, and black-backed *calcarata* in C & S Asia. Spends boreal winter mainly N of 10°N and rare in Africa (nominate). Vagrant Seychelles, Apr (unconfirmed race). Frequents freshwater margins and grassy areas. Behaviour like Yellow Wagtail.

Grey Wagtail *Motacilla cinerea* — 20 cm
From other wagtails by very *long tail, shorter pinkish (not black) legs, grey upperparts contrasting with black wings, single white wingbar visible above and below (in flight), and yellow undertail*. Non-breeding ♂ and ♀ have white chin, throat, supercilium and belly, with yellow vent and yellow-green rump. Breeding ♂ has black throat, entirely bright yellow underparts, white supercilium and white submoustachial stripe. First-winter is grey above with yellowish-green rump, buffish-white below. **Voice** A metallic, high-pitched, disyllabic *tzitzi*. **SHH** Breeds NW Africa and across Eurasia spending boreal winter to S including E Africa. Vagrant Seychelles and Comoros, mainly Oct–Nov. Generally favours fast-running clear streams in open woodland, although more widely observed on migration. Flight deeply undulating, long tail very evident. *Walks with more horizontal stance than other wagtails, constantly wagging tail.*

PLATE 119: VAGRANTS

Common Rock Thrush *Monticola saxatilis* 20 cm
Distinctive, thickset thrush, dull brown (ad) or buffy (imm) above, and paler below, from Malagasy rock thrushes (Plate 49) by larger size, intense blackish scaling throughout, shorter tail, longer wings; from other vagrant chats by blackish scaling, orangey-red tail above and below, some white on back, and rusty wash to breast and flanks; ♂ may show a blue wash on greyish crown. Breeding ♂ (unlikely in region) has blue-grey head, throat and upper mantle, with white central back and chestnut-orange underparts and underwing. **Voice** Low *chak-chak*. **SHH** Breeds S Europe to China, spending boreal winter mainly in sub-Saharan Africa S to E Africa. Vagrant Seychelles, Oct–Nov. Favours mountains and cultivation, but vagrants can turn up anywhere. Solitary and shy, quickly diving for cover. Flight low, often quivering tail on landing. Sometimes perches in open, bill uptilted, appearing slim and upright. Hops on the ground, appearing chat-like.

Common Redstart *Phoenicurus phoenicurus* 14 cm
Small, graceful chat, from other vagrant chats by *rusty-red rump and tail, and blackish central tail-feathers*. ♂ distinctive, with black face and throat, orangey-red breast, and white forehead extending as a short supercilium. ♀ is duller, grey-brown above and buffish below, whitish on throat and belly with a pale eye-ring. First-winter has no obvious supercilium; browner, lacking grey tones, orange breast much paler. Race *samamisicus* ♂ has *prominent white wing-panel*, and is slightly darker above, richer orange below; ♀ has pale grey or buff wing-panel. **Voice** Alarm call *twik* or *whee-tik-tik*. **SHH** Nominate breeds mainly Europe to C Asia, spending boreal winter mainly in tropical Africa S to Tanzania; *samamisicus* (recorded Seychelles) breeds Crimea to Iran, spending boreal winter to S. Vagrant Seychelles, Oct–Feb. Frequents woodland. Restless, constantly quivering tail and making short flights.

Whinchat *Saxicola rubetra* 13 cm
Small, stocky, short-tailed chat, from resident African and Reunion Stonechats (Plates 50 & 87) and vagrant Siberian Stonechat by *broad white or buff supercilium*. Blackish-brown tail with prominent white bases to sides shared with some Siberian, but not Reunion or African; also shows dark cheek-panel with pale lower border, dark rump, white wing-patches and buffy throat. ♂ has blackish-brown crown and cheeks, and a warm orangey-buff throat and chest. Imm warm buff, lacking white in wing. **Voice** Usual call a disyllabic *tu-tik*, sometimes *tu-tik-tik*. **SHH** Breeds Europe to Siberia, spending boreal winter in sub-Saharan Africa as far as E Africa. Vagrant Seychelles, Nov. Frequents open country, forest margins and cultivation. Flight is low, straight and rapid. Flicks wings and bobs tail. Perches on a low vantage point.

Siberian Stonechat *Saxicola maurus* 13 cm
See Whinchat for differences from that species. ♂ from African Stonechat (Plate 50) by *more white on rump and wing, more diffuse, paler chestnut on breast* (race *hemprichii*), and white in tail (race *variegatus*); ♀ and imm paler than African Stonechat, with paler throat contrasting with rufous-washed breast, and pale buffy supercilium. **Voice** Mainly silent. **SHH** Breeds temperate Asia, spending boreal winter mainly in S Asia, races *hemprichii* and *variegatus* (formerly known as *variegatus* and *armenicus*, respectively) reaching NE Africa. Vagrant Seychelles, Dec (race *hemprichii* or *variegatus*). Open habitats, often with some woody vegetation. Low, rapid flight. Hunts from a low exposed perch.

Spotted Flycatcher *Muscicapa striata* 14 cm
Small flycatcher, from others by *uniform grey-brown upperparts with dull whitish underparts streaked brown on sides of throat, breast and crown*, and whitish eye-ring around dark brown eye. First-winter has buff wash to face and forehead, and broad pale fringes to wing-feathers. Humblot's Flycatcher (Plate 76), resident in upland forest on Grande Comore) darker brown with bright orange-yellow bill. No other small flycatcher identified in region but indeterminate *Ficedula* sp. (Pied Flycatcher *F. hypoleuca*, Collared Flycatcher *F. albicollis* or Semi-collared Flycatcher *F. semitorquata*) vagrant to Seychelles; ♀/imm as Spotted Flycatcher but unstreaked below and on crown, and more striking white on tips of greater coverts, tertials and primaries; see specialist literature for details. **Voice** Usually silent. **SHH** Breeds mainly Europe to C Asia, spending boreal winter in sub-Saharan Africa to South Africa. Near-annual Seychelles, vagrant Comoros, Europa, Oct–Dec and Mar–Apr. Most likely races are nominate (recorded E Africa to South Africa), *neumanni* (commonest race in E Africa, paler grey above, whiter below), and *sarudnyi* (possibly also E Africa, paler sandy-brown above, streaking less distinct); *striata* and *sarudnyi* confirmed on Seychelles. Frequents woodland clearings and edges. Rapid, slightly undulating flight. Upright posture, occasionally flicking wings and tail.

PLATE 120: VAGRANTS

Northern Wheatear *Oenanthe oenanthe* — 15 cm
Breeding ♂ distinctive; ♀ and non-breeder similar to Isabelline Wheatear, but latter is larger and paler, lacking dark on wing-coverts, has clear whitish supercilium in front of eye, buffish and less distinct behind (opposite of Northern), ear-coverts not darker than crown, and more upright stance. Pied Wheatear is smaller, shorter-billed, darker with black underwing, more extensive white on outer-tail feathers and narrower black tail-band. ♀ Desert Wheatear has all-black tail. **Voice** A hard *chak* or a squeaky *weet*, sometimes combined. **SHH** Breeds across Eurasia, Alaska, Greenland and NE Canada, spending boreal winter entirely in sub-Saharan Africa. Near-annual throughout Seychelles mainly Dec–Mar (much later than other migrants from N) and by far the most likely wheatear to be encountered; also vagrant Madagascar and Comoros. Frequents open, often stony or sandy, habitats. Flight low and rapid, diving for cover or sweeping upward to perch. Makes short runs or hops, flicks wings and tail, bobs body or head.

Pied Wheatear *Oenanthe pleschanka* — 15 cm
Small, slim wheatear with **extensive white rump-patch and black underwing**; see Northern Wheatear for separation. Isabelline Wheatear is paler and larger with broader black tail-band. **Voice** Short monotonous song, *tri-tri-trees* (Feb onward). Calls include a harsh *chak*. **SHH** Breeds E Europe to NW India and Mongolia, spending boreal winter from SW Arabia to E Africa. Vagrant Seychelles, Nov–Feb. Frequents dry open areas, often perching on shrubs or low trees. Flight more buoyant, less flitting than Northern Wheatear, partly due to long tail.

Desert Wheatear *Oenanthe deserti* — 15 cm
In all plumages from other wheatears and chats by **almost entirely black tail** (a little white at the very base), and long primary projection. Tail longer than stonechats. C Asian race *oreophila* is larger than nominate and shows extensive white in outer webs of primaries. **Voice** Calls include a plaintive *sweeoo* and a clicking *chak*. **SHH** Nominate race breeds Middle East to C Asia, wintering S to NE Africa. Race *oreophila* breeds Kashmir to W China, spending boreal winter in Arabia, SW Asia and India, and also reaching NE Africa. Vagrant Seychelles, Nov (race *oreophila*). Favours dry open areas. Strong direct flight and long escape flight, then dives for cover.

Isabelline Wheatear *Oenanthe isabellina* — 17 cm
Large pale wheatear, from others **by larger size, lack of contrast above and below, pale cream underwing, black alula contrasting with pale wing-coverts**. Lores greyish, blacker in ♂, whitish supercilium broad in front of eye, tapering and more buffish behind it, and ear-coverts not darker than crown. **Voice** Non-breeders generally silent, but occasionally give a subdued song. Alarm call a harsh *chak*. **SHH** Breeds SE Europe to China, spending boreal winter to S including sub-Saharan Africa to Tanzania. Vagrant Seychelles (granitics and Amirantes), Oct–Jan. Frequents open terrain and short grass. Strong, direct flight. Upright stance with tail well off the ground. Wags tail and bobs head emphatically.

PLATE 121: VAGRANTS

Sedge Warbler *Acrocephalus schoenobaenus* 13 cm
Distinctive, fairly small warbler, from others by **boldly streaked upperparts, conspicuous creamy supercilium, and blackish-streaked crown with faint pale central crown-stripe**. Larger than resident Madagascar Cisticola (Plate 52) with slightly longer tail lacking white tips. First-winter may show faintly streaked or speckled breast. **Voice** Subdued song in winter quarters, a stream of grating, nasal or trilled notes and whistles. Call a short sharp *tuk* or muffled *churr*. **SHH** Breeds Europe to C Asia, spending boreal winter in sub-Saharan Africa. Vagrant Seychelles, Nov. Favours damp vegetation, sometimes drier habitats. Feeds low in dense vegetation, regularly flicking tail. Flitting, low flight often with tail depressed and spread.

Marsh Warbler *Acrocephalus palustris* 13 cm
Small unstreaked warbler, paler than larger Seychelles Warbler (Plate 64) and Madagascar Swamp Warbler (Plate 52), lacking streaking below and has shorter tail. Legs pale pinkish-brown usually with **pale claws**. From vagrant Icterine Warbler by lack of clear yellow on throat, shorter wings, and dark line from bill to behind eye. Several other unstreaked *Acrocephalus* may occur in the region and identification is a challenge. Sightings on Comoros include possible Lesser Swamp Warbler *A. gracilirostris* and Great Reed Warbler *A. arundinaceus*; Oriental Reed Warbler *A. orientalis*, Clamorous Reed Warbler *A. stentoreus*, Basra Reed Warbler *A. griseldis* and Eurasian Reed Warbler *A. scirpaceus* could all occur; see specialist literature. **Voice** Calls include a harsh *chaah*. **SHH** Breeds Europe to Kazakhstan, spending boreal winter to SE Africa. Vagrant Seychelles, Apr. Frequents tall vegetation, cultivation, reeds and edge of marshes. Low level flight.

Eurasian Blackcap *Sylvia atricapilla* 14 cm
Dark grey, robust warbler with long wings and legs, from others by **glossy crown to eye level, black (♂) or rusty-brown (♀/imm)**. **Voice** A harsh *tak-tak*. **SHH** Breeds Europe to Siberia, spending boreal winter S to C Africa. Vagrant Seychelles, Nov–Dec. Wide habitat choice, scrub to dense forest. Flight straight and rapid. Hops, sometimes raising cap, and may flick wings/tail.

Garden Warbler *Sylvia borin* 14 cm
Very plain, robust warbler, from others by grey-brown upperparts, paler below, faint blue-grey neck-patch, deep grey bill with slightly darker tip, thick grey legs, long primary projection. **Conspicuous dark eye**, narrow broken pale eye-ring and short, very indistinct, buff supercilium. Autumn ad often very worn when first-winter is fresh. **Voice** Call *chek-chek-chek*. **SHH** Breeds Europe to Siberia, spending boreal winter in E & S Africa. Vagrant Seychelles, Dec and Mar. Frequents scrub or woodland. Flight straight and rapid. Skulking, flitting between bushes.

Common Whitethroat *Sylvia communis* 14 cm
Medium-sized, long-tailed warbler, from others by **white throat contrasting with pale buff underparts, tinged pinkish on breast; also broad rufous fringes to tertials, secondaries and greater coverts forming a rufous wing-panel**, tertials have dark centres. Pale brown-orange eye and narrow white eye-ring sometimes conspicuous; tail dark brown with bold white edges. Races *icterops* and *rubicola* have rather long pointed wings, *rubicola* greyer above, whiter below and paler; *volgensis* is somewhat darker. **Voice** Call a repeated *chek* or harsh *charr*. **SHH** Breeds Europe to Eurasia, spending boreal winter in sub-Saharan Africa. Vagrant Seychelles, Oct and Mar. Vagrants may include races *volgensis* (E Europe to Siberia), *icterops* (Turkey to Turkmenistan) and *rubicola* (C Asia to Mongolia, palest race).

PLATE 122: VAGRANTS

Icterine Warbler *Hippolais icterina*   13 cm
Medium-sized, *long-winged* warbler, from other vagrant warblers by *long, broad bill with pale lower mandible, sloping forehead and narrow square tail, yellowish wing-panel and pale yellow underparts*. Short, diffuse yellow supercilium and pale lores give plain-faced appearance; blue-grey legs. *Phylloscopus* warblers and jeries (Plate 51) are smaller with slender bills and a dark eyestripe. Other *Hippolais* warblers are possible, including Eastern Olivaceous Warbler *H. pallida*, Upcher's Warbler *H. languida* and Olive-tree Warbler *H. olivetorum*. **Voice** Call note *tek*, often repeated, crown feathers raised. **SHH** Breeds Europe to Siberia, spending boreal winter in sub-Saharan Africa to South Africa. Vagrant Seychelles, Nov. Forest margins and glades, foraging restlessly mainly in high canopy. Flight dashing. Flicks tail and raises crown feathers when excited.

Willow Warbler *Phylloscopus trochilus*   11 cm
Small, slender warbler with long wings and usually pale legs, fairly strong pale supercilium and dark eyestripe. From jeries (Plate 51) by lack of grey on nape, dark eye, duller legs, and lack of streaking below. Wood Warbler is brighter, greener above with sharp contrast between yellow breast and white belly. Common Chiffchaff is more compact, olive-green above, olive-yellow (or sometimes buffish) below, head pattern weaker with shorter, less distinct supercilium but has distinct whitish eye-ring, legs usually blackish; dips tail. Icterine Warbler lacks dark eyestripe, bill mostly pale. Race *yakutensis* lacks olive tones, is more grey-brown above, dull whitish below with prominent white supercilium. Race *acredula* streaked greenish, with yellow underparts, greyish breast and white supercilium. **Voice** Call an almost disyllabic *hoo-eet*. **SHH** Breeds Europe to Siberia; race *acredula* from Scandinavia to Yenisey River, race *yakutensis* in Siberia, both (mainly *acredula*) spending boreal winter in E Africa. Vagrant Seychelles, Nov and Feb–Mar. Race(s) unknown. Frequents woodland, feeding at all levels. Dashing, agile flight. Sometimes flicks tail.

Common Chiffchaff *Phylloscopus collybita*   11 cm
Small, relatively short-winged warbler usually with dark legs; habitually dips tail downward when feeding. See Willow Warbler for separation. **Voice** Song a diagnostic *chiff-chiff-chaff* (Jan onward). Call a monosyllabic *hweet*. **SHH** Breeds Europe to Siberia, spending boreal winter from S Europe to Africa and India, mainly N of equator. Vagrant Seychelles, Jan, probably race *abietinus* which breeds N Europe and reaches E Africa. Frequents woodland at all levels, feeding actively.

Wood Warbler *Phylloscopus sibilatrix*   12 cm
Broad-chested, short-tailed warbler. From others by *long primary projection, long and prominent bright yellow supercilium extending to above rear ear-coverts, bright green upperparts, white belly with upper breast and throat washed lemon-yellow*. Some dull-plumaged birds lack yellow upperparts, but identified by prominent supercilium and very long wings. From below, tail has slight cleft and only a short projection beyond long undertail-coverts. Yellowish-brown legs, thicker than in Willow Warbler. **Voice** Usually silent; call a monotone *hweet*. **SHH** Breeds Europe to W Russia, spending boreal winter mostly in W Africa and rare in E Africa. Vagrant Seychelles, Nov–Dec. Frequents woodland, typically foraging high in trees. Flight fluent and agile, often gliding with wings half-closed. Active, but does not flick tail, sometimes holding wings low, tips below tail base.

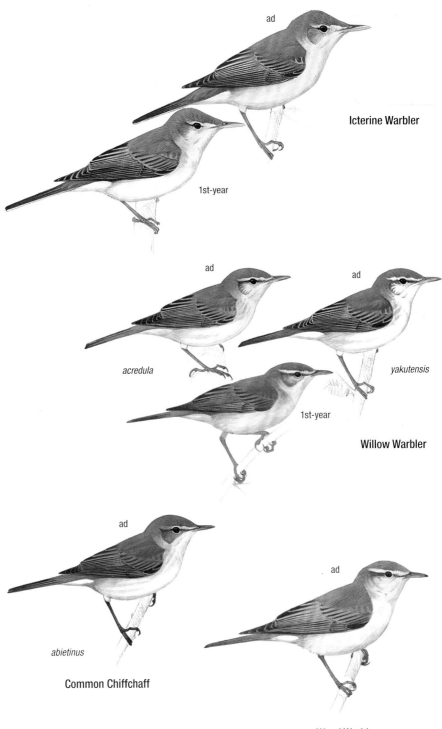

PLATE 123: VAGRANTS

Eurasian Golden Oriole *Oriolus oriolus* 22 cm
Very distinctive; ♂ unmistakable, ♀ from all other species of region by *yellowish rump, uppertail-coverts and flanks, breast streaked dull brown*. First-winter similar to ♀, but more heavily streaked below, median coverts usually tipped yellowish, fresh primaries tipped and edged whitish, eye dark brown or blackish and bill dark brown, later becoming pinkish. **Voice** A liquid, fluty whistle; call a harsh *kraar*. **SHH** Breeds N Africa and Europe E to Siberia and Mongolia, spending boreal winter in sub-Saharan Africa S to South Africa. Vagrant Madagascar, Seychelles, Comoros, Mauritius, mainly Oct–Nov but also Mar–Apr, usually singles but occasionally small numbers. Active but shy, often remaining concealed within vegetation. Frequents woodland, often near water. Rapid, powerful, undulating flight, wings set forward. Some reports do not rule out similar potential vagrants. ♂ Indian Golden Oriole *O. kundoo* (breeding C & S Asia) has larger black eye-patch, much more yellow in wing, including larger yellow carpal patch, and longer bill.

Red-backed Shrike *Lanius collurio* 18 cm
Small, compact, short-billed shrike. Adult ♂ distinctive with *blue-grey crown and nape* (extending to mantle in race *kobylini*), *black mask, chestnut back*. ♀ *dull rufous-brown above, buff below with heavy blackish scaling, dark brown ear-coverts*, brown or grey-brown crown, creamy supercilium and grey nape. Dark brown tail tipped and edged white. First-winter *rusty-brown and heavily scaled blackish above, pale whitish below with some scaling*, rufous tail tipped and edged white, lacking contrast with upperparts. Potential vagrant Isabelline Shrike *L. isabellinus* is a common Palearctic migrant to E Africa; first-winter from Red-backed by slightly paler, greyer and less heavily barred upperparts, and especially rufous tail above and below. **Voice** A harsh *chak-chak*. **SHH** Nominate race breeds S Europe to Asia, grading into *pallidifrons* in W Siberia and *kobylini* from Crimea to Iran; spends boreal winter in sub-Saharan Africa. Northbound passage typically further E than southbound, thus most Malagasy region records in boreal spring. Vagrant Seychelles, mainly Mar, a few Oct–Nov. Race unknown, but *kobylini* commonest in E Africa. Frequents bushes, hunting from an open perch. Flight rapid and agile, undulating over distance.

Lesser Grey Shrike *Lanius minor* 21 cm
Large-headed, grey-and-white shrike, ad from other species *by black mask on forehead, grey crown and upperparts*. Wings long and black with large white patch at base of primaries. First-winter tinged rufous-brown above, has brownish-black mask and faintly barred grey forehead with some white scaling on wing-coverts, pale edges and tips to primaries, and brown rump and tail. Southern Grey Shrike *L. meridionalis* is a potential Palearctic vagrant; it has white scapulars and whitish lores, shorter primary projection and no black frontal band. **Voice** Calls include a harsh piping chatter and a short *chek*. Song (Dec onward) a babbling chatter. **SHH** Breeds Spain to C Asia and Siberia, spending boreal winter in S Africa. Migrates further E on spring passage. Vagrant Seychelles, Nov and Mar–Apr. Rapid, direct flight, undulating over a distance.

Woodchat Shrike *Lanius senator* 18 cm
Small, stocky shrike, mainly black and white with *diagnostic chestnut rear crown and nape, and white uppertail-coverts* extending well onto tail base in race *niloticus*, which also has white primary patch larger than other races. ♀ duller and browner on back with dusky crescent-shaped bars on breast and flanks, forehead and mask flecked chestnut. First-winter is less 'tidy' and can show some scalloping on breast, indistinct mask and paler crown, forehead flecked brown-black, pale patch on scapulars and pale rump. First-winter Red-backed Shrike is darker, more rufous above with a brown rump. Masked Shrike *L. nubicus* is a potential vagrant: longer-tailed and smaller-billed; ad separated by black crown; imm mainly grey and scaly above, white below with pale face. **Voice** A sparrow-like chatter or a harsh *kwik-kwik*. **SHH** Nominate breeds S Europe and N Africa, spending boreal winter in W Africa to Uganda; race *niloticus* breeds Cyprus to Middle East, spending boreal winter mainly in NE Africa, a few S to Kenya. Vagrant Seychelles, Apr, race unknown. Frequents woodland edges and clearings with trees, perching on tops of shrubs. Flight rapid and direct.

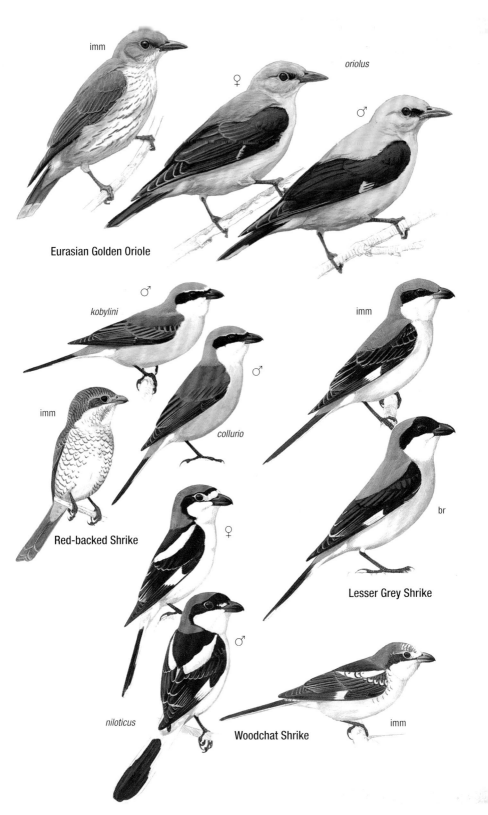

PLATE 124: VAGRANTS

Rosy Starling *Pastor roseus* — 21 cm
Ad very distinctive. Imm from Wattled Starling by indistinct pale supercilium, **sandy-brown upperparts**, palest on rump (but not white), **dark brown wings with buffish-white fringes**, bill yellowish with darker tip and legs pale pink. **Voice** A variety of loud twittering and chattering notes, but non-breeders generally silent. **SHH** Breeds SE Europe to *c.*50°N, E to NW China, spending boreal winter from Oman to Sri Lanka and Bangladesh. Vagrant Seychelles (granitics), mainly Oct–Dec. Open areas and woodland. Flight direct and rapid with some gliding.

Wattled Starling *Creatophora cinerea* — 21 cm
Breeding ad distinctive; non-breeder has **pale grey or grey-brown body contrasting with whitish rump, short black tail and black flight-feathers**, greater coverts whitish in ♂, black in ♀. Breeding ♂ has black wattles and extensive bare yellow skin from eye to hindcrown, and yellow bill. See Rosy Starling for differences from that species. Common Myna is darker brown, lacking grey tones and has conspicuous white wing-patches. **Voice** High, squeaky notes or in flight a harsh trisyllabic call. Alarm call a grating, nasal *graaaa*. **SHH** Breeds sub-Saharan Africa, mainly from Ethiopia and Sudan to South Africa; nomadic. Vagrant to Madagascar and Seychelles, recorded all months sometimes in small flocks and remaining for several months. Favours areas with scattered bushes or trees and cultivation. Highly gregarious. Direct, rapid flight.

Common Rosefinch *Carpodacus erythrinus* — 14 cm
Medium-sized finch with **stubby, bulbous bill**. Non-breeding ♂ and ♀/imm from House Sparrow and fodies by **greyish-white underparts with diffuse streaking, two indistinct narrow pale wingbars**; **dark eye stands out in featureless face**. Breeding ♂ has bright red head and breast. **Voice** Call a hoarse whistle. In flight or on take-off gives short *zik*. Alarm call a harsh *chay-eeee*. **SHH** Breeds Europe to Siberia, spending boreal winter from India to SE Asia. Vagrant Seychelles (granitic islands), Oct. Frequents undergrowth and bushes, forest edges and cultivation, often near water. Undulating flight.

Ortolan Bunting *Emberiza hortulana* — 16 cm
Distinctive in region; from other small seedeaters by **yellowish throat, moustachial stripe and eye-ring**, pinkish legs and bill. Breeding ♂ has olive-grey head, breast-band and malar stripe surrounding yellow throat and moustache, with orange-brown belly and flanks. First-winter grey-brown above, streaked on throat and breast. Cretzschmar's Bunting *E. caesia*, a potential vagrant (albeit less likely) not eliminated from some records; ♀/imm very similar but throat whiter, eye-ring lacking yellow tones and often has rufous rather than olive tinge to rump and vent. ♂ has rusty-brown throat and moustachial stripe. **Voice** Call a soft *tsip* or piping *tsoo*. **SHH** Breeds mainly Europe to Mongolia, spending boreal winter in N tropical Africa, mainly at 7–15°N in uplands. Vagrant Seychelles, mainly Oct–Nov. In scattered trees, woodland margins and clearings. Flight is light and undulating, diving for cover.

APPENDIX 1
CHECKLIST OF BIRDS OF THE MALAGASY REGION

Species name (English and scientific)

Background colour indicates native/endemic/introduced status at regional level

- native, non-endemic species
- endemic species (includes species which spend the non-breeding season outside the region)
- member of regionally endemic genus
- member of regionally endemic family (includes vangas Vangidae, although their status as an endemic family is dubious; see p. 44)
- member of regionally endemic order
- introduced

Subspecies

Yellow background colour indicates regional endemic subspecies of non-endemic species; a blank indicates that the species has no races (i.e. monotypic).

Distribution

X: Extinct within last c. 50 years (excludes exotics).
V: Vagrant. Records of vagrants other than on Seychelles have not been formally assessed.
?: Unconfirmed record, only included here where there are no confirmed records anywhere in the region (listing all 'unconfirmed' records per country or archipelago is beyond the scope of this book).

Colour codes are used as for species name, but refer to the status within the area described by each column (e.g. Madagascar Partridge is native to the region but introduced to Reunion).

English name	Scientific name	Subspecies	Madagascar	Granitic Seychelles	Coralline Seychelles	Grande Comore	Mohéli	Anjouan	Mayotte	Mauritius	Rodrigues	Reunion	Outer Islands
Rockhopper Penguin	Eudyptes chrysocome	filholi/moseleyi	V									V	
Little Grebe	Tachybaptus ruficollis	capensis				V							
Alaotra Grebe	Tachybaptus rufolavatus		X										
Madagascar Grebe	Tachybaptus pelzelnii												
Black-necked Grebe	Podiceps nigricollis	nigricollis		V									
		unconfirmed race			V								
Wandering Albatross	Diomedea exulans									V		V	
Black-browed Albatross	Thalassarche melanophris		V									V	
Shy Albatross	Thalassarche cauta	unconfirmed race											
		steadi											
Indian Yellow-nosed Albatross	Thalassarche carteri	bassi											
Sooty Albatross	Phoebetria fusca		?							?		?	
Light-mantled Albatross	Phoebetria palpebrata											V	
Southern Giant Petrel	Macronectes giganteus												
Northern Giant Petrel	Macronectes halli		V									V	
Southern Fulmar	Fulmarus glacialoides											?	
Cape Petrel	Daption capense	capense											
Antarctic Prion	Pachyptila desolata	desolata	V										
Slender-billed Prion	Pachyptila belcheri		V								V		

English name	Scientific name	Subspecies	Madagascar	Granitic Seychelles	Coralline Seychelles	Grande Comore	Mohéli	Anjouan	Mayotte	Mauritius	Rodrigues	Reunion	Outer Islands
Grey Petrel	Procellaria cinerea											V	
White-chinned Petrel	Procellaria aequinoctialis									V			
Spectacled Petrel	Procellaria conspicillata											V	
Mascarene Petrel	Pseudobulweria aterrima												
Bulwer's Petrel	Bulweria bulwerii												
Jouanin's Petrel	Bulweria fallax												
Streaked Shearwater	Calonectris leucomelas		V										
Cory's Shearwater	Calonectris diomedea	borealis											
Wedge-tailed Shearwater	Puffinus pacificus												
Flesh-footed Shearwater	Puffinus carneipes												
Short-tailed Shearwater	Puffinus tenuirostris									V			
Tropical Shearwater	Puffinus bailloni	bailloni											
		nicolae											
		temptator											
Little Shearwater	Puffinus assimilis	tunneyi ?								V			
Kerguelen Petrel	Aphrodroma brevirostris											V	
Barau's Petrel	Pterodroma baraui									V	V		
Trindade Petrel	Pterodroma arminjoniana												
Kermadec Petrel	Pterodroma neglecta			V									
Herald Petrel	Pterodroma heraldica			V									
Soft-plumaged Petrel	Pterodroma mollis	dubia ?											
Great-winged Petrel	Pterodroma macroptera	macroptera											
Black-winged Petrel	Pterodroma nigripennis									V			
Wilson's Storm-petrel	Oceanites oceanicus	oceanicus/exasperatus											
White-faced Storm-petrel	Pelagodroma marina	dulciae											
Black-bellied Storm-petrel	Fregetta tropica	tropica											
White-bellied Storm-petrel	Fregetta grallaria	leucogaster	?										?
Swinhoe's Storm-petrel	Oceanodroma monorhis												
Matsudaira's Storm-petrel	Oceanodroma matsudairae												
Red-billed Tropicbird	Phaethon aethereus	unconfirmed race		V	V								
		indicus		V									
Red-tailed Tropicbird	Phaethon rubricauda	rubricauda											
White-tailed Tropicbird	Phaethon lepturus	lepturus											
		europae											
Pink-backed Pelican	Pelecanus rufescens		V			X						V	
Cape Gannet	Morus capensis		V									V	
Australasian Gannet	Morus serrator									V			
Masked Booby	Sula dactylatra	melanops										V	
Red-footed Booby	Sula sula	rubripes								V			
Brown Booby	Sula leucogaster	plotus											
Great Cormorant	Phalacrocorax carbo	lucidus/sinensis		V									
Reed Cormorant	Phalacrocorax africanus	unconfirmed race		V	V	V							
		pictilis											

English name	Scientific name	Subspecies	Madagascar	Granitic Seychelles	Coralline Seychelles	Grande Comore	Mohéli	Anjouan	Mayotte	Mauritius	Rodrigues	Reunion	Outer Islands
African Darter	Anhinga rufa	vulsini											
Great Frigatebird	Fregata minor	unconfirmed race								V	V	V	
		aldabrensis											
Lesser Frigatebird	Fregata ariel	unconfirmed race											
		iredalei											
Grey Heron	Ardea cinerea	cinerea								V		V	
		firasa											
Black-headed Heron	Ardea melanocephala		V										
Madagascar Heron	Ardea humbloti								V				
Goliath Heron	Ardea goliath		V										
Purple Heron	Ardea purpurea	purpurea				V							
		madagascariensis											
Great Egret	Ardea alba	melanorhynchos	V	V									
Intermediate Egret	Ardea intermedia	intermedia	V	V									
Cattle Egret	Ardea ibis	ibis											
Black Heron	Egretta ardesiaca												
Little Egret	Egretta garzetta	dimorpha											
		garzetta	V										
Squacco Heron	Ardeola ralloides		V	V	V	V	V	V	V				
Indian Pond Heron	Ardeola grayii		V	V									
Madagascar Pond Heron	Ardeola idae		V									V	
Striated Heron	Butorides striata	rutenbergi											
		rhizophorae											
		javanica											
		degens											
		crawfordi											
Black-crowned Night Heron	Nycticorax nycticorax	nycticorax											
Yellow Bittern	Ixobrychus sinensis				V								
Little Bittern	Ixobrychus minutus	minutus	V										
		podiceps											
Cinnamon Bittern	Ixobrychus cinnamomeus		V										
Eurasian Bittern	Botaurus stellaris	stellaris	V	V									
Hamerkop	Scopus umbretta	tenuirostris											
Yellow-billed Stork	Mycteria ibis												
African Openbill	Anastomus lamelligerus	madagascariensis											
White Stork	Ciconia ciconia	ciconia/asiatica	V	V									
African Sacred Ibis	Threskiornis aethiopicus				V								
Madagascar Sacred Ibis	Threskiornis bernieri	bernieri											
		abbotti											
		unconfirmed race								V			
Glossy Ibis	Plegadis falcinellus		V									V	

English name	Scientific name	Subspecies	Madagascar	Granitic Seychelles	Coralline Seychelles	Grande Comore	Mohéli	Anjouan	Mayotte	Mauritius	Rodrigues	Reunion	Outer Islands
Madagascar Crested Ibis	Lophotibis cristata	cristata											
		urschi											
African Spoonbill	Platalea alba												
Greater Flamingo	Phoenicopterus roseus			V									V
Lesser Flamingo	Phoeniconaias minor										V	V	V	V
Fulvous Whistling Duck	Dendrocygna bicolor													
White-faced Whistling Duck	Dendrocygna viduata					V		V		V			V	
White-backed Duck	Thalassornis leuconotus	insularis											
Comb Duck	Sarkidiornis melanotos	melanotos				V							
African Pygmy Goose	Nettapus auritus													
Eurasian Wigeon	Anas penelope			V									
Madagascar Teal	Anas bernieri												
Mallard	Anas platyrhynchos	platyrhynchos		V									
Meller's Duck	Anas melleri												
Northern Pintail	Anas acuta			V									
Red-billed Teal	Anas erythrorhyncha				V								
Hottentot Teal	Anas hottentota												
Garganey	Anas querquedula										V	V		
Northern Shoveler	Anas clypeata			V	V								
Ferruginous Duck	Aythya nyroca			V	V								
Madagascar Pochard	Aythya innotata												
Tufted Duck	Aythya fuligula				V								
Osprey	Pandion haliaetus	haliaetus		V			V		V				
Madagascar Cuckoo-hawk	Aviceda madagascariensis												
European Honey-buzzard	Pernis apivorus			V	V								
Bat Hawk	Macheiramphus alcinus	anderssoni											
Black-winged Kite	Elanus caeruleus	caeruleus	V										
Black Kite	Milvus migrans	migrans		V	V								
		parasitus			V	X							V
Madagascar Fish Eagle	Haliaeetus vociferoides										V			
Madagascar Serpent Eagle	Eutriorchis astur												
Western Marsh Harrier	Circus aeruginosus	aeruginosus		V									
Reunion Harrier	Circus maillardi												V	
Madagascar Harrier	Circus macrosceles									V			
Pallid Harrier	Circus macrourus				V								
Madagascar Harrier-hawk	Polyboroides radiatus												
Frances's Sparrowhawk	Accipiter francesiae	francesiae											
		griveaudi											
		pusillus											
		brutus											
Madagascar Sparrowhawk	Accipiter madagascariensis												
Henst's Goshawk	Accipiter henstii												

English name	Scientific name	Subspecies	Madagascar	Granitic Seychelles	Coralline Seychelles	Grande Comore	Mohéli	Anjouan	Mayotte	Mauritius	Rodrigues	Reunion	Outer Islands
Madagascar Buzzard	*Buteo brachypterus*		■										
Booted Eagle	*Hieraaetus pennatus*		V										
Long-crested Eagle	*Lophaetus occipitalis*		V										
Lesser Kestrel	*Falco naumanni*		V	V					V				
Common Kestrel	*Falco tinnunculus*	tinnunculus?		V									
Madagascar Kestrel	*Falco newtoni*	newtoni	■										
		aldabranus			■								
Mauritius Kestrel	*Falco punctatus*									■			
Seychelles Kestrel	*Falco araeus*			■									
Banded Kestrel	*Falco zoniventris*		■										
Red-footed Falcon	*Falco vespertinus*		V										
Amur Falcon	*Falco amurensis*								V				
Eleonora's Falcon	*Falco eleonorae*		V	V									V
Sooty Falcon	*Falco concolor*		V	V						V	V		
Eurasian Hobby	*Falco subbuteo*	subbuteo	V	V									
Saker Falcon	*Falco cherrug*	cherrug/milvipes	V										
Peregrine Falcon	*Falco peregrinus*	radama	■										
		calidus	V							V			
Grey Francolin	*Francolinus pondicerianus*	pondicerianus											
Madagascar Partridge	*Margaroperdix madagarensis*		■										
Common Quail	*Coturnix coturnix*	africana	V				V		V	V	V		
Harlequin Quail	*Coturnix delegorguei*	delegorguei				V							
King Quail	*Coturnix chinensis*	palmeri?	■										
Jungle Bush Quail	*Perdicula asiatica*	asiatica?											
Red Junglefowl	*Gallus gallus*	gallus											
Helmeted Guineafowl	*Numida meleagris*	mitratus	■										
Brown Mesite	*Mesitornis unicolor*		■										
White-breasted Mesite	*Mesitornis variegatus*		■										
Subdesert Mesite	*Monias benschi*		■										
Madagascar Buttonquail	*Turnix nigricollis*		■										
Madagascar Flufftail	*Sarothrura insularis*		■										
Slender-billed Flufftail	*Sarothrura watersi*		■										
Madagascar Wood Rail	*Mentocrex kioloides*	kioloides	■										
		berliozi	■										
Tsingy Wood Rail	*Mentocrex beankensis*		■										
Buff-banded Rail	*Gallirallus philippensis*	unconfirmed race								V			
Madagascar Rail	*Rallus madagascariensis*		■										
White-throated Rail	*Dryolimnas cuvieri*	cuvieri								V	V		
		aldabranus			■								
Corn Crake	*Crex crex*		V										
White-breasted Waterhen	*Amauromis phoenicurus*	phoenicurus	V										
Sakalava Rail	*Amauromis olivieri*		■										

English name	Scientific name	Subspecies	Madagascar	Granitic Seychelles	Coralline Seychelles	Grande Comore	Mohéli	Anjouan	Mayotte	Mauritius	Rodrigues	Reunion	Outer Islands
Little Crake	Porzana parva		V										
Baillon's Crake	Porzana pusilla	intermedia											
Spotted Crake	Porzana porzana		V	V									
Striped Crake	Aenigmatolimnas marginalis				V								
Purple Swamphen	Porphyrio porphyrio	madagascariensis											
Allen's Gallinule	Porphyrio alleni		V	V	V				V				
Common Moorhen	Gallinula chloropus	pyrrhorrhoa											
		orientalis				V							
Lesser Moorhen	Gallinula angulata					V							
Red-knobbed Coot	Fulica cristata												
Madagascar Jacana	Actophilornis albinucha												
Greater Painted-snipe	Rostratula benghalensis	benghalensis											
Crab-plover	Dromas ardeola									V		V	
Eurasian Oystercatcher	Haematopus ostralegus	longipes?	V	V				V					
Black-winged Stilt	Himantopus himantopus	himantopus	V	V									
Pied Avocet	Recurvirostra avosetta		V	V									
Eurasian Stone-curlew	Burhinus oedicnemus	oedicnemus/harterti	V	V									
Collared Pratincole	Glareola pratincola	pratincola	V	V									
Oriental Pratincole	Glareola maldivarum		V	V	V				V		V		
Black-winged Pratincole	Glareola nordmanni		V	V									
Madagascar Pratincole	Glareola ocularis		V	V	V		V	V			V	V	
Blacksmith Lapwing	Vanellus armatus												V
Spur-winged Lapwing	Vanellus spinosus		V										
Sociable Lapwing	Vanellus gregarius		V										
Pacific Golden Plover	Pluvialis fulva		V					V	V				V
Grey Plover	Pluvialis squatarola												
Common Ringed Plover	Charadrius hiaticula	tundrae											
Little Ringed Plover	Charadrius dubius	curonicus	V	V									
Madagascar Plover	Charadrius thoracicus												
Kittlitz's Plover	Charadrius pecuarius												
Three-banded Plover	Charadrius tricollaris	unconfirmed race				V							
		bifrontatus											
White-fronted Plover	Charadrius marginatus	tenellus											
Lesser Sand Plover	Charadrius mongolus	pamirensis	V						V				
Greater Sand Plover	Charadrius leschenaultii	scythicus/leschenaultii											
Caspian Plover	Charadrius asiaticus		V	V									
Oriental Plover	Charadrius veredus		V	V									
Jack Snipe	Lymnocryptes minimus		V										
Pin-tailed Snipe	Gallinago stenura		V	V									
Madagascar Snipe	Gallinago macrodactyla												
Great Snipe	Gallinago media		V	V									
Common Snipe	Gallinago gallinago	gallinago	V	V									

English name	Scientific name	Subspecies	Madagascar	Granitic Seychelles	Coralline Seychelles	Grande Comore	Mohéli	Anjouan	Mayotte	Mauritius	Rodrigues	Reunion	Outer Islands
Black-tailed Godwit	Limosa limosa	limosa	V	V	V								
Bar-tailed Godwit	Limosa lapponica	taymyrensis											
		menzbieri		V								V	
Little Curlew	Numenius minutus		V										
Whimbrel	Numenius phaeopus	phaeopus											
Eurasian Curlew	Numenius arquata	orientalis											
Terek Sandpiper	Xenus cinereus									V	V		V
Common Sandpiper	Actitis hypoleucos												
Green Sandpiper	Tringa ochropus		V	V									
Grey-tailed Tattler	Tringa brevipes			V	V					V	V		
Spotted Redshank	Tringa erythropus			V									
Common Greenshank	Tringa nebularia												
Marsh Sandpiper	Tringa stagnatilis			V						V			
Wood Sandpiper	Tringa glareola									V			
Common Redshank	Tringa totanus	totanus/ussuriensis	V										
Ruddy Turnstone	Arenaria interpres	interpres											
Great Knot	Calidris tenuirostris			V						V			
Red Knot	Calidris canutus	canutus?			V								
Sanderling	Calidris alba												
Little Stint	Calidris minuta						V		V				
Temminck's Stint	Calidris temminckii			V									
Long-toed Stint	Calidris subminuta			V									
Pectoral Sandpiper	Calidris melanotos		V	V	V					V			
Sharp-tailed Sandpiper	Calidris acuminata		V	V	V								
Curlew Sandpiper	Calidris ferruginea												
Broad-billed Sandpiper	Limicola falcinellus	falcinellus/sibirica		V									
Buff-breasted Sandpiper	Tryngites subruficollis		V	V	V								
Ruff	Philomachus pugnax		V	V	V	V				V	V		
Red-necked Phalarope	Phalaropus lobatus		V	V									
South Polar Skua	Stercorarius maccormicki			V									
Subantarctic Skua	Stercorarius antarcticus	lonnbergi											
Pomarine Skua	Stercorarius pomarinus			V			V						V
Arctic Skua	Stercorarius parasiticus		V	V									
Long-tailed Skua	Stercorarius longicaudus	longicaudus/pallescens							V		V		
Sooty Gull	Ichthyaetus hemprichii		V	V									
Kelp Gull	Larus dominicanus	unconfirmed race										V	
		melisandae											
Lesser Black-backed Gull	Larus fuscus	fuscus	V	V									
		heuglini	V	V									
Grey-headed Gull	Chroicocephalus cirrocephalus	poiocephalus											
Common Black-headed Gull	Chroicocephalus ridibundus		V	V									

English name	Scientific name	Subspecies	Madagascar	Granitic Seychelles	Coralline Seychelles	Grande Comore	Mohéli	Anjouan	Mayotte	Mauritius	Rodrigues	Reunion	Outer Islands
Sabine's Gull	Xema sabini		V										
Gull-billed Tern	Gelochelidon nilotica	nilotica	V	V									
Caspian Tern	Hydroprogne caspia			V		V							
Lesser Crested Tern	Thalasseus bengalensis	bengalensis								V	V	V	
Sandwich Tern	Thalasseus sandvicensis	sandvicensis	V	V	V								
Greater Crested Tern	Thalasseus bergii	thalassinus								V		V	
		velox	V	V									
Roseate Tern	Sterna dougallii	dougallii?							V	V		V	
		arideensis											
Black-naped Tern	Sterna sumatrana	mathewsi	V	V		V	V	V	V				
Common Tern	Sterna hirundo	hirundo									V		
White-cheeked Tern	Sterna repressa			V	V								
Antarctic Tern	Sterna vittata	unconfirmed race	V										
Little Tern	Sternula albifrons	albifrons?		V	V								
Saunders's Tern	Sternula saundersi									V		V	
Bridled Tern	Onychoprion anaethetus	antarcticus										V	
Sooty Tern	Onychoprion fuscatus	nubilosus											
Whiskered Tern	Chlidonias hybrida	delalandii											
		unconfirmed race		V	V								
White-winged Tern	Chlidonias leucopterus				V					V	V	V	
Black Tern	Chlidonias niger	niger	V										
Lesser Noddy	Anous tenuirostris	tenuirostris											
Brown Noddy	Anous stolidus	pileatus											
Fairy Tern	Gygis alba	candida	V									V	
Madagascar Sandgrouse	Pterocles personatus												
Feral Pigeon	Columba livia												
Comoro Olive Pigeon	Columba pollenii												
Pink Pigeon	Nesoenas mayeri												
Madagascar Turtle Dove	Nesoenas picturata	picturata											
		coppingeri											
		comorensis											
		rostrata											
European Turtle Dove	Streptopelia turtur	turtur		V	V								
Ring-necked Dove	Streptopelia capicola	tropica											
Spotted Dove	Spilopelia chinensis	tigrina											
Laughing Dove	Spilopelia senegalensis	cambayensis/senegalensis											
Tambourine Dove	Turtur tympanistria												
Namaqua Dove	Oena capensis	aliena											
		unconfirmed race		V	V								
Zebra Dove	Geopelia striata												
Madagascar Green Pigeon	Treron australis	australis											
		xenius											
		griveaudi											

English name	Scientific name	Subspecies	Madagascar	Granitic Seychelles	Coralline Seychelles	Grande Comore	Mohéli	Anjouan	Mayotte	Mauritius	Rodrigues	Reunion	Outer Islands
Madagascar Blue Pigeon	*Alectroenas madagascariensis*		■										
Comoro Blue Pigeon	*Alectroenas sganzini*	sganzini				■	■	■					
		minor							■				
Seychelles Blue Pigeon	*Alectroenas pulcherrimus*			■									
Rose-ringed Parakeet	*Psittacula krameri*	borealis/manillensis								■			
Echo Parakeet	*Psittacula echo*									■			
Grey-headed Lovebird	*Agapornis canus*	ablectaneus											
		canus	■										
Greater Vasa Parrot	*Coracopsis vasa*	vasa	■										
		drouhardi	■										
		comorensis				■	■	■					
Lesser Vasa Parrot	*Coracopsis nigra*	nigra	■										
		libs	■										
		sibilans				■	■	■					
Seychelles Black Parrot	*Coracopsis barklyi*			■									
Madagascar Coucal	*Centropus toulou*	toulou	■										
		insularis											■
Crested Coua	*Coua cristata*	cristata	■										
		dumonti	■										
		pyropyga	■										
		maxima	■										
Verreaux's Coua	*Coua verreauxi*		■										
Blue Coua	*Coua caerulea*		■										
Red-capped Coua	*Coua ruficeps*	ruficeps	■										
		olivaceiceps	■										
Red-fronted Coua	*Coua reynaudii*		■										
Coquerel's Coua	*Coua coquereli*		■										
Running Coua	*Coua cursor*		■										
Giant Coua	*Coua gigas*		■										
Red-breasted Coua	*Coua serriana*		■										
Great Spotted Cuckoo	*Clamator glandarius*			V	V								
Jacobin Cuckoo	*Clamator jacobinus*	jacobinus/pica/serratus		V	V								
Thick-billed Cuckoo	*Pachycoccyx audeberti*	audeberti	■										
Lesser Cuckoo	*Cuculus poliocephalus*		V	V	V								
Madagascar Cuckoo	*Cuculus rochii*		■										
Common Cuckoo	*Cuculus canorus*	canorus/subtelephonus		V	V								
Madagascar Red Owl	*Tyto soumagnei*		■										
Barn Owl	*Tyto alba*	poensis	■										
Eurasian Scops Owl	*Otus scops*	scops/turanicus/pulchellus				V							
Seychelles Scops Owl	*Otus insularis*			■									

English name	Scientific name	Subspecies	Madagascar	Granitic Seychelles	Coralline Seychelles	Grande Comore	Mohéli	Anjouan	Mayotte	Mauritius	Rodrigues	Reunion	Outer Islands
Karthala Scops Owl	Otus pauliani					■							
Anjouan Scops Owl	Otus capnodes							■					
Mohéli Scops Owl	Otus moheliensis						■						
Madagascar Scops Owl	Otus rutilus	rutilus	■										
		madagascariensis	■										
Mayotte Scops Owl	Otus mayottensis								■				
Brown Fish Owl	Ketupa zeylonensis	unconfirmed race	V										
White-browed Owl	Ninox superciliaris		■										
Madagascar Long-eared Owl	Asio madagascariensis		■										
Marsh Owl	Asio capensis	hova	■										
Collared Nightjar	Gactornis enarratus		■										
European Nightjar	Caprimulgus europaeus	europaeus?	V										
Madagascar Nightjar	Caprimulgus madagascariensis	madagascariensis	■										
		aldabrensis			■								
Seychelles Swiftlet	Aerodramus elaphrus			■									
Mascarene Swiftlet	Aerodramus francicus									■		■	
Madagascar Spinetail	Zoonavena grandidieri	grandidieri	■										
		mariae				■							
White-throated Needletail	Hirundapus caudacutus	caudacutus	V	V						V			
African Palm Swift	Cypsiurus parvus	unconfirmed race											V
		gracilis			V								
		griveaudi				■							
Alpine Swift	Tachymarptis melba	unconfirmed race	V		V			V					
		willsi	■										
Common Swift	Apus apus	apus	V	V									
		pekinensis			V								
African Black Swift	Apus barbatus	balstoni	■										
		mayottensis							■				
Pacific Swift	Apus pacificus	pacificus?	V	V									
Little Swift	Apus affinis	affinis?	V	V									
Madagascar Malachite Kingfisher	Corythornis vintsioides	vintsioides	■										
		johannae				■	■	■	■				
		unconfirmed race			V								
Madagascar Pygmy Kingfisher	Corythornis madagascariensis	madagascariensis	■										
		dilutus	■										
Grey-headed Kingfisher	Halcyon leucocephala	unconfirmed race	V	V									
Blue-cheeked Bee-eater	Merops persicus	persicus	V	V									
Olive Bee-eater	Merops superciliosus	superciliosus			V								
European Bee-eater	Merops apiaster		V	V									V
European Roller	Coracias garrulus	garrulus/semenowi								V	V		
Broad-billed Roller	Eurystomus glaucurus	glaucurus	■	V						V		V	
Short-legged Ground-roller	Brachypteracias leptosomus		■										

English name	Scientific name	Subspecies	Madagascar	Granitic Seychelles	Coralline Seychelles	Grande Comore	Mohéli	Anjouan	Mayotte	Mauritius	Rodrigues	Reunion	Outer Islands
Scaly Ground-roller	*Geobiastes squamiger*		■										
Pitta-like Ground-roller	*Atelornis pittoides*		■										
Rufous-headed Ground-roller	*Atelornis crossleyi*		■										
Long-tailed Ground-roller	*Uratelornis chimaera*		■										
Cuckoo-roller	*Leptosomus discolor*	discolor	■				■	■					
		gracilis							■				
		intermedius				■							
Madagascar Hoopoe	*Upupa marginata*		■										
Common Hoopoe	*Upupa epops*	africana		V	V								
		epops/senegalensis			V								
Velvet Asity	*Philepitta castanea*		■										
Schlegel's Asity	*Philepitta schlegeli*		■										
Common Sunbird Asity	*Neodrepanis coruscans*		■										
Yellow-bellied Sunbird Asity	*Neodrepanis hypoxantha*		■										
Madagascar Lark	*Mirafra hova*		■										
Bimaculated Lark	*Melanocorypha bimaculata*	unconfirmed race	V										
Greater Short-toed Lark	*Calandrella brachydactyla*	unconfirmed race	V	V									
Mascarene Martin	*Phedina borbonica*	madagascariensis	V	V					V				
		borbonica								■	■	■	
Brown-throated Martin	*Riparia paludicola*	cowani	■										
		paludicola/chinensis		V									
Sand Martin	*Riparia riparia*	riparia/others?	V	V	V	V		V					V
Barn Swallow	*Hirundo rustica*	rustica											
Wire-tailed Swallow	*Hirundo smithii*	smithii				V							
Common House Martin	*Delichon urbicum*	urbicum		V	V	V							V
Lesser Striped Swallow	*Cecropis abyssinica*	unconfirmed race	V										V
Richard's Pipit	*Anthus richardi*	unconfirmed race				V							
Tree Pipit	*Anthus trivialis*	trivialis											V
Red-throated Pipit	*Anthus cervinus*			V	V								V
White Wagtail	*Motacilla alba*	alba		V	V						V		
Yellow Wagtail	*Motacilla flava*	unconfirmed race		V	V								V
		flava			V								
		lutea			V								
		beema	V										
		feldegg			V								
Citrine Wagtail	*Motacilla citreola*	unconfirmed race	V										
Madagascar Wagtail	*Motacilla flaviventris*		■										
Grey Wagtail	*Motacilla cinerea*	cinerea		V	V	V							
Madagascar Cuckoo-shrike	*Coracina cinerea*	cinerea	■										
		pallida	■										
		cucullata				■							
		moheliensis					■						

English name	Scientific name	Subspecies	Madagascar	Granitic Seychelles	Coralline Seychelles	Grande Comore	Mohéli	Anjouan	Mayotte	Mauritius	Rodrigues	Reunion	Outer Islands
Mauritius Cuckoo-shrike	Coracina typica									■			
Reunion Cuckoo-shrike	Coracina newtoni											■	
Red-whiskered Bulbul	Pycnonotus jocosus	emeria								■		■	
Reunion Bulbul	Hypsipetes borbonicus											■	
Mauritius Bulbul	Hypsipetes olivaceus									■			
Madagascar Bulbul	Hypsipetes madagascariensis	madagascariensis	■					■					
		rostratus				■							
Mohéli Bulbul	Hypsipetes moheliensis						■						
Grande Comore Bulbul	Hypsipetes parvirostris					■							
Seychelles Bulbul	Hypsipetes crassirostris			■									
Comoro Thrush	Turdus bewsheri	bewsheri						■					
		comorensis				■							
		moheliensis					■						
Forest Rock Thrush	Monticola sharpei	sharpei	■										
		erythronotus	■										
Littoral Rock Thrush	Monticola imerinus		■										
Common Rock Thrush	Monticola saxatilis			V	V								
Seychelles Magpie-robin	Copsychus sechellarum			■									
Madagascar Magpie-robin	Copsychus albospecularis	albospecularis	■										
		inexspectatus	■										
		pica	■										
Common Redstart	Phoenicurus phoenicurus	phoenicurus?		V	V								
		samamisicus		V									
Whinchat	Saxicola rubetra			V	V								
African Stonechat	Saxicola torquatus	sibilla	■										
		ankaratrae	■										
		tsaratananae	■										
		voeltzkowi	■										
Siberian Stonechat	Saxicola maurus	hemprichii/variegatus		V									
Reunion Stonechat	Saxicola tectes											■	
Northern Wheatear	Oenanthe oenanthe	oenanthe	V	V	V				V				
Pied Wheatear	Oenanthe pleschanka			V	V								
Desert Wheatear	Oenanthe deserti	unconfirmed race		V									
Isabelline Wheatear	Oenanthe isabellina			V	V								
Spotted Flycatcher	Muscicapa striata	striata		V	V								
		sarudnyi?			V								
Humblot's Flycatcher	Humblotia flavirostris					■							
Madagascar Paradise Flycatcher	Terpsiphone mutata	mutata	■										
		comorensis				■							
		voeltzkowiana					■						
		vulpina						■					
		pretiosa							■				

English name	Scientific name	Subspecies	Madagascar	Granitic Seychelles	Coralline Seychelles	Grande Comore	Mohéli	Anjouan	Mayotte	Mauritius	Rodrigues	Reunion	Outer Islands
Seychelles Paradise Flycatcher	*Terpsiphone corvina*			■									
Mascarene Paradise Flycatcher	*Terpsiphone bourbonnensis*	*bourbonnensis*										■	
		desolata								■			
Common Jery	*Neomixis tenella*	*tenella*	■										
		orientalis	■										
		debilis	■										
		decaryi	■										
Green Jery	*Neomixis viridis*	*viridis*	■										
		delacouri	■										
Stripe-throated Jery	*Neomixis striatigula*	*striatigula*	■										
		sclateri	■										
		pallidior	■										
Madagascar Cisticola	*Cisticola cherina*		■										■
Brown Emu-tail	*Bradypterus brunneus*		■										
Grey Emu-tail	*Amphilais seebohmi*		■										
Madagascar Yellowbrow	*Crossleyia xanthophrys*		■										
Dusky Tetraka	*Xanthomixis tenebrosa*		■										
Spectacled Tetraka	*Xanthomixis zosterops*	*zosterops*	■										
		fulvescens	■										
		andapae	■										
		maroantsetrae	■										
		ankafanae	■										
Appert's Tetraka	*Xanthomixis apperti*		■										
Grey-crowned Tetraka	*Xanthomixis cinereiceps*		■										
Wedge-tailed Tetraka	*Hartertula flavoviridis*		■										
Rand's Warbler	*Randia pseudozosterops*		■										
Thamnornis	*Thamnornis chloropetoides*		■										
Cryptic Warbler	*Cryptosylvicola randrianasoloi*		■										
White-throated Oxylabes	*Oxylabes madagascariensis*		■										
Long-billed Tetraka	*Bernieria madagascariensis*	*madagascariensis*	■										
		inceleber	■										
Aldabra Brush Warbler	*Nesillas aldabrana*				X								
Madagascar Brush Warbler	*Nesillas typica*	*typica*	■										
		ellisii	■										
		obscura	■										
		moheliensis					■						
		longicaudata							■				
Subdesert Brush Warbler	*Nesillas lantzii*		■										
Grande Comore Brush Warbler	*Nesillas brevicaudata*					■							
Mohéli Brush Warbler	*Nesillas mariae*						■						
Sedge Warbler	*Acrocephalus schoenobaenus*			V									

English name	Scientific name	Subspecies	Madagascar	Granitic Seychelles	Coralline Seychelles	Grande Comore	Mohéli	Anjouan	Mayotte	Mauritius	Rodrigues	Reunion	Outer Islands
Marsh Warbler	Acrocephalus palustris				V								
Madagascar Swamp Warbler	Acrocephalus newtoni		■										
Rodrigues Warbler	Acrocephalus rodericanus										■		
Seychelles Warbler	Acrocephalus sechellensis			■									
Icterine Warbler	Hippolais icterina			V	V								
Willow Warbler	Phylloscopus trochilus	unconfirmed race		V	V								
Common Chiffchaff	Phylloscopus collybita	abietinus?			V								
Wood Warbler	Phylloscopus sibilatrix			V	V								
Red-billed Leiothrix	Leiothrix lutea	unconfirmed race										■	
Eurasian Blackcap	Sylvia atricapilla	unconfirmed race		V	V								
Garden Warbler	Sylvia borin				V								
Common Whitethroat	Sylvia communis	icterops/volgensis		V	V								
Madagascar Green Sunbird	Nectarinia notata	notata	■										
		moebii				■							
		voeltzkowi											■
Humblot's Sunbird	Nectarinia humbloti	humbloti				■							
		mohelica					■						
Seychelles Sunbird	Nectarinia dussumieri			■									
Mayotte Sunbird	Nectarinia coquerellii								■				
Anjouan Sunbird	Nectarinia comorensis							■					
Souimanga Sunbird	Nectarinia souimanga	souimanga	■										
		apolis											■
		buchenorum											■
		aldabrensis											■
		abbotti											■
Karthala White-eye	Zosterops mouroniensis					■							
Mauritius Olive White-eye	Zosterops chloronothos									■			
Reunion Olive White-eye	Zosterops olivaceus											■	
Reunion Grey White-eye	Zosterops borbonicus											■	
Mauritius Grey White-eye	Zosterops mauritianus									■			
Seychelles White-eye	Zosterops modestus			■									
Madagascar White-eye	Zosterops maderaspatanus	maderaspatanus	■										
		kirki						■					
		comorensis				■							
		anjuanensis						■					
		mayottensis							■				
		voeltzkowi											■
		aldabrensis											■
Eurasian Golden Oriole	Oriolus oriolus	oriolus	V	V	V				V				
Red-backed Shrike	Lanius collurio	unconfirmed race		V	V								
Lesser Grey Shrike	Lanius minor	unconfirmed race		V	V								
Woodchat Shrike	Lanius senator	unconfirmed race		V									

English name	Scientific name	Subspecies	Madagascar	Granitic Seychelles	Coralline Seychelles	Grande Comore	Mohéli	Anjouan	Mayotte	Mauritius	Rodrigues	Reunion	Outer Islands
Archbold's Newtonia	*Newtonia archboldi*		●										
Common Newtonia	*Newtonia brunneicauda*	*brunneicauda*	●										
		monticola	●										
Dark Newtonia	*Newtonia amphichroa*		●										
Red-tailed Newtonia	*Newtonia fanovanae*		●										
Red-tailed Vanga	*Calicalicus madagascariensis*		●										
Red-shouldered Vanga	*Calicalicus rufocarpalis*		●										
Tylas Vanga	*Tylas eduardi*	*eduardi*	●										
		albigularis	●										
Nuthatch Vanga	*Hypositta corallirostris*		●										
Blue Vanga	*Cyanolanius madagascarinus*	*madagascarinus*	●										
		comorensis				●							
		bensoni					●						
Crossley's Vanga	*Mystacornis crossleyi*		●										
Chabert Vanga	*Leptopterus chabert*	*chabert*	●										
		schistocercus	●										
Ward's Vanga	*Pseudobias wardi*		●										
Hook-billed Vanga	*Vanga curvirostris*	*curvirostris*	●										
		cetera	●										
Helmet Vanga	*Euryceros prevostii*		●										
Rufous Vanga	*Schetba rufa*	*rufa*	●										
		occidentalis	●										
Bernier's Vanga	*Oriolia bernieri*		●										
Pollen's Vanga	*Xenopirostris polleni*		●										
Van Dam's Vanga	*Xenopirostris damii*		●										
Lafresnaye's Vanga	*Xenopirostris xenopirostris*		●										
White-headed Vanga	*Artamella viridis*	*viridis*	●										
		annae	●										
Sickle-billed Vanga	*Falculea palliata*		●										
Grande Comore Drongo	*Dicrurus fuscipennis*					●							
Mayotte Drongo	*Dicrurus waldenii*								●				
Aldabra Drongo	*Dicrurus aldabranus*												●
Crested Drongo	*Dicrurus forficatus*	*forficatus*	●										
		potior							●				
House Crow	*Corvus splendens*	unconfirmed race		●									
Pied Crow	*Corvus albus*		●										
Madagascar Starling	*Hartlaubius auratus*		●										
Common Myna	*Acridotheres tristis*	*tristis*	●	●						●	●	●	●
Rosy Starling	*Pastor roseus*			V									
Wattled Starling	*Creatophora cinerea*			V	V	V							
House Sparrow	*Passer domesticus*	unconfirmed race				V							
Village Weaver	*Ploceus cucullatus*	*spilonotus*											●

322

English name	Scientific name	Subspecies	Madagascar	Granitic Seychelles	Coralline Seychelles	Grande Comore	Mohéli	Anjouan	Mayotte	Mauritius	Rodrigues	Reunion	Outer Islands
Nelicourvi Weaver	*Ploceus nelicourvi*		✓										
Sakalava Weaver	*Ploceus sakalava*	sakalava	✓										
		minor							✓				
Red-billed Quelea	*Quelea quelea*	unconfirmed race										✓	
Seychelles Fody	*Foudia sechellarum*			✓	✓								
Rodrigues Fody	*Foudia flavicans*										✓		
Mauritius Fody	*Foudia rubra*									✓			
Comoro Fody	*Foudia eminentissima*	eminentissima				✓							
		consobrina					✓						
		anjuanensis						✓					
		algondae							✓				
Aldabra Fody	*Foudia aldabrana*												✓
Madagascar Fody	*Foudia madagascariensis*		✓	✓	✓	✓	✓	✓	✓	✓	✓	✓	✓
Forest Fody	*Foudia omissa*		✓										
Common Waxbill	*Estrilda astrild*	astrild		✓						✓	✓	✓	
Red Avadavat	*Amandava amandava*	amandava										✓	
Madagascar Mannikin	*Lepidopygia nana*		✓										
Bronze Mannikin	*Spermestes cucullatus*	scutatus							✓				
Scaly-breasted Munia	*Lonchura punctulata*	unconfirmed race										✓	
Pin-tailed Whydah	*Vidua macroura*											✓	
Common Rosefinch	*Carpodacus erythrinus*	unconfirmed race	V										
Cape Canary	*Serinus canicollis*	canicollis										✓	
Yellow-fronted Canary	*Crithagra mozambica*	granti								✓	✓	✓	
Ortolan Bunting	*Emberiza hortulana*			V	V								

APPENDIX 2
A NEW CLASSIFICATION OF BIRD ORDERS AND FAMILIES NATIVE TO THE MALAGASY REGION

This classification follows a detailed high-level classification of all extant birds as part of the fourth edition of the *Howard and Moore' Checklist of Birds of the World* (Dickinson & Remsen 2013, Dickinson & Christidis 2014; see References), prepared by J. Cracraft. All living birds (class Aves) were placed in an order and family; other units (between class and order, between order and family, and between family and genus) were also proposed, but not for all species, and are not shown here, although subfamilies for endemic species, also proposed for some but not all families, are mentioned in the notes column. Shortly thereafter, another high-level phylogenetic tree of birds was published by Jarvis *et al.* (2014), producing an almost identical set and sequence of bird orders, differing only in the placement of a very few groups.

Exotics and vagrants are not included. Families known only as vagrants in the Malagasy region are Spheniscidae (penguins), Oriolidae (orioles), Laniidae (shrikes), Phylloscopidae (leaf warblers), Sylviidae (*Sylvia* warblers, parrotbills and allies) and Emberizidae (buntings). Families represented only by exotics are Viduidae (whydahs), Passeridae (sparrows) and Leiothricidae (babblers). Fringillidae (finches) includes both exotic and vagrant species.

Mesites, Cuckoo-roller, ground-rollers, asities and tetrakas are agreed to be endemic families, the first two also being endemic orders (and thus not members of the Gruiformes or Coraciiformes, respectively). Vangas are an endemic subfamily of the larger family Vangidae, which now includes Asian and African genera.

Other notable findings, listed according to position in the traditional sequence, include:
- ducks, geese and swans and gamebirds are the most ancient bird groups among species extant in the region (the extinct elephant birds would have come first), and so are now placed first in the sequence;
- grebes and flamingos are each other's closest relatives, and are not close to any other families;
- the two traditional large waterbird assemblages, the pelicans and allies (former Pelecaniformes), and storks, herons and allies (former Ciconiiformes), belong in a single order, the Pelecaniformes, which excludes tropicbirds;
- tropicbirds are not Pelecaniformes, and thus unrelated to boobies, frigatebirds and cormorants; they are highly distinct and placed in a separate order;
- flufftails are so well separated from other rails that they merit a separate family (subsequent research suggests Malagasy wood rails belong there too);
- buttonquails are members of the Charadriiformes, not the Gruiformes;
- three Malagasy raptors, Madagascar Serpent Eagle, Cuckoo-hawk and Harrier-hawk, belong in the raptor subfamily Gypaetinae, previously thought to be limited to certain vultures such as Bearded, Egyptian and Palmnut Vultures;
- all Malagasy parrots belong in the Old World family Psittaculidae, representing all three subfamilies; most African parrots (*Poicephalus* and *Psittacus*, but not *Agapornis* or *Psittacula*) are in a different family (Psittacidae);
- the Malagasy region has only one native bulbul genus (*Hypsipetes*) and no native babblers;
- four families of the former 'Old World warbler' assemblage are present: Acrocephalidae, Locustellidae, Bernieridae and Cisticolidae; Sylviidae (*sensu stricto*) occur only as vagrants;
- chats, including rock thrushes, are transferred from the true thrushes (Turdidae) to the flycatchers (Muscicapidae), leaving just one true thrush (on the Comoros).

Orders in Malagasy region	Families in Malagasy region	Notes
Anseriformes ducks, geese and swans	Anatidae ducks	All extant Malagasy species in subfamilies Anatinae (*Anas*, *Aythya*, extinct shelducks) or Dendrodygninae (remainder)
Galliformes gamebirds	Numididae guineafowl	
	Phasianidae pheasants, quails and partridges	
Phoenicopteriformes flamingos and grebes	Phoenicopteridae flamingos	Separated into two orders by Jarvis *et al.* (2014), but remain each other's closest relatives
	Podicipedidae grebes	
Columbiformes pigeons and doves	Columbidae pigeons and doves	Malagasy native species in subfamilies Columbinae or Raphinae
Pterocliformes sandgrouse	Pteroclidae sandgrouse	
Mesitornithiformes mesites	Mesitornithidae mesites	Endemic order and family
Phaethontiformes tropicbirds	Phaethontidae tropicbirds	
Caprimulgiformes nightjars and allies	Caprimulgidae nightjars	Both Malagasy species in subfamily Caprimulginae
	Apodidae swifts	All Malagasy swifts in subfamily Apodinae
Cuculiformes cuckoos	Cuculidae cuckoos	Couas and coucals as separate 'tribes' in subfamily Centropodinae; remaining Malagasy species in subfamily Cuculinae
Gruiformes cranes, rails and allies	Rallidae rails	Order excludes mesites and buttonquails; all Malagasy Rallidae in subfamily Rallinae
	Sarothruridae flufftails and wood rails	New family, includes flufftails (and probably also wood rails)
Procellariiformes tubenoses	Oceanitidae southern storm-petrels	
	Diomedeidae albatrosses	
	Hydrobatidae northern storm-petrels	Includes Swinhoe's and Matsudaira's Storm-petrels (other storm-petrels in Oceanitidae)
	Procellariidae petrels and shearwaters	
Pelecaniformes pelicans, cormorants, storks, herons and allies	Ciconiidae storks	
	Pelecanidae pelicans	
	Scopidae Hamerkop	
	Ardeidae herons	

Orders in Malagasy region	Families in Malagasy region	Notes
	Threskiornithidae ibises	
	Fregatidae frigatebirds	
	Sulidae gannets and boobies	
	Phalacrocoracidae cormorants	
	Anhingidae darters	
Charadriiformes shorebirds, skuas, gulls and terns	Recurvirostridae stilts, avocets, thick-knees, oystercatchers	
	Charadriidae plovers	*Charadrius* plovers in subfamily Charadriinae, extinct endemic lapwing in subfamily Vanellinae
	Rostratulidae painted-snipes	
	Jacanidae jacanas	
	Scolopacidae sandpipers and allies	Madagascar Snipe in subfamily Scolopacinae
	Turnicidae buttonquails	
	Dromadidae crab-plover	
	Glareolidae pratincoles and coursers	Madagascar Pratincole in subfamily Glareolinae
	Stercorariidae skuas	
	Laridae gulls and terns	
Accipitriformes raptors	Accipitridae raptors	Madagascar Serpent Eagle, Cuckoo-hawk and Harrier-hawk in subfamily Gypaetinae, remainder in subfamily Accipitrinae
Strigiformes owls	Tytonidae barn owls	
	Strigidae true owls	White-browed Owl in subfamily Surniinae (assuming true relationship to *Athene*), remainder in subfamily Striginae
Leptosomiformes Cuckoo-roller	Leptosomidae Cuckoo-roller	Endemic order and family
Bucerotiformes hornbills, hoopoes and allies	Upupidae hoopoes	
Coraciiformes bee-eaters, rollers, kingfishers and allies	Meropidae bee-eaters	
	Coraciidae rollers	
	Brachypteraciidae ground-rollers	Endemic family
	Alcedinidae kingfishers	Malagasy species in subfamily Alcedininae

Orders in Malagasy region	Families in Malagasy region	Notes
Falconiformes falcons and allies	Falconidae falcons	Malagasy species in subfamily Falconinae
Psittaciformes parrots	Psittaculidae parrots	Vasa parrots in subfamily Psittricasinae, parakeets in Psittaculinae, lovebirds in Loriinae
Passeriformes passerines	Philepittidae asitios	Endemic family; only Malagasy family in suborder Tyranni (suboscines); remaining passerines in suborder Passeri (oscines)
	Campephagidae cuckoo-shrikes and orioles	Malagasy species in subfamily Campephaginae
	Vangidae vangas and allies	Malagasy species in endemic subfamily Vanginae
	Dicruridae drongos	
	Corvidae crows	
	Monarchidae monarchs	
	Nectariniidae sunbirds	
	Ploceidae weavers	
	Estrildidae waxbills	Madagascar Mannikin in subfamily Lonchurinae
	Motacillidae wagtails and pipits	
	Alaudidae larks	
	Cisticolidae African warblers	Jeries in endemic subfamily Neomixinae, Madagascar Cisticola in subfamily Cisticolinae
	Locustellidae grass and bush warblers	Includes *Amphilais* (provisionally) and *Bradypterus*
	Bernieridae tetrakas	Endemic family (recognised in text)
	Acrocephalidae reed warblers	Includes *Nesillas*, *Acrocephalus*
	Hirundinidae swallows and martins	Mascarene Martin in subfamily Hirundininae
	Pycnonotidae bulbuls	Only Malagasy genus is *Hypsipetes*
	Zosteropidae white-eyes	
	Sturnidae starlings and mynas	Madagascar Starling in subfamily Sturninae
	Muscicapidae flycatchers and chats	Magpie-robins and Humblot's Flycatcher in subfamily Muscicapinae, stonechats and rock thrushes in subfamily Saxicolinae
	Turdidae thrushes	Includes (only) *Turdus* in subfamily Turdinae

REFERENCES

Bird Studies Canada. *Avibase—The World Bird Database*. http://avibase.bsc-eoc.org/avibase.jsp. Ottawa: Bird Studies Canada. Accessed in 2012.
BirdLife International. *Marine IBA e-atlas*. http://maps.birdlife.org/marineIBAs/. Cambridge, UK: BirdLife International. Accessed in 2013
Cheke, A. & Hume, J. 2008. *Lost Land of the Dodo*. London, UK: T & AD Poyser.
del Hoyo, J. & Collar, N. J. 2014. *HBW and BirdLife International Illustrated Checklist of the Birds of the World*. Vol. 1. Non-passerines. Barcelona: Lynx Edicions.
del Hoyo, J. Elliott, A., Sargatal, J. & Christie, D. A. (eds.) 1992–2011. *Handbook of the Birds of the World*. Vols. 1–16. Barcelona: Lynx Edicions.
Dickinson, E. C. & Christidis, L. (eds.) 2014. *The Howard and Moore Complete Checklist of the Birds of the World*. Fourth edn. Vol. 2. Eastbourne: Aves Press.
Dickinson, E. C. & Remsen, J. V. (eds.) 2013. *The Howard and Moore Complete Checklist of the Birds of the World*. Fourth edn. Vol. 1. Eastbourne: Aves Press.
Fry, C. H., Keith, S. & Urban, E. K. (eds.) 1982–2004. *The Birds of Africa*. Vols. 1–7. London, UK: Academic Press (vols. 1–6) and Christopher Helm (vol. 7).
Gill, F. & Wright, M. 2006. *Birds of the World: Recommended English Names*. Princeton, NJ: Princeton University Press (with updates published online).
Jarvis, E. D. & 104 authors. 2014. Whole-genome analyses resolve early branches in the tree of life of modern birds. *Science* 346: 1320–1331.
Safford, R. J. & Hawkins, A. F. A. (eds.) 2013. *The Birds of Africa. Volume 8: The Malagasy Region*. London, UK: Christopher Helm.
Warren, B. H., Safford, R. J., Strasberg, D. & Thébaud, C. 2013. Bird biogeography and evolution. Pp. 35–40 in Safford, R. J. & Hawkins, A. F. A. (eds.) *The Birds of Africa. Volume 8: The Malagasy Region*. London, UK: Christopher Helm.

Sound recordings

Herremans, M. 2001. *Guide Sonore des Oiseaux Nicheurs des Comores*. Tervuren: Royal Museum of Central Africa.
Huguet, P. 2001. *Voix de la Nature de Mayotte*. Mamoudzou: Naturalistes de Mayotte.
Huguet, P. & Chappuis, C. 2003. *Birds Sounds/Oiseaux de Madagascar, Mayotte, Comores, Seychelles, Réunion, Maurice*. Paris: Société d'Etudes Ornithologiques de France.
Hawkins, A. F. A. & Ranft, R. 2007. *Bird Sounds of Madagascar*. London, UK: British Library.
Randrianary, V., Rifflet, S. & Roché, J. C. 1997. *Madagascar Soundscapes*. Le Verdier: Sittelle.
Rocamora, G. & Solé, A. 2001. *Sounds of Seychelles. Fauna of the Granitic Islands*. Ministry of Environment and Transport/Sounds of Nature Pty Ltd (Seychelles).

INDEX

A

Accipiter henstii 94
Accipiter madagascariensis 94
Accipter francesiae 94
Acridotheres tristis 178, 188, 214, 224, 228, 238
Acrocephalus newtoni 162
Acrocephalus palustris 300
Acrocephalus rodericanus 222
Acrocephalus schoenobaenus 300
Acrocephalus sechellensis 186
Actitis hypoleucos 120
Actophilornis albinucha 110
Aenigmatolimnas marginalis 266
Acrodramus elaphrus 184
Aerodramus franciscus 222, 230
Agapornis canus 134, 200
Albatross, Black-browed 244
    Indian Yellow-nosed 60
    Light-mantled 244
    Shy 60
    Sooty 244
    Wandering 244
Alectroenas madagascariensis 132
Alectroenas pulcherrimus 182
Alectroenas sganzini 190, 198
Amandava amandava 236
Amaurornis olivieri 106
Amaurornis phoenicurus 266
Amphilais seebohmi 162
Anas acuta 256
Anas bernieri 88
Anas clypeata 256
Anas erythrorhyncha 88
Anas hottentota 88
Anas melleri 88
Anas penelope 256
Anas platyrhynchos 88
Anas querquedula 88
Anastomus lamelligerus 80
Anhinga rufa 72
Anous stolidus 130
Anous tenuirostris 130
Anthus cervinus 292
Anthus richardi 292
Anthus trivialis 292
Aphrodroma brevirostris 248
Apus affinis 144
Apus apus 286
Apus balstoni 144
Apus barbatus 144, 204
Apus pacificus 286
Ardea alba 78
Ardea cinerea 74
Ardea goliath 254
Ardea humbloti 74
Ardea ibis 78
Ardea intermedia 254
Ardea melanocephala 254
Ardea purpurea 74
Ardeola grayii 254
Ardeola idae 74
Ardeola ralloides 74
Arenaria interpres 120
Artamella viridis 170
Asio capensis 142
Asio madagascariensis 142
Asity, Common Sunbird 152
    Schlegel's 146
    Velvet 146
    Yellow-bellied Sunbird 152
Atelornis crossleyi 150
Atelornis pittoides 150
Avadavat, Red 236
Aviceda madagascariensis 90
Avocet, Pied 268
Aythya fuligula 256
Aythya innotata 86
Aythya nyroca 256

B

Bee-eater, Blue-cheeked 288
    European 288
    Madagascar 148, 204, 240
    Olive 148, 204, 240
Bernieria madagascariensis 164
Bittern, Cinnamon 254
    Eurasian 252
    Little 76, 254
    Yellow 76
Blackcap, Eurasian 300
Booby, Brown 70
    Masked 70
    Red-footed 70
Botaurus stellaris 252
Brachypteracias leptosomus 150
Bradypterus brunneus 166
Bulbul, Grande Comore 208
    Madagascar 174, 192, 208, 240
    Mauritius 220
    Mohéli 208
    Red-whiskered 220, 230, 242
    Reunion 230
    Seychelles 186
Bulweria bulweria 62
Bulweria fallax 62
Bunting, Ortolan 306
Burhinus oedicnemus 268
Buteo brachypterus 90
Butorides striata 76
Buttonquail, Madagascar 102
Buzzard, Madagascar 90

C

Calandrella brachydactyla 292
Calicalicus madagascariensis 170
Calicalicus rufocarpalis 170
Calidris acuminata 276
Calidris alba 120
Calidris canutus 276
Calidris ferruginea 120
Calidris melanotos 276
Calidris minuta 120
Calidris subminuta 278
Calidris temminckii 278
Calidris tenuirostris 276
Calonectris diomedea 66
Calonectris leucomelas 248
Canary, Cape 236
    Yellow-fronted 224, 236, 238
Caprimulgus europaeus 286

Caprimulgus madagascariensis 140, 192
Carpodacus erythrinus 306
Cecropis abyssinica 290
Centropus toulou 140, 190
Charadrius asiaticus 270
Charadrius dubius 270
Charadrius bifrontatus 116
Charadrius hiaticula 114
Charadrius leschenaultii 114
Charadrius marginatus 116
Charadrius mongolus 114
Charadrius pecuarius 116
Charadrius thoracicus 116
Charadrius tricollaris 116
Charadrius veredus 270
Chiffchaff, Common 302
Chlidonias hybrida 128
Chlidonias leucopterus 128
Chlidonias niger 282
Chroicocephalus cirrocephalus 124
Chroicocephalus ridibundus 280
Ciconia ciconia 252
Circus aeruginosus 260
Circus macrosceles 90
Circus macrourus 260
Circus maillardi 90
Cisticola cherina 162, 194, 242
Cisticola, Madagascar 162, 194, 242
Clamator glandarius 284
Clamator jacobinus 284
Columba livia 132, 182, 198, 218, 228
Columba pollenii 198
Coot, Red-knobbed 110
Copsychus albospecularis 158
Copsychus sechellarum 186
Coracias garrulus 288
Coracina cinerea 174, 208
Coracina newtoni 230
Coracina typica 220
Coracopsis barklyi 184
Coracopsis nigra 134, 200
Coracopsis vasa 134, 200
Cormorant, Great 252
    Long-tailed 72
    Reed 72

Corvus albus 178, 192, 214, 240
Corvus splendens 224
Corythornis madagascariensis 146
Corythornis vintsioides 146, 206
Coturnix chinensis 102
Coturnix coturnix 102
Coturnix delegorguei 102
Coua, Blue 136
    Coquerel's 138
    Crested 136
    Giant 138
    Red-breasted 136
    Red-capped 138
    Red-fronted 136
    Running 138
    Verreaux's 136
Coua caerulea 136
Coua coquereli 138
Coua cristata 136
Coua cursor 138
Coua gigas 138
Coua reynaudii 136
Coua ruficeps 138
Coua serriana 136
Coua verreauxi 136
Coucal, Madagascar 140, 190
Crab-plover 112
Crake, Baillon's 106
    Corn 266
    Little 266
    Spotted 266
    Striped 266
Creatophora cinerea 306
Crex crex 266
Crithagra mozambica 224, 236, 238
Crossleyia xanthophrys 166
Crow, House 224
    Pied 178, 192, 214, 240
Cryptosylvicola randrianasoloi 160
Cuckoo, Common 284
    Great Spotted 284
    Jacobin 284
    Lesser 284
    Madagascar 140
    Thick-billed 140
Cuckoo-hawk, Madagascar 90

Cuckoo-roller 148, 206
Cuckoo-shrike, Madagascar 174, 208
    Mauritius 220
    Reunion 230
Cuculus canorus 284
Cuculus poliocephalus 284
Cuculus rochii 140
Curlew, Eurasian 118
    Little 274
Cyanolanius madagascarinus 170, 214
Cypsiurus parvus 144, 204

D

Daption capense 60
Darter, African 72
Delichon urbicum 290
Dendrocygna bicolor 86
Dendrocygna viduata 86
Dicrurus aldabranus 192
Dicrurus forficatus 178, 214
Dicrurus fuscipennis 214
Dicrurus waldenii 214
Diomedea exulans 244
Dove, Barred Ground 182, 190, 218, 228, 238
    Cape Turtle 200
    European Turtle 284
    Laughing 218
    Madagascar Turtle 132, 182, 190, 198, 218, 228, 238
    Namaqua 132
    Ring-necked 200
    Spotted 218
    Tambourine 200
    Zebra 182, 190, 218, 228, 238
Dromas ardeola 112
Drongo, Aldabra 192
    Crested 178, 214
    Grande Comore 214
    Mayotte 214
Dryolimnas cuvieri 108
Duck, Comb 86
    Ferruginous 256
    Fulvous Whistling 86
    Knob-billed 86

Meller's 88
Tufted 256
White-backed 86
White-faced Whistling 86

E

Eagle, Booted 258
    Long-crested 258
    Madagascar Fish 92
    Madagascar Serpent 94
Egret, Black 78
    Cattle 78
    Dimorphic 78
    Great 78
    Intermediate 254
    Little 78, 254
Egretta ardesiaca 78
Egretta garzetta 78, 254
Elanus caeruleus 260
Emberiza hortulana 306
Emu-tail, Brown 166
    Grey 162
Estrilda astrild 188, 196, 224, 236, 242
Eudyptes chrysocome 250
Euryceros prevostii 172
Eurystomus glaucurus 148, 192, 206, 240
Eutriorchis astur 94

F

Falco amurensis 264
Falco araeus 96
Falco cherrug 262
Falco concolor 98
Falco eleonorae 98
Falcon, Amur 264
    Eleonora's 98
    Peregrine 98
    Red-footed 264
    Saker 262
    Sooty 98
    'Tundra' Peregrine 262
Falco naumanni 262
Falco newtoni 96
Falco peregrinus 98, 262
Falco punctatus 96

Falco subbuteo 264
Falco tinnunculus 262
Falco vespertinus 264
Falco zoniventris 96
Falculea palliata 172
Finch, Spice 224, 236
Flamingo, Greater 80
    Lesser 80
Flufftail, Madagascar 106
    Slender-billed 106
Flycatcher, Humblot's 210
    Madagascar Paradise 176, 210
    Mascarene Paradise 222, 232
    Seychelles Paradise 186
    Spotted 296
Fody, Aldabra 196
    Comoro 216
    Forest 180
    Madagascar 180, 188, 196, 216, 226, 234, 238
    Mauritius 226
    Rodrigues 226
    Seychelles 188, 196
Foudia aldabrana 196
Foudia eminentissima 216
Foudia flavicans 226
Foudia madagascariensis 180, 188, 196, 216, 226, 234, 238
Foudia omissa 180
Foudia rubra 226
Foudia sechellarum 188, 196
Francolin, Grey 100
Francolinus pondicerianus 100
Fregata ariel 72
Fregata minor 72
Fregetta grallaria 68
Fregetta tropica 68
Frigatebird, Great 72
    Lesser 72
Fulica cristata 110
Fulmar, Southern 246
Fulmarus glacialoides 246

G

Gactornis enarratus 140
Gallinago gallinago 272

Gallinago macrodactyla 112
Gallinago media 272
Gallinago stenura 272
Gallinula angulata 266
Gallinula chloropus 110
Gallinule, Allen's 110
Gallirallus philippensis 266
Gallus gallus 100
Gannet, Australasian 250
    Cape 250
Garganey 88
Gelochelidon nilotica 282
Geobiastes squamiger 150
Geopelia striata 182, 190, 218, 228, 238
Glareola maldivarum 268
Glareola nordmanni 268
Glareola ocularis 112
Glareola pratincola 268
Godwit, Bar-tailed 118
    Black-tailed 274
Goose, African Pygmy 84
Goshawk, Henst's 94
Grebe, Alaotra 84
    Black-necked 256
    Little 84
    Madagascar 84
Greenshank, Common 118
Ground-roller, Long-tailed 150
    Pitta-like 150
    Rufous-headed 150
    Scaly 150
    Short-legged 150
Guineafowl, Helmeted 100
Gull, Common Black-headed 280
    Grey-headed 124
    Kelp 124
    Lesser Black-backed 280
    Sabine's 280
    Sooty 280
Gygis alba 126

H

Haematopus ostralegus 268
Halcyon leucocephala 288
Haliaeetus vociferoides 92
Hamerkop 80

Harrier-hawk, Madagascar 92
Harrier, Madagascar 90
    Pallid 260
    Reunion 90
    Western Marsh 260
Hartertula flavoviridis 166
Hartlaubius auratus 178
Hawk, Bat 92
Heron, Black 78
    Black-crowned Night 76
    Black-headed 254
    Goliath 254
    Green-backed 76
    Grey 74
    Indian Pond 254
    Madagascar 74
    Madagascar Pond 74
    Purple 74
    Squacco 74
    Striated 76
Hieraaetus pennatus 258
Himantopus himantopus 112
Hippolais icterina 302
Hirundapus caudacutus 286
Hirundo rustica 154, 184, 194, 204, 242
Hirundo smithii 290
Hobby, Eurasian 264
Honey-buzzard, European 258
Hoopoe, Common 288
    Madagascar 148
Humblotia flavirostris 210
Hydroprogne caspia 124
Hypositta corallirostris 176
Hypsipetes borbonicus 230
Hypsipetes crassirostris 186
Hypsipetes madagascariensis 174, 192, 208, 240
Hypsipetes moheliensis 208
Hypsipetes olivaceus 220
Hypsipetes parvirostris 208

I

Ibis, African Sacred 252
    Glossy 82
    Madagascar Crested 82
    Madagascar Sacred 82

Ichthyaetus hemprichii 280
Ixobrychus cinnamomeus 254
Ixobrychus minutus 76, 254
Ixobrychus sinensis 76

J

Jacana, Madagascar 110
Jaeger, Long-tailed 122
    Parasitic 122
    Pomarine 122
Jery, Common 160
    Green 160
    Stripe-throated 160
Junglefowl, Red 100

K

Kestrel, Banded 96
    Common 262
    Lesser 262
    Madagascar 96
    Mauritius 96
    Seychelles 96
Ketupa zeylonensis 286
Kingfisher, Grey-headed 288
    Madagascar Malachite 146, 206
    Madagascar Pygmy 146
Kite, Black 92, 260
    Black-shouldered 260
    Black-winged 260
    Yellow-billed 92
Knot, Great 276
    Red 276

L

Lanius collurio 304
Lanius minor 304
Lanius senator 304
Lapwing, Blacksmith 270
    Sociable 270
    Spur-winged 270
Lark, Bimaculated 292
    Greater Short-toed 292
    Madagascar 154
Larus dominicanus 124
Larus fuscus 280
Leiothrix lutea 230

Leiothrix, Red-billed 230
Lepidopygia nana 180
Leptopterus chabert 170
Leptosomus discolor 148, 206
Limicola falcinellus 278
Limosa lapponica 118
Limosa limosa 274
Lonchura punctulata 224, 236
Lophaetus occipitalis 258
Lophotibis cristata 82
Lovebird, Grey-headed 134, 200
Lymnocryptes minimus 272

M

Macheiramphus alcinus 92
Macronectes giganteus 60
Macronectes halli 60
Magpie-robin, Madagascar 158
    Seychelles 186
Mallard 88
Mannikin, Bronze 216
    Madagascar 180
Margaroperdix madagarensis 100
Martin, Brown-throated 154
    Common House 290
    'Grey-throated' 290
    Mascarene 154, 222, 230
    Sand 290
Melanocorypha bimaculata 292
Mentocrex beankaensis 108
Mentocrex kioloides 108
Merops apiaster 288
Merops persicus 288
Merops superciliosus 148, 204, 240
Mesite, Brown 104
    Subdesert 104
    White-breasted 104
Mesitornis unicolor 104
Mesitornis variegatus 104
Milvus migrans 92, 260
Mirafra hova 154
Monias benschi 104
Monticola imerinus 156
Monticola saxatilis 296
Monticola sharpei 156
Moorhen, Common 110

Lesser 266
Morus capensis 250
Morus serrator 250
Motacilla alba 294
Motacilla cinerea 294
Motacilla citreola 294
Motacilla flava 294
Motacilla flaviventris 154
Munia, Scaly-breasted 224, 236
Muscicapa striata 296
Mycteria ibis 80
Myna, Common 178, 188, 214, 224, 228, 238
Mystacornis crossleyi 176

N

Nectarinia comorensis 212
Nectarinia coquerellii 212
Nectarinia dussumieri 186
Nectarinia humbloti 212
Nectarinia notata 152, 212
Nectarinia souimanga 152, 194, 242
Needletail, White-throated 286
Neodrepanis coruscans 152
Neodrepanis hypoxantha 152
Neomixis striatigula 160
Neomixis tenella 160
Neomixis viridis 160
Nesillas aldabrana 194
Nesillas brevicaudata 210
Nesillas lantzii 162
Nesillas mariae 210
Nesillas typica 162, 210
Nesoenas mayeri 218
Nesoenas picturata 132, 182, 190, 198, 218, 228, 238
Nettapus auritus 84
Newtonia amphichroa 168
Newtonia archboldi 168
Newtonia brunneicauda 168
Newtonia fanovanae 168
Newtonia, Archbold's 168
  Common 168
  Dark 168
  Red-tailed 168
Nightjar, Collared 140

European 286
  Madagascar 140, 192
Ninox superciliaris 142
Noddy, Brown 130
  Lesser 130
Numenius arquata 118
Numenius minutus 274
Numenius phaeopus 118
Numida meleagris 100
Nycticorax nycticorax 76

O

Oceanites oceanicus 68
Oceanodroma matsudairae 68
Oceanodroma monorhis 68
Oena capensis 132
Oenanthe deserti 298
Oenanthe isabellina 298
Oenanthe oenanthe 298
Oenanthe pleschanka 298
Onychoprion anaethetus 130
Onychoprion fuscatus 130
Openbill, African 80
Oriole, Eurasian Golden 304
Oriolia bernieri 172
Oriolus oriolus 304
Osprey 258
Otus capnodes 202
Otus insularis 184
Otus mayottensis 202
Otus moheliensis 202
Otus pauliani 202
Otus rutilus 142
Otus scops 286
Owl, Anjouan Scops 202
  Barn 142, 184, 202, 240
  Brown Fish 286
  Eurasian Scops 286
  Karthala Scops 202
  Madagascar Long-eared 142
  Madagascar Red 142
  Madagascar Scops 142
  Marsh 142
  Mayotte Scops 202
  Mohéli Scops 202
  Seychelles Scops 184

White-browed 142
Oxylabes madagascariensis 166
Oxylabes, White-throated 166
Oystercatcher, Eurasian 268

P

Pachycoccyx audeberti 140
Pachyptila belcheri 246
Pachyptila desolata 246
Painted-snipe, Greater 112
Pandion haliaetus 258
Parakeet, Echo 220
  Mauritius 220
  Ring-necked 184, 220, 228
  Rose-ringed 184, 220, 228
Parrot, Greater Vasa 134, 200
  Lesser Vasa 134, 200
  Seychelles 184
  Seychelles Black 184
Partridge, Madagascar 100
Passer domesticus 180, 196, 216, 226, 234
Pastor roseus 306
Pelagodroma marina 68
Pelecanus rufescens 252
Pelican, Pink-backed 252
Penguin, Rockhopper 250
Perdicula asiatica 102
Pernis apivorus 258
Petrel, Barau's 62
  Black-winged 248
  Bulwer's 62
  Cape 60
  Great-winged 62
  Grey 246
  Herald 64
  Jouanin's 62
  Kerguelen 248
  Kermadec 64
  Mascarene 62
  Northern Giant 60
  Round Island 64
  Soft-plumaged 62
  Southern Giant 60
  Spectacled 246
  Trindade 64

White-chinned 246
Phaethon aethereus 250
Phaethon lepturus 70
Phaethon rubricauda 70
Phalacrocorax africanus 72
Phalacrocorax carbo 252
Phalarope, Red-necked 278
Phalaropus lobatus 278
Phedina borbonica 154, 230
Philepitta castanea 146
Philepitta schlegeli 146
Philomachus pugnax 276
Phoebetria fusca 244
Phoebetria palpebrata 244
Phoeniconaias minor 80
Phoenicopterus roseus 80
Phoenicurus phoenicurus 296
Phylloscopus collybita 302
Phylloscopus sibilatrix 302
Phylloscopus trochilus 302
Pigeon, Comoro Blue 190, 198
    Comoro Green 198
    Comoro Olive 198
    Feral 132, 182, 198, 218, 228
    Madagascar Blue 132
    Madagascar Green 132, 198
    Pink 218
    Seychelles Blue 182
Pintail, Northern 256
Pipit, Red-throated 292
    Richard's 292
    Tree 292
Platalea alba 82
Plegadis falcinellus 82
Ploceus cucullatus 226, 234
Ploceus nelicourvi 180
Ploceus sakalava 180
Plover, Caspian 270
    Common Ringed 114
    Greater Sand 114
    Grey 114
    Kittlitz's 116
    Lesser Sand 114
    Little Ringed 270
    Madagascar 116
    Madagascar Three-banded 116

    Mongolian 114
    Oriental 270
    Pacific Golden 114
    Three-banded 116
    White-fronted 116
Pluvialis fulva 114
Pluvialis squatarola 114
Pochard, Madagascar 86
Podiceps nigricollis 256
Polyboroides radiatus 92
Porphyrio porphyrio 110
Porphyrio alleni 110
Porzana parva 266
Porzana porzana 266
Porzana pusilla 106
Pratincole, Black-winged 268
    Collared 268
    Madagascar 112
    Oriental 268
Prion, Antarctic 246
    Slender-billed 246
Procellaria aequinoctialis 246
Procellaria cinerea 246
Procellaria conspicillata 246
Pseudobias wardi 176
Pseudobulweria atterima 62
Psittacula eques 220
Psittacula krameri 184, 220, 228
Pterocles personatus 132
Pterodroma arminjoniana 64
Pterodroma baraui 62
Pterodroma heraldica 64
Pterodroma macroptera 62
Pterodroma mollis 62
Pterodroma neglecta 64
Pterodroma nigripennis 248
Puffinus assimilis 248
Puffinus bailloni 66
Puffinus carneipes 66
Puffinus pacificus 66
Puffinus tenuirostris 248
Pycnonotus jocosus 220, 230, 242

Q

Quail, Common 102
    Harlequin 102

    Jungle Bush 102
    King 102
Quelea quelea 234
Quelea, Red-billed 234

R

Rail, Aldabra 108
    Buff-banded 266
    Madagascar 108
    Madagascar Wood 108
    Sakalava 106
    Tsingy Wood 108
    White-throated 108
Rallus madagascariensis 108
Randia pseudozosterops 160
Recurvirostra avosetta 268
Redshank, Common 274
    Spotted 274
Redstart, Common 296
Riparia paludicola 154, 290
Riparia riparia 290
Roller, Broad-billed 148, 192, 206, 240
    European 288
Rosefinch, Common 306
Rostratula benghalensis 112
Ruff 276

S

Sanderling 120
Sandgrouse, Madagascar 132
Sandpiper, Broad-billed 278
    Buff-breasted 276
    Common 120
    Curlew 120
    Green 274
    Marsh 118
    Pectoral 276
    Sharp-tailed 276
    Terek 120
    Wood 118
Sarkidiornis melanotos 86
Sarothrura insularis 106
Sarothrura watersi 106
Saxicola maurus 296
Saxicola rubetra 296
Saxicola tectes 232

Saxicola torquatus 158, 210
Schetba rufa 172
Scopus umbretta 80
Serinus canicollis 236
Shearwater, Cory's 66
   Flesh-footed 66
   Little 248
   Persian 66
   Short-tailed 248
   Streaked 240
   Tropical 66
   Wedge-tailed 66
Shoveler, Northern 256
Shrike, Lesser Grey 304
   Red-backed 304
   Woodchat 304
Skua, Arctic 122
   Brown 122
   Long-tailed 122
   Pomarine 122
   South Polar 122
   Subantarctic 122
Snipe, Common 272
   Great 272
   Jack 272
   Madagascar 112
   Pin-tailed 272
Sparrowhawk, Frances's 94
   Madagascar 94
Sparrow, House 180, 196, 216, 226, 234
Spermestes cucullatus 216
Spilopelia chinensis 218
Spilopelia senegalensis 218
Spinetail, Madagascar 144, 204
Spoonbill, African 82
Starling, Madagascar 178
   Rosy 306
   Wattled 306
Stercorarius antarcticus 122
Stercorarius longicaudus 122
Stercorarius maccormicki 122
Stercorarius parasiticus 122
Stercorarius pomarinus 122
Sterna dougallii 126
Sterna hirundo 126

Sterna repressa 282
Sterna sumatrana 126
Sterna vittata 282
Sternula albifrons 128
Sternula saundersi 128
Stilt, Black-winged 112
Stint, Little 120
   Long-toed 278
   Temminck's 278
Stonechat, African 158, 210
   Reunion 232
   Siberian 296
Stone-curlew, Eurasian 268
Stork, White 252
   Yellow-billed 80
Storm-petrel, Black-bellied 68
   Matsudaira's 68
   Swinhoe's 68
   White-bellied 68
   White-faced 68
   Wilson's 68
Streptopelia capicola 200
Streptopelia turtur 284
Sula dactylatra 70
Sula leucogaster 70
Sula sula 70
Sunbird, Anjouan 212
   Humblot's 212
   Madagascar Green 152, 212
   Mayotte 212
   Seychelles 186
   Souimanga 152, 194, 242
Swallow, Barn 154, 184, 194, 204, 242
   Lesser Striped 290
   Mascarene 222, 230
   Wire-tailed 290
Swamphen, African 110
   Purple 110
Swift, African Black 144, 204
   African Palm 144, 204
   Alpine 144
   Common 286
   Little 144
   Madagascar Black 144
   Pacific 286
Swiftlet, Mascarene 222, 230

Seychelles 184
Sylvia atricapilla 300
Sylvia borin 300
Sylvia communis 300

T

Tachybaptus pelzelnii 84
Tachybaptus ruficollis 84
Tachybaptus rufolavatus 84
Tachymarptis melba 144
Tattler, Grey-tailed 274
Teal, Hottentot 88
   Madagascar 88
   Red-billed 88
Tern, Antarctic 282
   Black 282
   Black-naped 126
   Bridled 130
   Caspian 124
   Common 126
   Fairy 126
   Greater Crested 124
   Gull-billed 282
   Lesser Crested 124
   Little 128
   Roseate 126
   Sandwich 282
   Saunders's 128
   Sooty 130
   Swift 124
   Whiskered 128
   White 126
   White-cheeked 282
   White-winged 128
   White-winged Black 128
Terpsiphone bourbonnensis 222, 232
Terpsiphone corvina 186
Terpsiphone mutata 176, 210
Tetraka, Appert's 164
   Dusky 164
   Grey-crowned 164
   Long-billed 164
   Spectacled 164
   Wedge-tailed 166
Thalassarche carteri 60
Thalassarche cauta 60

Thalassarche melanophris 244
Thalasseus bengalensis 124
Thalasseus bergii 124
Thalasseus sandvicensis 282
Thalassornis leuconotus 86
Thamnornis 162
Thamnornis chloropetoides 162
Threskiornis aethiopicus 252
Threskiornis bernieri 82
Thrush, Common Rock 296
    Comoro 208
    Forest Rock 156
    Littoral Rock 156
Treron australis 132, 198
Treron griveaudi 198
Tringa brevipes 274
Tringa erythropus 274
Tringa glareola 118
Tringa nebularia 118
Tringa ochropus 274
Tringa stagnatilis 118
Tringa totanus 274
Tropicbird, Red-billed 250
    Red-tailed 70
    White-tailed 70
Tryngites subruficollis 276
Turdus bewsheri 208
Turnix nigricollis 102
Turnstone, Ruddy 120
Turtur tympanistria 200
Tylas eduardi 174
Tyto alba 142, 184, 202, 240
Tyto soumagnei 142

U

Upupa epops 288
Upupa marginata 148
Uratelornis chimaera 150

V

Vanellus armatus 270
Vanellus gregarius 270
Vanellus spinosus 270
Vanga, Bernier's 172

    Blue 170, 214
    Chabert 170
    Crossley's 176
    Helmet 172
    Hook-billed 172
    Lafresnaye's 174
    Nuthatch 176
    Pollen's 174
    Red-shouldered 170
    Red-tailed 170
    Rufous 172
    Sickle-billed 172
    Tylas 174
    Van Dam's 174
    Ward's 176
    White-headed 13
Vanga curvirostris 172
Vidua macroura 236

W

Wagtail, Citrine 294
    Grey 294
    Madagascar 154
    White 294
    Yellow 294
Warbler, Aldabra Brush
    Cryptic 160
    Garden 300
    Grande Comore Brush 210
    Icterine 302
    Madagascar Brush 162, 210
    Madagascar Swamp 162
    Marsh 300
    Mohéli Brush 210
    Rand's 160
    Rodrigues 222
    Sedge 300
    Seychelles 186
    Subdesert Brush 162
    Willow 302
    Wood 302
Waterhen, White-breasted 266
Waxbill, Common 188, 196, 224, 236, 242

Weaver, Nelicourvi 180
    Sakalava 180
    Village 226, 234
Wheatear, Desert 298
    Isabelline 298
    Northern 298
    Pied 298
Whimbrel 118
Whinchat 296
White-eye, Karthala 212
    Madagascar 168, 194, 212, 242
    Mauritius Grey 222
    Mauritius Olive 222
    Reunion Grey 232
    Reunion Olive 232
    Seychelles 186
Whitethroat, Common 300
Whydah, Pin-tailed 236
Wigeon, Eurasian 256

X

Xanthomixis apperti 164
Xanthomixis cinereiceps 164
Xanthomixis tenebrosa 164
Xanthomixis zosterops 164
Xema sabini 280
Xenopirostris damii 174
Xenopirostris polleni 174
Xenopirostris xenopirostris 174
Xenus cinereus 120

Y

Yellowbrow, Madagascar 166

Z

Zoonavena grandidieri 144, 204
Zosterops borbonicus 232
Zosterops chloronothos 222
Zosterops maderaspatanus 168, 194, 212, 242
Zosterops mauritianus 222
Zosterops modestus 186
Zosterops mouroniensis 212
Zosterops olivaceus 232